T0222934

Advances in Intelligent Systems and Computing

Volume 1305

The series "Advances in Intelligent Systems and Computing" contains publications on theory, applications, and design methods of Intelligent Systems and Intelligent Computing. Virtually all disciplines such as engineering, natural sciences, computer and information science, ICT, economics, business, e-commerce, environment, healthcare, life science are covered. The list of topics spans all the areas of modern intelligent systems and computing such as: computational intelligence, soft computing including neural networks, fuzzy systems, evolutionary computing and the fusion of these paradigms, social intelligence, ambient intelligence, computational neuroscience, artificial life, virtual worlds and society, cognitive science and systems, Perception and Vision, DNA and immune based systems, self-organizing and adaptive systems, e-Learning and teaching, human-centered and human-centric computing, recommender systems, intelligent control, robotics and mechatronics including human-machine teaming, knowledge-based paradigms, learning paradigms, machine ethics, intelligent data analysis, knowledge management, intelligent agents, intelligent decision making and support, intelligent network security, trust management, interactive entertainment, Web intelligence and multimedia.

The publications within "Advances in Intelligent Systems and Computing" are primarily proceedings of important conferences, symposia and congresses. They cover significant recent developments in the field, both of a foundational and applicable character. An important characteristic feature of the series is the short publication time and world-wide distribution. This permits a rapid and broad dissemination of research results.

Indexed by Scopus, DBLP, EI Compendex, INSPEC, WTI Frankfurt eG, zbMATH, Japanese Science and Technology Agency (JST), SCImago.

All books published in the series are submitted for consideration in Web of Science.

More information about this series at http://www.springer.com/series/11156

Yupeng Li · Quanmin Zhu · Feng Qiao ·
Zhiping Fan · Yinong Chen
Editors

Advances in Simulation and Process Modelling

Proceedings of the Second International
Symposium on Simulation and Process
Modelling (ISSPM 2020)

 Springer

Editors
Yupeng Li
Shenyang Jianzhu University
Shenyang, Liaoning, China

Feng Qiao
Faculty of Information and Control
Engineering
Shenyang Jianzhu University
Shenyang, Liaoning, China

Yinong Chen
School of Computing, Informatics,
and Decision Systems Engineering
Arizona State University
Tempe, AZ, USA

Quanmin Zhu
Department of Engineering Design
and Mathematics
The University of the West of England
Bristol, UK

Zhiping Fan
School of Business Administration
Northeastern University
Shenyang, Liaoning, China

ISSN 2194-5357   ISSN 2194-5365   (electronic)
Advances in Intelligent Systems and Computing
ISBN 978-981-33-4574-4   ISBN 978-981-33-4575-1   (eBook)
https://doi.org/10.1007/978-981-33-4575-1

This Springer imprint is published by the registered company Springer Nature Singapore Pte Ltd.
The registered company address is: 152 Beach Road, #21-01/04 Gateway East, Singapore 189721,
Singapore

The Second International Symposium on Simulation and Process Modelling

ISSPM 2020

August 29-30, 2020, Shenyang, China

Convened by

Liaoning Association of Automation, China

Organized by

Shenyang Jianzhu University, Shenyang, China
Northeastern University, Shenyang, China

Sponsored by

University of the West of England, Bristol, UK
Arizona State University, Tempe, AZ, USA

Supported by

Henryton Science and Technology Co. Ltd. Shenyang, China

Organisors, Sponsors and Supporter

Organisors

Sponsors

Supporter

Committees

Advisory Committee

Advisory Chair

Prof. Quanmin Zhu, the University of the West of England, UK

Members of Advisory Committee

Prof. Bernard P. Zeigler, University of Arizona, USA
Prof. Chun-Hung Chen, George Mason University, USA
Prof. Agostino Bruzzone, University of Genoa, Italy
Prof. Libero Nigro, Universit della Calabria, Italy
Prof. Alexander Verbraeck, TU Delft, Netherlands

Organising Committee

General Chairs

Prof. Yupeng Li, Shenyang Jianzhu University, China
Prof. Peter Ball, University of York, UK

General Co-chairs

Prof. Zhijun Gao, Shenyang Jianzhu University, China
Prof. Kunyuan Hu, Shenyang Institute of Automation, CAS, China
Prof. Zhiping Fan, Northeastern University, China

Executive Chair

Prof. Feng Qiao, Shenyang Jianzhu University, China

Organising Chairs

Prof. Liangliang Sun, Shenyang Jianzhu University, China
Prof. Shaowen Lu, Northeastern University, China

Publication Chair

Prof. Jing Na, Kunming University of Science and Technology, China

Publicity Chairs

Prof. Li Xia, Sun Yat-sen University, China
Prof. Weicun Zhang, University of Science and Technology Beijing, China

Special and Invited Session Chairs

Dr. Franco Cicirelli, National Research Council of Italy, Italy
Prof. Ruozhen Qiu, Northeastern University, China
Dr. Lingzhong Guo, Sheffield University, UK
Dr. Lai Xu, Bournemouth University, UK

Student Activities Chair

Mr. Hongwei Jiang, Shenyang Jianzhu University, China

General Secretary

Ms. Yifan Chen, Shenyang Jianzhu University, China
Mr. Xiaohan Wang, Shenyang Institute of Automation, CAS, China

International Program Committee

IPC Chairs

Dr. Yinong Chen, Arizona State University, USA
Prof. John Wang, Montclair State University, USA

IPC Members

Ahmad Taher Azar, Benha University, Egypt
Fernando Barros, Universidade de Coimbra, Plo II, Portugal
Franco Cicirelli, National Research Council of Italy, Italy
Michael Devetsikiotis, University of New Mexico, USA
Lingzhong Guo, University of Sheffield, UK
Zhonghua Han, Shenyang Jianzhu University, China
In Lee, Western Illinois University, USA
Jian Liu, Shenyang Jianzhu University, China
Yong Ma, Wuhan University, China
Roger McHaney, Kansas State University, USA
Yuri Merkuryev, Riga Technical University, Latvia

Keynote Speeches and Invited Talks

Keynote Speeches

Prof. Kornel F. Ehmann, Department of Mechanical Engineering, Northwestern University, Evanston, IL, USA

Prof. Emeritus and Dr. Bernard P. Zeigler, Electrical and Computer Engineering, University of Arizona, Tucson, AZ, USA and RTSync Corp. USA

Prof. Qianchuan Zhao, Department of Automation, Tsinghua University, Beijing, China.

Invited Talks

Dr. Mengchu Huang, Applied Sciences Books, Springer Nature, Shanghai, China

Dr. Shan Bai, Karlsruher Institut für Technologie, Eggenstein-Leopoldshafen, Germany

Preface

In recent decades, more and more attentions are paid to modelling and simulation (M and S) to investigate complex systems in various sectors and areas for the requirements of the quick reaction to innovations and improvements. As for the increasing competition, higher demand of quality, lower cost, shorter lifecycles, more complexity and variety of the systems, it is obvious that M and S has been a powerful tool in identifying the bottlenecks and hidden potentials of the systems, verifying the effectiveness of the proposed strategies and evaluating the managerial, operational and control plans.

This book is the collection of 46 papers from the Second International Symposium on Simulation and Process Modelling (ISSPM 2020) held online on 29 and 30 August 2020 in Shenyang, China.

ISSPM 2020 was convened by Liaoning Association of Automation, China and jointly organized by Shenyang JianZhu University (SJZU) and the Northeastern University (NEU), Shenyang, China. It was sponsored by the University of the West of England, UK, and Arizona State University, USA. It was supported by the International Journal of Simulation and Process Modelling (IJSPM), International Journal of Information Systems and Supply Chain Management (IJISSCM) and Henryton Science and Technology Company Limited, Shenyang, China.

ISSPM 2020 provided a forum for scholars, researchers and practitioners interested in the modelling and simulation of business processes, production and industrial processes, service and administrative processes, and public sector processes to develop the theory and practice of simulation and process modelling.

The symposium invited professors and scholars from well-known universities to exchange and share their experiences, present research results, explore collaborations and to spark new ideas with the aim of developing new projects and exploiting new technology in these fields, and bridge theoretical studies and practical applications in all science and engineering branches.

At the symposium, three distinguished professors were invited to give keynote speeches; they were Prof. Kornel Ehmann, from the Northwestern University, USA;

Emeritus Prof. Bernard Zeigler, from the University of Arizona, USA; and Prof. Qianchuan Zhao, from Tsinghua University, China. They gave speeches on "Modelling of Hybrid Multi-scale Manufacturing Processes", "Theory of Modelling and Simulation: computational support for systems of systems design and testing", and "Energy Efficient Building Control Strategies in a Multi-Agent Framework", respectively. And also, at the symposium, Dr. Mengchu Huang, from Springer Nature, Shanghai, China, was invited to give a talk on "Publishing from a Springer Book Editor's Perspective"; and Dr. Shan Bai, from Karlsruhe Institute of Technology, Germany, was invited to give invited talk on "Modelling a Decision Support System for Risk Management of COVID-19". Their speeches and talks addressed the state-of-the-art development and the cutting-edge research topics in both theory and practical application.

This edition of ISSPM 2020 covers a wide range of research areas in modelling and simulation of manufacturing and production processes, supply chains, transportation and traffic systems, built environment, smart city, smart building and smart home, energy systems and automation, COVID-19 transmission and impact, performance optimization through simulation.

The papers in this volume are categorized into five tracks in simulation and process modelling as follows:

1. Theory, Methodology and Application of Modelling and Simulation,
2. Modelling and Simulation of Manufacturing and Production Processes,
3. Transportation and Traffic Systems,
4. Smart City, Smart Building and Smart Home, and
5. Automation, Identification and Robotics.

For the well-known reason, ISSPM 2020 was greatly affected by COVID-19, it was postponed from August 1 and 2, 2020 to August 29 and 30, 2020, and it was forced to be moved from onsite to online. The organization committee of the symposium made a great effort and tried every means to promote the event, and we received tremendous support from various sources. Therefore, on behalf of the organizing committee of ISSPM 2020, I am very thankful to the Editor-in-Chief of the Springer Series on Advances in Intelligent Systems and Computing (AISC) and the staff from the Springer who have supported and helped to bring out the Proceedings of the Second International Symposium on Simulation and Process Modelling; I am grateful to the keynote speakers, Profs. Kornel Ehmann, Bernard Zeigler and Qianchuan Zhao, and the invited talkers, Drs. Mengchu Huang and Shan Bai, who shared their research and professional experience at the symposium; Last but not least, I would like to express my heartfelt thanks to the authors to contribute the results of their research work to the symposium and the reviewers who anonymously gave their full support in reviewing the papers submitted to the symposium in time.

It is expected that the papers in this volume will stimulate further research and development in simulation and process modelling, and the readers will get great help from this volume of AISC series on advancements in simulation and process modelling.

Symposium Website: http://conf.neu.edu.cn/isspm/index.html

Shenyang, China   Feng Qiao
                                                                        Executive Chair of ISSPM 2020

Contents

Theory, Methodology and Application of Modeling and Simulation

Modeling a Decision Support System for Risk Management
of COVID-19 3
Shan Bai

Research on the Numerical Representation of Stock Linkage
Prediction 13
Chi Ma, Shaofan Wang, Shengliang Lu, and Guojing Han

Research on Teaching Method of EDA Simulation Design Based
on Numerical Calculation 21
Yuanyuan Deng, Zhijun Gao, and Zhuhan Wang

Sub-pixel Edge Contour Detection Algorithm Based on Cubic
B-Spline Interpolation 29
Jianzhao Cao, Ruwei Ma, Renning Pang, and Yuanwei Qi

Spectrum Sensing Algorithm Based on Random Forest in Dynamic
Fading Channel 39
Zhijun Gao and Xin Wang

A Study on Escape Path Planning of Multi-Source/Multi-Sink
for Public Buildings 47
Yi Zhang, Tianqi Liu, Chi Wang, and Chenlei Xie

Design of 3D Building Model Based on Airborne LiDAR
Point Cloud 59
Maohua Liu, Manwen Li, and Jiahua Liang

Research on the Connection Performance of SMA Pipe Coupling
for Sports Equipment 71
Xudong Yang, Fan Gu, Yuyu Zhang, and Yuhan Jiang

A New Shingling Similar Text Detection Algorithm   83
Peng Li, Tianling Qiao, Yongxing Guang, and Lan Zhang

Kalman Consensus Filtering Algorithm Based on Update Scheduling
Scheme for Estimating the Concentrations of Pollutants   93
Rui Wang and Yahui Li

Modeling and Simulation of Manufacturing and Production Processes

Numerical Simulation of Welding Quality of Reinforcement
Framework Under Different Welding Sequence 103
Shuwen Ren, Shizhong Chen, Zijin Liu, Zhongxian Xia, Yonghua Wang,
and Songhua Li

Theoretical Study and Experimental Test on Solenoid Actuator
of Active Control Mount 119
Fang-Hua Yao, Rang-Lin Fan, and Song-Qiang Qi

Research on the Problems and Countermeasures of Sustainable
Development of Green Building Economy in China 131
Xuefeng Li and Lijuan Song

Research on the Balance of Automobile Mixed-Flow Assembly Line . . . 139
Qi Ge, Qianqian Shao, Yunfeng Zhang, and Siqi Zhang

Research on Scheduling Method for Uncertainty of Hit Rate of Molten
Steel Based on Q Learning 149
Liangliang Sun, Tianyi Lu, Shuya Sha, Wanying Zhu, Qiuxia Qu,
and Baolong Yuan

Research on Obstacle Factors of Project Operation and Maintenance
Based on BIM Technology 159
Shengkai Zhao, Haiyi Sun, Zhe Huang, Ning Li, Mingze Guo,
and Xiaohu Li

Simulation of Shenyang Pork Supply Chain System Based on System
Dynamics Model. 169
Qianqian Shao, Xiaojing Zhang, Yunfeng Zhang, and Siqi Zhang

Experimental Study on Grinding Surface Roughness of Full-Ceramic
Bearing Ring 181
Songhua Li, Kechong Wang, and Jian Sun

Transportation and Traffic Systems

Research on Traffic Impact Assessment of Project Under Construction
Based on TransCAD 193
Lei Zhang and Zhu Bai

Research on Influencing Factors of Combined Transportation System
Based on Analytic Hierarchy Process 205
Yan Xing, Yingwen Xing, Weidong Liu, Zongze Li, and Shanshan Fan

Simulation on Two Types of Improved Displaced Left-Turn
Intersections Based on VISSIM 213
Yingcheng Zheng, Qianqian Shao, Yunfeng Zhang, and Siqi Zhang

Simulation and Optimization of Multi-UAV Route Planning Based
on Hybrid Particle Swarm Optimization 223
Xuezhao Peng, Feng Guan, Zhongpu Wang, and Shushida Gao

Simulation Research of Arctic Route and Traditional Route Based
on the Logit Model 233
Feng Guan, Yi Cao, Xuezhao Peng, and Yongbao Wang

Airport Taxi Driver Decision and Ride Area Scheme Design Based
on Hybrid Strategy 243
Jingxuan Yang, Longsheng Bao, Qicheng Xu, Junjie Liu, and Yang Wang

Dynamic Route Optimization Problem Based on Variable Range
Short-Term Traffic Flow Forecast 253
Guanghui Dai, Qianqian Shao, Yunfeng Zhang, and Siqi Zhang

GAPSO-Based Traffic Signal Control in Isolated Intersection
with Multiple Objectives 263
Yifan Chen, Feng Qiao, Lingzhong Guo, and Tao Liu

Multi-objective Optimization of Traffic Signal Systems on Urban
Arterial Roads 277
Tao Liu, Feng Qiao, Lingzhong Guo, and Yifan Chen

Study on Traffic Volume Transferred by Bohai Strait Tunnel 287
Bing Wang, Fei Liu, and Yu Peng Li

Smart City, Smart Building and Smart Home

Effects of Cap Gap and Spiral-Welded Seam Composite Defects
on Concrete-Filled Steel Tubes 303
Zhengran Lu, Chao Guo, and Guochang Li

Fuzzy Sliding Mode Control of a VAV Air-Conditioning Terminal
Temperature System 315
Ziyang Li, Lijian Yang, Zhengtian Wu, Baoping Jiang, and Baochuan Fu

Study on the Treatment of Cutting Fluid Wastewater Chromaticity
by Multiphase Fenton System 325
Nana Wu, Yang He, Qiang Liu, Youchen Tan, Licheng Zhang, Na Huang,
and Yulin Gan

**Optimization for Finite Element Model of a Steel Ring Restrainer
with Sectional Defect** 333
Qiang Zhang, Yue Ma, Jin Feng, and Zhanfei Wang

**An Epidemic Prevention Robot System Based on RoboMaster
Technology** 341
Tao Li, Lei Cheng, Huanlin Li, Yanjie Wu, and Guang Li

**Research on the Fire Resistance of Grout Sleeve Splicing Joint
for Precast Stadium** 351
Xudong Yang, Fan Gu, Liqiang Liu, and Qinghe Li

**Mechanical Property of Grout Sleeve Splicing Joint Under Reversed
Cyclic Loading for Precast Stadium** 365
Xudong Yang, Fan Gu, Qinghe Li, and Liqiang Liu

**Self-test of Athletic Ability for the Elderly Using Inertia Motion
Capture Device** 375
Jun Sun, Donghua Li, and Lianjie Lv

**A Method Based on CNN + SVM for Classifying Abnormal
Audio Indoors** 385
Jian Liu, Shuyan Ning, Sanmu Wang, Jiarui Yi, and Mingrui Zhao

**A Simplified Method of Radiator to Improve the Simulation Speed
of Room Temperature Distribution** 395
Zhenqiang Cao, Tong Niu, Haiyi Sun, and Xia Lu

Automation, Identification and Robotics

**An Optimal Maintenance Cycle Decision of Relay Protection Device
Based on Weibull Distribution Model** 409
Qiuyu Zhuang and Meiju Liu

**Hysteretic Behavior Analysis of Concrete-Filled Double-Skin Steel
Tubular Column Under the Constraint of Mortise and Tenon Joint
with Low-Cycle Reciprocating Load** 419
Wei Sun, Junshan Yang, and Bing Li

Face Mask Recognition Based on MTCNN and MobileNet 433
Jianzhao Cao, Renning Pang, Ruwei Ma, and Yuanwei Qi

**Quadcopter UAV Finite Time Sliding Mode Control Based
on Super-Twisting Algorithm** 443
Jianhua Zhang, Wenbo Fei, and Yang Li

**Cartesian Admittance Control Based on Maxwell Model for Human
Robot Interaction** 451
Xin Wang, Jia Sun, Zhijun Gao, Languang Zhao, Jianshun Liu,
and Naifeng He

**Bearing Fault Detection Method Based on Improved
Convolution Network** 459
Pengyu Cheng, Binbin Li, and Bin Jiao

**Path Planning of AFM-Based Manipulation Using Virtual
Nano-hand** 467
Shuai Yuan, Tianshu Chu, and Jing Hou

**Automatic Reading Algorithm of Pointer Water Meter Based
on Deep Learning and Double Centroid Method** 477
Hongqing Li, Juan Wang, Bing Bai, and Chuang Lu

Author Index 487

Theory, Methodology and Application of Modeling and Simulation

Modeling a Decision Support System for Risk Management of COVID-19

Shan Bai

Abstract A decision support system that performs a comprehensive analysis on risks of the COVID-19 pandemic and its subsequent medium- and long-term impacts is highly desirable, especially in view of the still ongoing pandemic in the world. This paper focuses on the modeling process of such a decision support system. The modeling process includes the following steps: generating appropriate strategies that suit the specific situation concerned, where each strategy is composed of measures and each measure consists of a static (textual) part relating to the contexts abstracted from existing literature via techniques inspired by natural language processing and a dynamic part adapted by using scenario planning method and agent-based simulation; generating utility-based recommender based on soft systems methodology and recommending the Top-N strategies to the group of stakeholders based on the scores of the strategies using multiple-criteria decision analysis method. The proposed model is highly flexible in that it recommends appropriate strategies in respect of the various scenarios concerned and may thus be applied to various types of risk management.

Keywords Decision making · Risk management · Ontology-based text summarization · Scenario planning · Soft system thinking · Agent-based simulation · Recommender system · COVID-19

1 Introduction

The COVID-19 pandemic has plunged the whole world into a crisis of unprecedented scope and scale, so that most countries engage in a collective endeavor to tackle the dramatic impacts of the pandemic. To mitigate relative damages to public health, economy, and society, adequate strategies shall be adopted at each stage of the pandemic. In the past months, extensive progress has been made in predictions

S. Bai (✉)

Karlsruher Institut für Technologie, Hermann-von-Helmholtz Platz 1, 76344 Eggenstein-Leopoldshafen, Germany

e-mail: shan.bai@kit.edu

© The Author(s), under exclusive license to Springer Nature Singapore Pte Ltd. 2021
Y. Li et al. (eds.), *Advances in Simulation and Process Modelling*,
Advances in Intelligent Systems and Computing 1305,
https://doi.org/10.1007/978-981-33-4575-1_1

of infected cases and in assessments on the effectiveness of intervention strategies. Artificial intelligence (AI) has made significant contributions to coping with the COVID-19 pandemic [1], including modeling its spread process and forecasting the infected cases. For example, a modified stacked auto-encoder model was used in [2] to forecast in real time the COVID-19 confirmed cases across China, and a hybrid AI model for COVID-19 infection rate forecasting was proposed in [3], which combined the epidemic susceptible infected model, natural language processing (NLP), and deep learning tools. For further applications of AI against the COVID-19 pandemic, please refer to [1] and references cited therein.

This paper relates to a general framework for establishing the decision-making system that performs a comprehensive analysis on risks of the COVID-19 pandemic and its subsequent medium- and long-term impacts. The current work focuses on the modeling process of such a system. The process involves gathering appropriate information for creating possible measures, and searching from the so created possible measures for the proper combination of measures to figure out the most efficient intervention or mitigation strategies against the outbreak of the pandemic.

Section 2 discusses gathering appropriate information, i.e., summarizing existing measures from available literature, which amounts to collecting and analyzing the corresponding textual resources, using information technology. In this paper, we focus on how to make use of document summarizing, which is an NLP technique, to prepare the measures for a special target from hundreds of documents, wherein similarity search is performed in order to retrieve similar measures to the queries. Section 3 is devoted to determining the most efficient intervention or mitigation strategies from the measures prepared in Sect. 2, wherein soft systems methodology (SSM) is used to describe the problem situation that participants in a group play decision-making roles and rank various scenario-based strategies. Moreover, agent-based modeling is applied to predict the performance of measures which reflects how practical needs are met, wherein the parameters in the corresponding solver are inspired by those in the compartmental model in epidemiology. Finally, the conclusions of the paper are presented in Sect. 4.

2 Generation of New Strategies

This section tackles the generation of new strategies for concrete situations in which no existing appropriate strategies are available. Generally speaking, one strategy consists of a plurality of measures, where each measure may be divided into a static/textual part and a dynamic part. The static part relates to contexts that are abstracted from documents, while the dynamic part is adjustable to meet the requirements in view of the situation concerned. In this paper, the dynamic part will be adapted using scenario planning method and agent-based simulation.

2.1 Textual Description of Measure

Since February, many institutions have developed lots of measures to prepare for and respond to the pandemics and published them in a large volume of documents. However, it is not so easy for a stakeholder confronted with a specific scenario to identify the optimal measures from hundreds of documents due to various writing styles and structures. Therefore, how to efficiently extract information from texts in multiple resources is the first important task for constructing appropriate measures specifically designed for the stakeholder. This task can be completed in the following steps: (a) extracting sentences from the original text by identifying important sections thereof and generating them verbatim; (b) based on the extraction, interpreting, and examining the text using advanced techniques from NLP to generate a new shorter text that conveys the most critical information from the original text.

Here, the technique from automatic text summarization [4] is utilized, which can give an overview of all relevant literature data needed. The key points of all the documents are comprehended to deliver summaries to each document. The ontology of the measures in all the documents is concluded, and thus, a graph database for text searching is generated.

We take one example from Ref. [5], where the recommendations in the case of mild severity are given as follows:

- Patients should be instructed to self-isolate and contact COVID-19 information line for advice on testing and referral.
- Test suspected COVID-19 cases according to diagnostic strategy, isolation/cohorting in:

  - Health facilities, if resources allow;
  - Community facilities (e.g., stadiums, gymnasiums, hotels) with access to rapid health advice.

- Self-isolation at home according to WHO guidance.

The presence of a domain ontology is key to summarization, which represents all concepts of the organization. The ontology includes entities and relations. The semantic representations are not only simple words, but also relations that can be semantically typed, e.g., "isolated by," in order to express interrelations between concepts.

As illustrated in Fig. 1, the abstractive summarization of measures in documents can be graphically represented such that the nodes of the graph represent entities of these sentences, while the edges, i.e., the lines that connect the nodes, indicate the relation (or e.g., action) between the entities. The nodes, the edges, and properties of the nodes that are information associated with the nodes constitute ingredients of the graph database. One goal of the research in the generation of the textual part of measures is to create an onto base, which is a repository that comprises Web ontology language (OWL) ontological entities, so that domain-relevant OWL

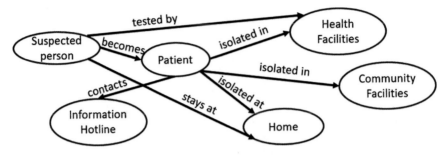

Fig. 1 Ontology-based concept map of the example introduced in Sect. 2.1

ontologies present in the onto base can be extracted and transformed into their equivalent resource description framework (RDF) structures [6]. More details about the ontology model will be reported in the future.

Usually, the measures stored in the database may not be appropriate in view of a concrete application scenario, which requires that the system shall be capable of generating new measures based on experience data. To fulfill this requirement, the efficient searching machine shall be introduced, which can return a proper set of measures that are relevant to the query for new measures. Furthermore, to improve the quality of the search, it is necessary to define a suitable similarity function between query and entities. The query is described by the domain ontology for matching measures and the search aims to identify the sets of measures stored in the graph database that contains nodes similar to the cluster of query words. The semantic similarity function can be divided into two categories, local and global similarity, where local similarities are aggregated to a global similarity. Various definitions of similarity functions were reviewed and categorized in both syntactic and semantic relationships in [7]. The searched similar measures are essential to solving the retrieval problem, one of pattern recognition problems. In addition, a domain adaptation framework for the knowledge retrieval system is under development. The new measures can be used in the next step as the textual part and then enlarged in view of the specific scenario by adding dynamic attributes.

2.2 Scenario Planning Method for Generating Dynamical Measures

The dynamic attributes of the measures can be generated by, e.g., scenario planning method [8], which concerns planning based on systematic examination of the future by picturing plausible and consistent images of that future, while scenario describes a certain topic or issue about the future. The process of developing and using a number of contrasting scenarios explores the consequences of future uncertainty surrounding a decision. In the present work, instead of reducing to 2–3 scenarios like the standard

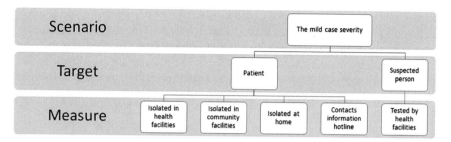

Fig. 2 Tree structure of the example introduced in Sect. 2.1

approach, the number of scenarios will not be limited because of the processing capability of computer simulation.

Given a scenario and several (protection) targets in query, similar measures can be retrieved from the database by the method introduced in the previous subsection, and revised so as to generate new proper measures. The comprehensive list of new measures is utilized to create the respective scenario-based baseline strategies. The tree structure of the example introduced in Sect. 2.1 is shown in Fig. 2. Other structures, e.g., network structure, can also be implemented in accordance with the specified requirements.

Up to now, the measures are described statically, that is, in terms of textual illustrations of entities, relationships, and actions. In the next step, we introduce dynamical structures into the measures by modeling mathematically. For example, "Patient isolated at home" can be interpreted to be "The number of contact persons for each patient reduces to be no more than the number of his/her family, the average of which being, e.g., 2." Therefore, in one epidemic model, the modified SEIRD model in [9], the parameter, "the number of contact persons", can be set to be a random number in the range of, e.g., [0, 5] obeying the Gaussian distribution.

The dynamic attributes can be adapted according to the requirements of the scenario. For example, one category of the mitigation measures, which is considered efficient and has been widely adopted, is to reduce the number of contact persons. In the scenario of mass gathering, the measure can be to reduce the number of participants to be less than 1000, or 500, or 100, depending on the risk scores [10].

With a view to epidemic models, there are other dynamic attributes that can be introduced into the measures. It may relate to the affected people, for example, the population of the affected people, the number of the contact persons for each affected people, and so on.

2.3 Estimator of Scenario-Based Strategies

As mentioned above, the scenario-based baseline strategy is a comprehensive combination of measures. However, the effect of one strategy cannot be always simply a linear combination of the effect of each constituent measure, which is a characterizing feature of a complex system due to the potential interdependence of the constituents, i.e., measures in the context. Taking this characterizing feature into account means that we shall create the estimation model for the effectiveness of the whole strategy instead of each constituent measure, which can be realized by agent-based modeling (ABM) [11, 12]. ABM is currently used in social science to describe and understand the dynamics of social and economic systems because the system under consideration is composed of interacting agents; and the system exhibits emergent properties, that is, properties arising from the interactions of the agents that are driven by a set of rules which govern the agents' behavior [13]. ABM can be considered as a bottom-up approach that relies on interacting agents to build system behavior or properties. It consists of an environment in which interactions take place and a number of agents whose behavior is defined by a basic set of rules and by characteristic parameters. The states of agents and their relations with one another are changing through time. These collective changes in state can be considered a process, which can interact with any relevant external component and alter the environment. An example of the estimator of the intervention strategy was given in [9]. The spatial ABM was introduced in the background of a small city where the agents refer to citizens. Compared to those existing ordinary differential equation models, the spatial ABM model specially describes the willingness for population to move and introduces a sub-model to illustrate the occupancy of the hospital beds. The intervention strategies can act directly on individuals, and the effectiveness of these strategies could be assessed by comparing the simulation results with/without these strategies. The idea in [9] can be generalized so as to accommodate more complicated situations.

3   Generating a Utility-Based Recommender

In order to implement such a decision-making process that the strategies prepared in Sect. 2 are selected as recommendations to a group of stakeholders, the underlying paradigm for system thinking that is intrinsically related to the modeling process has to be determined, wherein the paradigm is characterized by the degree to which there is agreement on the different purposes among various stakeholders, tolerance of uncertainty, and uncertainty in the amount of data required [14, 15].

In this section, we consider the design of the recommender, which is the system for the recommendation of strategy, in view of soft systems methodology (SSM). The SSM aims to solve the soft problem, which is involved in complex situations such that there are divergent views about the definition of the problem. It is a process-oriented approach for channeling debate about situations characterized by messy ill-structured problems with multiple perspectives [16, 17]. Furthermore, effective use of SSM requires a situation where participants have sufficient time to share with and learn from each other and have a high enough level of trust to discuss their preferences and requirements openly. To intervene in such situations, the debate is undertaken among participants. After initial appreciation of the problem situation, the evaluation of the strategy for participants is performed individually. By discussions and exploration of these concerning issues, the participants can arrive at common decisions based on consensus over what kind of changes may be systemically desirable and feasible in the situation.

Since the recommendation for proper strategy is often requested by a group of decision makers, many aspects related to the group, for example, the size of the group, the relationships among the members, etc., shall be considered in order to choose and design optimal group recommender approach. On the other hand, the group of decision makers can actively choose strategies instead of only passively accepting those, which is also a remarkable feature of soft system thinking. The decision makers in a group may have conflicting preferences, so that after generating recommendations to every group member individually, these recommendation candidates are aggregated together into a recommendation list intended for the whole group [18].

The aggregation method can be based on the scores of strategies by using multiple-criteria decision analysis (MCDA) method [19]. Let $S = (S_j), j \in \{1, 2, \ldots, m\}$ be the strategy with m attributes, $P = (P_{ik})(i \in \{1, 2, \ldots, n\}, k \in \{1, 2, \ldots, r\})$ be the preference values of n group member w.r.t. r criteria. Therefore, the member-based utility is the score of the strategy for each member, defined as a kind of weighted sums of evaluations of alternatives $U_i = \sum_{j=1}^{m} w'_j \cdot \sum_{k=1}^{r} w_k \cdot E(P_{ik}, S_j)$, where the weights of criteria are $w_k > 0$ and satisfy $\sum_{k=1}^{r} w_k = 1$. The weights of attributes are $w'_j > 0$ and satisfy $\sum_{j=1}^{m} w'_j = 1$. The definition of the functional E, which acts on the preference values and the attributes of strategies, depends on the detailed requirements from the problem to be solved. The scores of the strategy for individual decision makers U_i are aggregated in order to solve conflicting preferences and obtain one score for the group, which is the utility for the group. The utility-based ranking will be taken, so that the final recommended strategies should represent interest of the whole group. In detail, the group ranks the final scores of the strategies and then gets the ranking of the strategies. The Top-N strategies in the ranking list are the ultimate recommendations.

4 Conclusions

This paper presented the modeling process for developing a decision support system for strategy recommendation. The modeling process consisted of generating new measures by extracting relevant existing strategies from the literature using techniques inspired by NLP and creating the scenario-based baseline strategies based on the collections of these measures to meet the stakeholders' particular requirements and the addition of dynamical attributes with the aid of agent-based simulation, and constructing a utility-based recommender that selects the Top-N recommended strategies from the generated new strategies. The proposed model is highly flexible in view of the diverse situations that the decision makers are confronted with, as it will help them figure out the appropriate strategies that meet their concrete requirements in the same framework. Therefore, the proposed model may be applied to various types of risk management.

In the recommender proposed in Sect. 3, the strategies were selected by ranking. Alternatively, they may also be voted, which has a structure that is similar to polls. There is a decisive aspect in voting since a group or a community decides on which alternative should be chosen, i.e., there is clear pragmatics of the decision outcome. One implementation for modeling and simulation of a voting process was shown in [20] in the situation that after negotiation, there is no strategy agreed on by most negotiation participants as the best strategy, which is the special case of $N = 1$ for Top-N strategy. Therefore, the strategies, e.g., which get high total scores (the sum of the scores overall participants) were used for voting.

The code implementation of the proposed decision support system, wherein the recommender may be designed to select the strategies by either ranking or voting, is under investigation. We hope to report on the progress along this direction in the near future.

References

1. Nguyen, T.T (2020): Artificial intelligence in the battle against coronavirus (COVID-19): a survey and future research directions. arXiv preprint arXiv:2008.07343
2. Hu, Z., Ge, Q., Jin, L., and Xiong, M. (2020): Artificial intelligence forecasting of Covid-19 in China. arXiv preprint arXiv:2002.07112
3. Du, S., Wang, J., Zhang, H., Cui, W., Kang, Z., Yang, T., … Yuan, Q.: Predicting COVID-19 using hybrid AI model. Available at (2020) http://dx.doi.org/10.2139/ssrn.3555202
4. Torres-Moreno, J. M.: Automatic Text Summarization. Wiley. ISBN 978-1-848-21668-6: 320 (2014)
5. World Health Organization: Operational considerations for case management of COVID-19 in health facility and community: interim guidance (2020). https://apps.who.int/iris/handle/10665/331492
6. Deepak, G., Ahmed, A., Skanda, B.: An intelligent inventive system for personalised webpage recommendation based on ontology semantics. Int. J. Intell. Syst. Technol. Appl. **18**(1–2), 115–132 (2019)
7. Cha, S.H.: Comprehensive survey on distance/similarity measures between probability density functions. City **1**(2), 1 (2007)
8. Bradfield, R., Wright, G., Burt, G., Cairns, G., Van Der Heijden, K.: The origins and evolution of scenario techniques in long range business planning. Futures **37**(8), 795–812 (2005)
9. Bai, S.: Simulations of COVID-19 spread by spatial agent-based model and ordinary differential equations, Int. J. Simulation Process Modell. **15**(3), 268–277 (2020). https://doi.org/10.1504/ijspm.2020.107334
10. World Health Organization: How to use WHO risk assessment and mitigation checklist for mass gatherings in the context of COVID-19: interim guidance (2020). https://apps.who.int/iris/handle/10665/331536
11. Gilbert, N.: Agent-Based Models. SAGE Publications, Inc. (2008)
12. Wooldridge, M.: An Introduction to Multiagent Systems. Wiley, Chichester, UK (2008)
13. Axelrod, R.: Advancing the art of simulation in the social sciences. In: Rennard, J.-P. (ed.) Handbook of Research on Nature Inspired Computing for Economics and Management. Hershey, IGR (2006)
14. Jackson, M.C.: Systems Approaches to Management. Kluwer/Plenum, New York (2000)

15. Jackson, M.C.: Systems Thinking: Creative Holism for Managers. Wiley, Chichester (2003)
16. Checkland, P.B.: Systems Thinking, Systems Practice. Wiley, Chichester (1999)
17. Checkland, P., Poulter, J.: Soft systems methodology. In: Reynolds, M., Holwell, S. (eds.) Systems Approaches to Making Change: A Practical Guide, 2nd edn, pp. 201–254. Springer, London (2020)
18. Kaššák, O., Kompan, M., Bieliková, M.: Personalized hybrid recommendation for group of users: Top-N multimedia recommender. Inf. Processing Manag. 52(3), 459–477. ISSN: 0306-4573 (2016). https://doi.org/10.1016/j.ipm.2015.10.001
19. Müller, T., Bai, S., Raskob, W.: MCDA handling uncertainties. Radioprotection 2020, 55 (HS1), S181–S185 (2020)
20. Bai, S., Raskob W., Müller T.: Agent based model, Radioprotection 2020, 55 (HS1), S187–S191 (2020)

Research on the Numerical Representation of Stock Linkage Prediction

Chi Ma, Shaofan Wang, Shengliang Lu, and Guojing Han

Abstract In this paper, time series of numerical correlations and morphological similarities are analyzed. It is proposed to combine the correlation coefficient with a time-weighted dynamic time warping (DTW) distance to emphasize timeliness as a stock linkage numerical formula. Therefore, the problem of finding the connection relationship between stocks can be converted into a numerical representation problem of stock linkage, and a stock linkage prediction optimized model based on long short-term memory (LSTM) can be established. At the same time, in order to improve the prediction performance of the LSTM model for the time series of stock interconnection values, wavelet transform and denoising autoencoder are used to denoise and reconstruct the input samples, thereby achieving more accurate prediction and analysis of stock linkage.

Keywords Stock linkage · Long short-term memory · Numerical representation · Dynamic time warping · Optimized model

1 Introduction

The stock linkage [1] common phenomenon is that stock market trends are highly correlated over time and have similar stock price fluctuation curves. Effectively finding stock linkages helps investors avoid certain investment risks and improve portfolio efficiency [2]. Because there are many factors that affect the stock price

C. Ma · S. Wang (✉) · S. Lu
School of Computer Science and Engineering, Huizhou University, Huizhou 516007, China
e-mail: wsf19961230@163.com

S. Wang · S. Lu
School of Computer Science and Software Engineering, University of Science and Technology Liaoning, Anshan 114051, China

G. Han
Meteorological Administration, Anshan, China

© The Author(s), under exclusive license to Springer Nature Singapore Pte Ltd. 2021   13
Y. Li et al. (eds.), *Advances in Simulation and Process Modelling*,
Advances in Intelligent Systems and Computing 1305,
https://doi.org/10.1007/978-981-33-4575-1_2

trend, it is difficult to explain the stock price-related phenomena, and it is even more difficult to formulate a unified analysis and evaluation standard [3].

In general, the mining can be divided into related networks based on the textual information of related mining [4] or based on volume and price time series data [5, 6]. Scholars have awarded that financial time series data exhibit certain time-varying characteristics [7]. In the financial field, such complex data cannot be handled by traditional data mining models [8]. However, the development of neural network technology, especially deep learning, can alleviate these problems to some extent [9, 10]. Cheng S H settled the inventory classification problem through neural networks and decision trees [11]. Dixon M et al. described the application of deep neural networks to predict the direction of financial market movements [12]. Akita et al. put forward long short-term memory (LSTM) and paragraph vectors to predict financial time series [13]. Deep learning models for different application scenarios and research tasks are currently being designed in different areas [14, 15].

2 Numerical Representation of Stack Linkage

Stock linkage is a numerical standard used to describe the extent of a stock link. In the actual application, the linkage between stocks is reflected not only in the numerical correlation of the stock series, but also in the tendency similarity. Therefore, we put forward a new formula. This equation shows improved continuity and delay consideration of the stock linkage phenomenon.

2.1 Stock Relevance Based on Pearson Correlation Coefficient

Pearson correlation coefficient is used to quantitatively represent the possible linear correlation between fixed-distance continuous random variables. It is given by Eq. (1).

$$\rho_{x,y} = \frac{\text{cov}(X, Y)}{\sigma_x \sigma_y} = \frac{E((X - \mu_X)(Y - \mu_Y))}{\sigma_x \sigma_y} = \frac{E(XY) - E(X)E(Y)}{\sqrt{E(X^2) - E^2(X)}\sqrt{E(Y^2) - E^2(Y)}}$$

(1)

where $\text{cov}(\mathbf{X}, \mathbf{Y})$ is the covariance between variables X and Y. σ_x, σ_y are the standard deviation of variables x and y, respectively. μ_X, μ_Y are the means of variables x and y, respectively. $E(\mathbf{X})$ is the expectation for variable x.

The coefficient is usually used to calculate the correlation of the time series of stock prices and describe the stock links. In the short-term analysis of stock price fluctuations, it can achieve good results. Considering the stock market volatility, we chose the time series data of the Shanghai and Shenzhen 300 Index as market environment variables. This partial correlation coefficient is given by Eq. (2) as follows.

$$\gamma_{ij(h)} = \frac{\gamma_{ij} - \gamma_{ih}\gamma_{jh}}{\sqrt{1 - \gamma_{ih}^2}\sqrt{1 - \gamma_{jh}^2}} \tag{2}$$

where $\gamma_{ij(h)}$ represents the partial correlation coefficient of stocks i and j after controlling market environment variable, h. γ_{ih} represents the simple correlation coefficient between stock and environmental variables.

2.2 Stock Linkage Based on DTW

DTW minimize the cumulative distance between two time series to seek the optimal alignment path between them. Not only can DTW process time series of different lengths, it can also fold and twist times to align corresponding peaks or valleys at different times [16].

The DTW algorithm calculates the distance $d_{i,j}$ between different elements of the time series of length m and n and then calculates the minimum cumulative distance $D_{i,j}$ to establish the distance map D. It is given by Eq. (3) as follows.

$$D_{i,j} = d_{i,j} + \min\{D_{i,j-1}, D_{i-1,j}, D_{i-1,j-1}\} \tag{3}$$

$$i = 2, \ldots, m$$
$$j = 2, \ldots, n$$

The initial conditions in the equation are set as follows:

$$D_{1,1} = d_{1,1} \tag{4}$$

$$D_{1,j} = \sum_{p-1}^{j} d_{1,p} \quad j = 1, \ldots, n \tag{5}$$

$$D_{i,1} = \sum_{q-1}^{i} d_{q,1} \quad i = 1, \ldots, m \tag{6}$$

This study adds the first exponential smoothing method to the DTW algorithm. It is given by Eq. (7) as follows.

$$D_{i,j} = ad_{i,j} + (1-a)\min\{D_{i,j-1}, D_{i-1,j}, D_{i-1,j-1}\} \tag{7}$$

$$i = 2, \ldots, m$$
$$j = 2, \ldots, n$$

As shown in Eq. 7, the influence of the weighting coefficient and the recent shape on the calculation of the similarity between the stock price time series is proportional. In this study, $a = 0.98$.

2.3 Numerical Algorithm of Stock Linkage

Considering the properties of Pearson partial correlation coefficient and time-weighted DTW similarity, the linear combination of the above two numerical expressions yielded more stock linkage indicators.

By Eq. (8), the time-weighted DTW distance of the time series is converted into DTW similarity, $s_{i,j}$. Then, it can be linearly combined with the Pearson partial correlation coefficient, $\gamma_{i,j(h)}$, to obtain the numerical expression of the linkage that considers the stock market environmental factor h, as shown in Eq. (9).

$$s_{i,j} = \frac{1}{1 + d_{i,j}} \tag{8}$$

$$c_{i,j} = \alpha_1 \cdot s_{i,j} + \alpha_2 \cdot \gamma_{i,j(h)} \tag{9}$$

3 Establishment of Stock Linkage Prediction Model Based on Optimized LSTM

In this study, we use trading scales and stock prices as characteristics of stocks to build the attributes of the input time series based on the LSTM model.

3.1 Construction of Training Samples

Combine the time series data of two stocks by using the difference construction method. The time series structure of the sample is shown in Fig. 1.

In Fig. 1, the stock feature time series is listed as columns, with each row being an input sample at a particular time. Each sample contains multiple feature dimensions that correspond to linked values. The training output is the linkage value at the next moment after the end of the period, which has dimension 1.

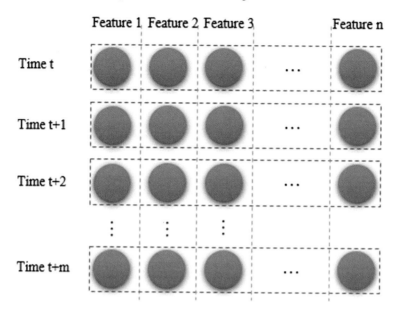

Fig. 1 Structural diagram of input sample time series

3.2 *Structural Design of Optimized LSTM Model*

The stock linkage forecasting task designs a model structure that reflects relevance and independence. To deal with noise in financial time series, after input, the noise reduction processing layer is configured, including the corresponding noise reduction autoencoder module or wavelet transform module. It is as shown in Fig. 2.

4 Experimental Results and Analysis

4.1 *Experimental Preparation*

Considering the influence of the number of layers and hidden units in the LSTM module on the prediction results of the model, in the following experiments of stock link prediction, all prediction models are based on the two-layer LSTM module. Meanwhile, this study selects the time window size of 20 days and the time-weighted DTW distance with the largest co-directional volatility as the numerical expression of the stock linkage.

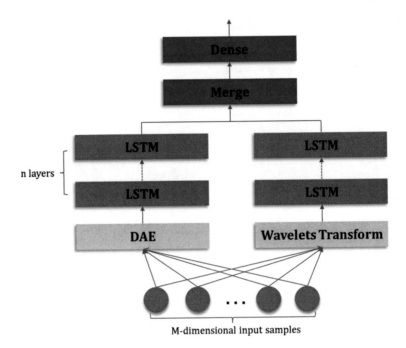

M-dimensional input samples

Fig. 2 Prediction model network structure diagrammable

4.2 Experimental Results

Three different types of comparative experiments were conducted in this study to verify the predictive performance of the optimized LSTM model for numerical time series of inter-stock linkage.

The first set of experiments used basic reference models such as autoregressive integrated moving average (ARIMA) statistical models and traditional LSTM models.

The second set of experiments includes comparative models that use wavelet transform technology. It includes the deep bidirectional LSTM (DB-LSTM) models and deep bidirectional ARIMA (DB-ARIMA). Among them, the DB4 wavelet is used as the basis.

The third set of experiments uses automatic encoder to reconstruct input sequence. It includes the stacked denoising autoencoder (SDAE)-LSTM model [17].

In this study, the dataset was split by year, and 20% of the end-of-year trading day data was used as a test set to validate the performance of the predictive model. The final experimental result is the average of five repeated experiments each year. Table 1 shows the results of comparative experiments on the prediction models.

Table 1 shows that the performance of the predictive model is improved by the noise reduction of the input samples. By the DB noise reduction, the prediction RMSE of ARIMA and LSTM models are reduced by 35.86% and 41.67%, respectively.

Table 1 Comparison of errors in prediction models

Models	Root mean square error (RMSE)	Mean square error (MSE)	Mean absolute error (MAE)
LSTM	0.156	0.028	0.123
ARIMA	0.555	0.308	0.493
DB-ARIMA	0.356	0.127	0.317
DB-LSTM	0.091	0.009	0.075
SDAE-LSTM	0.138	0.019	0.113
Optimized LSTM	0.074	0.006	0.059

Through the SDAE noise reduction, the RMSE of the LSTM model is reduced by 11.54%. Although the quality of the SDAE-LSTM model has improved, the generalization ability has declined due to the increase in network complexity. In the optimized LSTM model, the input samples are reconstructed through a simplified structure. The smoothing trend features and the new reconstructed feature sequence acquired by the wavelet transform are used as training material for the model, which gives relatively improved prediction performance. Compared with the RMSE of the DB-LSTM and SDAE-LSTM model, the RMSE errors of the optimized LSTM model are reduced by 18.68% and 46.38%, respectively.

5 Conclusion

This paper proposed an optimized LSTM model based on the LSTM model and the numerical sequence of stock linkage, combining noise reduction autoencoder and wavelet transform module as the noise reduction processing layer. It could be seen from the comparison experiment that the optimized model had better prediction.

Acknowledgements This paper is supported in the part by the Foundation of Guangdong Educational Committee under Grant 2018KTSCX218, in the part by the Professorial and Doctoral Scientific Research Foundation of Huizhou University, China, under Grant 2018JB020.

References

1. Arshanapalli, B., Doukas, J. et al.: International stock market linkages: evidence from the pre- and post-October 1987 period. J. Bank. Financ. **17**(1), 193–208 (1993)
2. Pan, M.S., Fok, C.W., Liu, Y.A.: Dynamic linkages between exchange rates and stock prices: evidence from East Asian markets. Int. Rev. Econ. Financ. **16**(4), 503–520 (2007)
3. Thomas, S.: Dynamic Linkages between Fundamental Drivers and Stock Market: An Empirical Analysis from India. Social Science Electronic Publishing (2011)

4. Al-Augby, S., Majewski, S., Nermend, K. et al.: Proposed investment decision support system for stock exchange using text mining method. In: Al-Sadeq International Conference on Multidisciplinary in IT and Communication Science & Applications (2016)

5. Belen, N., Rosa, R.: Correlation Between Individual Stock and Corporate Bond Returns. Social Science Electronic Publishing

6. Okamoto, T., Lida, D., Toge, K., et al.: Optical correlation domain reflectometry based on coherence synchronization: theoretical analysis and proof-of-concept. J. Lightw. Technol. **34**(18), 4259–4265 (2016)

7. Patton, A.J.: A review of copula models for economic time series (2016)

8. Bao, W., Yue, J., Rao, Y. et al.: A deep learning framework for financial time series using stacked autoencoders and long-short term memory. PLOS ONE **12**(7), e0180944 (2017)

9. Chen, J. F., Chen, W. L., Huang, C. P., et al. Financial Time-Series Data Analysis Using Deep Convolutional Neural Networks. In: 2016 7th International Conference on Cloud Computing and Big Data (CCBD) (2016)

10. Heaton, J.B., Polson, N.G., Witte, J.H.: Deep Learning in Finance (2016)

11. Cheng, S.H.: Predicting Stock Returns by Decision Tree Combining Neural Network. Intelligent Information and Database Systems, LNCP (2014)

12. Dixon, M.F, Klabjan, D., Bang, J.H.: Classification-based Financial Markets Prediction using Deep Neural Networks. Social Science Electronic Publishing (2016)

13. Akita, R., Yoshihara, A., Matsubara, T. et al.: Deep learning for stock prediction using numerical and textual information. In: IEEE/ACIS International Conference on Computer & Information Science IEEE (2016)

14. Day, M.Y., Lee, C.: Deep learning for financial sentiment analysis on finance news providers. In: 2016 IEEE/ACM International Conference on Advances in Social Networks Analysis and Mining (ASONAM) ACM (2016)

15. Sahar, S., Wang, D., Anna, P., et al.: Big Data: deep learning for financial sentiment analysis. J. Big Data **5**(1), 3 (2018)

16. Marion, M., Catherine, A., Richard, K., et al.: Time-series averaging using constrained dynamic time warping with tolerance. Pattern Recogn. **74**, 77–89 (2018)

17. Shao, L., Cai, Z., Liu, L., Lu, K.: Performance evaluation of deep feature learning for RGB-D image/video classification. Inf. Sci. **385–386**, 266–283 (2017)

Research on Teaching Method of EDA Simulation Design Based on Numerical Calculation

Yuanyuan Deng, Zhijun Gao, and Zhuhan Wang

Abstract Nowadays, the globally integrated circuit industry is developing rapidly. With the progress of information technology, an integrated circuit has become a new pillar industry in China. EDA technique is the development of modern electronic design, which must be grasped by a modern electronic engineer. More and more electronic information courses choose EDA simulation design as the course content. Numerical analysis theory and computational geometry are the most widely used theoretical and technical methods among various EDA technology tools. In this paper, the numerical analysis method is applied to the teaching research of EDA design method. By applying the numerical calculation teaching content into the EDA simulation design teaching content, students' ideas are broadened, their innovation ability will be improved, and China's talent building process of in the integrated circuit industry can be completed. It will also cultivate students' practical ability and train them to have a strong level of innovation and scientific research.

Keywords EDA simulation design · Numerical calculation · Teaching method

1 Introduction

Integrated circuit industry is the foundation of the information technology industry, and it is also a national strategic, fundamental, and leading industry. Nowadays, electronic design automation (EDA) simulation design method [1] is one of the most

Y. Deng (✉)
School of Science, Shenyang Jianzhu University, Shenyang 110168, China
e-mail: dyy_0124@163.com

Z. Gao
Faculty of Information and Control Engineering, Shenyang Jianzhu University, Shenyang 110168, China

Z. Wang
China Overseas Holdings Limited, Shenyang 110005, China

© The Author(s), under exclusive license to Springer Nature Singapore Pte Ltd. 2021   21
Y. Li et al. (eds.), *Advances in Simulation and Process Modelling*,
Advances in Intelligent Systems and Computing 1305,
https://doi.org/10.1007/978-981-33-4575-1_3

important cornerstones of the IC industry. The teaching content of information technology speciality in colleges and universities has strong theoretical and engineering practicality [2]. Its task is to provide the necessary basic theory and experimental skills for cultivating applied qualified students. Now the EDA technology has become an important means of digital electronic technology system design [3]. Almost all science and Engineering (especially electronic information) colleges and universities have set up EDA simulation design method courses. The teaching of technology, electronic technology, and numerical science calculation method should be better integrated to improve students' ability of critical thinking and expansion, which requires construction, reform, and practice in many aspects.

Due to the late start of China's integrated circuit industry, the independent innovation ability of integrated circuit design of enterprises is relatively weak. Traditional digital electronic technology experiments use small- and medium-sized integrated circuits to carry out experiments, but there are often some technical and technological problems in the operation process. In the teaching process of EDA technology, it can make students understand the related principle and working process of the circuit visually by simulating the experimental process and then letting the students carry out the experiment after the simulation verification. Through the deep integration of numerical calculation modeling technology in college teaching, students' independent thinking ability can be cultivated. In addition to that, the diversity of teaching methods can be improved and the independent innovation ability of China's integrated circuit design can be fundamentally improved enhanced. As a result, the development of China's integrated circuit industry can be accelerated. Strengthening the development of core technology provides strong support and guarantee.

2 The Reason for Integrating Numerical Calculation Method into EDA Teaching

The numerical calculation method is an instrumental, methodological, and marginal subject, and it has become an important means of modern scientific development with theoretical research. It focuses on the close combination of theory and practice. It has the characteristics of pure mathematical abstraction along with extensive application and highly technical practical experiment [4]. The mathematical model derived in the practical application cannot easily get an accurate solution. So we will ignore some simple factors and transform the complex nonlinear model into a linear model which can obtain an accurate solution, which makes the development of electronic technology play a greater role. Especially in the research of EDA development direction, numerical calculation method has played a great role.

The numerical calculation method is the basis of applied mathematics. Through the establishment of a mathematical model, it is closely related to mathematics. The mathematical calculation method is applied in various fields of science and technology in various forms. The numerical calculation method and its theory and software realization for solving various mathematical problems by computer are studied. Generally, we guide students to solve problems in three steps in teaching:

(1) Establish the mathematical model according to the actual problems;
(2) The numerical calculation method is given by the mathematical model; and
(3) Program the algorithm according to the calculation method.

Nowadays, EDA development covers a wide range of technologies, including circuit diagram simulation and debugging [5], design rule checking/layout versus schematic (DRC/LVS) physical verification, logic programming and simulation, the establishment of a basic unit library, logic synthesis and automatic layout and routing, timing analysis and optimization design, etc. They all involve the application of numerical analysis modeling method. In Fig. 1, we can see the EDA development technology involving numerical calculation. In the teaching content of EDA design technology, there is a lot of principle analysis, formula derivation, data calculation, and the content is relatively abstract and difficult to understand. At the same time, it also needs certain mathematical knowledge as the basis. But it is often because of the lack of knowledge of numerical calculation methods that students generally feel more difficult in building methods, and teachers and students often spend a lot of time and energy, but the result of students' learning is not very ideal. One of the most important reasons is that students do not have a good grasp and understanding of numerical calculation methods.

In order to solve the above problems in the teaching of electronic technology and to better apply the advanced EDA to the teaching of electronic technology at the same time, we need to start from many aspects, combining the numerical calculation method with EDA. The study of technology [6] infiltrates into all aspects of teaching. In the process of simulation design, the theoretical knowledge of numerical analysis modeling is introduced to optimize the design of the model continuously, so as

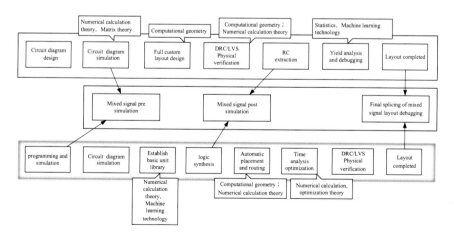

Fig. 1 EDA involved technologies

to cultivate students' independent thinking and independent innovation ideas. It is helpful for students to understand the EDA simulation design method thoroughly.

3 Teaching Method of Integrating Numerical Calculation into EDA Simulation Design

EDA simulation design method is a subject that has a wide range of disciplines. Taking analog/digital /hybrid IC design as an example, in the traditional teaching methods, circuit simulation is the most important link in circuit design. It can help us analyze the circuit performance [7], to adjust the circuit parameters according to the simulation results to meet the design requirements. But at present, simulation program with integrated circuit emphasis (SPICE) method is used for most simulation software of analog circuit, based on the functions and usages of various SPICE simulators that are similar. Therefore, with the increasing complexity of modern analog circuit devices and the increasingly advanced technology, the number of resistor capacitance (RC) devices encountered in post-simulation will increase exponentially. Under this condition, the teaching of numerical calculation method is discussed [8]. It is particularly important to bring in the content, because with the rapid increase of the scale of the matrix to be solved, the simulation time is significantly prolonged. As a result, in the SPICE simulation process, many advanced commercial SPICE software, such as Alps, appear. Usually, various methods are used to improve the running speed, and a variety of numerical calculation [9] acceleration technologies is integrated to improve the simulation performance. For example:

(1) Parallel computing technology
   The calculation of different submatrices is allocated to multiple CPUs or even integrated into multiple servers for parallel computing. The simulation performance is improved by comprehensively utilizing the computing power of multi-core and even distributed systems. However, this method often requires very high specification background backend hardware.

(2) Merge Technology
   Using simple circuit combination, linear correction, and other technologies to reduce circuit nodes and matrix size.

(3) RC reduction technology
   For the post-simulation circuit, RC device reduction is carried out under the premise of satisfying the accuracy, including frequency domain reduction and time domain reduction.

(4) Bypass technology
   When the matrix elements are not updated after a tran time step, it is not necessary to calculate the matrix repeatedly, but to modify the results by linear interpolation.

(5) Homomorphism Technology

Several modules with the same structure and similar voltage share the same Jacobian matrix of the submatrix, which is solved only once. Small differences between other modules are corrected by linear interpolation.

(6) Multiple rate

According to the working frequency, the circuit is divided into high-frequency module and low-frequency module. The high-frequency module has less time step, while the low-frequency module has a larger time step.

In the teaching process of EDA method, we found that SPICE simulation is the process of establishing and solving a series of partial differential equations according to the circuit netlist. These partial differential equations [10] are established by the device parameters and basic equations defined in the model. Through the above simulation process, a large number of partial differential equations can be established. In SPICE simulation, tran simulation(transient simulation) is the most time consuming and the most critical. The main purpose of simulation is to calculate all net voltage and node current at each time. When the nonlinear equations solved by students are too large, then in the teaching method, we introduce the method of numerical calculation to intervene in the tran simulation process. Generally, the steps of solving tran simulation are shown in Fig. 2.

(1) Because there are many parameters of partial differential equations, it can be regarded as a large matrix with high dimension. Then, the large matrix can be cut into several small matrices by using matrix calculation method.

(2) Although the size of each small matrix is reduced, the dimension is still very high, but the approximate solution cannot be obtained. We call this process convergence.

(3) The solutions of the small matrix are synthesized to generate the solution of the total large matrix at each time point.

In the teaching of circuit simulation, it only involves the process of circuit construction and analog design, while the concept of numerical calculation is the method and process of solving approximate solution of mathematical problems by digital computer, and the discipline composed of related theories. The generalized numerical calculation includes matrix numerical calculation, numerical calculus, solution of partial differential equations, optimization theory, computational geometry, calculation, etc. However, it is well known that the results obtained by numerical calculation are discrete, so there is a certain error with the accurate value, which is the main difference between the numerical calculation method and the traditional analytical method. In introducing EDA technology to students, we usually use Newton iteration method, interpolation method, and Rayleigh difference quotient method [11] to solve partial differential equations in SPICE simulation.

In solving the convergence of matrix, we often use Newton iteration method in SPICE simulation design teaching [12]. Generally, Newton iteration method is used to solve the differential equation considering one-dimensional matrix.

$F(x) = y$, solving the approximate solution x_k

the function $f(x)$ will unfold at x_k, we will get

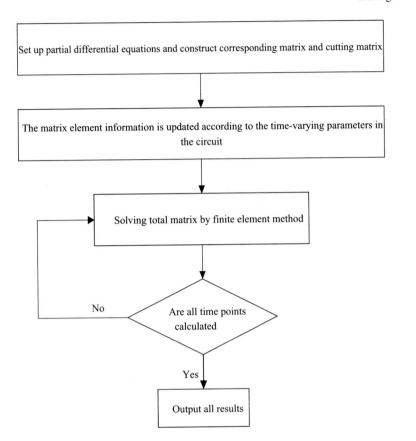

Fig. 2 Tran simulation steps

$$f(x) \approx f(x_k) + f'(x_k)(x - x_k) \tag{1}$$

then the approximate solution

$$x_{k+1} = x_k - \frac{f(x_k)}{f'(x_k)}, \quad k = 0, 1, \ldots, n \tag{2}$$

Finally, the excellent results are obtained x_{k+1}, $|f(x_{k+1})| < \text{tol}$, the above results are applied to SPICE residuals, the residual error of net voltage and node current in the circuit is expressed as:

$$V_{\text{tol}} = \text{absvtol} + \text{reltol} * V_{\text{max}} \tag{3}$$

$$I_{tol} = absitol + reltol * I_{max} \qquad (4)$$

As mentioned above, absvtol, absitol, and reltol are set outside the program (the accuracy of SPICE simulation will correspond to different setting values), while V_{max} and I_{max} refer to the maximum net voltage and the maximum node current of all nets in the circuit at each time.

In the correction of numerical calculation, the approximate value obtained in each calculation may be different. For example, in 32B system, the maximum bit width that a double byte DWORD variable can quantify is 32B. When it is used to represent a continuous physical variable, it will cause quantization error at least less than the absolute value represented by the last one bit, which is in a large number of complex iteration process when the error is amplified, it will be infinitely enlarged, and the final result will be affected. One of the most typical examples is that after the error is amplified, the parameter will fluctuate continuously with time.

The process of oscillation may affect the accuracy of circuit simulation design, because when the test circuit has external oscillation excitation, this small numerical oscillation phenomenon will be amplified by the circuit phenomenon, resulting in more obvious oscillation results. But through the tran simulation under the optimized numerical calculation method, we find that after the introduction of the Newton iteration method, we get better results in the output voltage and current.

4 Conclusion

In the new generation of integrated circuit design, students are required to have a stronger ability of multidisciplinary, comprehensive application, and practical innovation. From the introduction of this article, we can understand that the learning of numerical calculation methods provides important ideas and methods for the construction of EDA design methods and plays an important role in students' use and in-depth understanding of EDA design methods. It has become the core of modern electronic design. The current situation has put forward deeper and higher requirements for the teaching and scientific research of electronic courses in colleges and universities. Then, the teaching content and emphasis of traditional electronic technology should be changed accordingly. Development is the future direction of teaching development, but also the focus and breakthrough of electronic teaching. Therefore, it is necessary to continuously update the teaching mode and method in the teaching content. Integrate more advanced and cutting-edge knowledge and technology into the teaching practice. Cultivate students' self-learning ability, practical ability, and innovative ability, so as to make the trained students to have stronger social competitiveness, truly achieve the social needs and use for the society.

Acknowledgements The authors also wish to thank Mr. Han Yu and Zhengyan Chen, the research staff of Empyrean Software, Beijing, China, for their valuable suggestions and encouragement to this study.

The research work was financially supported by the project of Industry University Cooperation of the Ministry of Education, China, with project No. 201902024016.

References

1. Wang, L., Shen, L.: The application of EDA technology in education reform of system of hardware courses for electric information. Specialties. In: International Conference on Future Information Technology and Management Engineering, pp. 268–271, Guangzhou (2010)
2. Qi, H.: Application of EDA technology on professional teaching for electronic information engineering. In: 2010 International Conference on E-Health Networking, pp. 203–206, Shenzhen (2010)
3. Zhao, H., Jing, Y.: Application of EDA technology in designing of digital system. In: Proceedings of 2015 International Industrial Informatics and Computer Engineering Conference (IIICEC 2015), pp. 1981–1984, Shanxi (2015)
4. Li, Q., Wang, N., Yi, D.: Numerical Analysis, 5th edn. Tsinghua University Press, Beijing (2008). (in Chinese)
5. Chen, Q.: EDA Technology and Application, 1st edn. Machinery Industry Press, Beijing (2013). (in Chinese)
6. Jiang, D.: Integrate Teaching of Electronic Information Courses Based on EDA. High. Educ. Forum 2(2), 30–32 (2009)
7. Sargentm, A.: Neuroergonomic assessment of hot beverage preparation and consumption: an EEG and EDA study. Front. Hum. Neurosci., p. 175 (2020)
8. Li, J.: A study on the multi-curriculum integrated teaching mode based on the EDA technology. J. Chongqing Electr. Power Coll. 25(1), 37–39 (2020)
9. Jiang, H.: FPGA design flow based on multi-EDA tools. Microcomput. Inf. 23(11–2), 201–203 (2007)
10. Gu, S.: Exploration of teaching method of EDA course innovative experiment. Exp. Technol. Manag. 32(3), 40–46 (2015)
11. Zhang, X., Wang, S.: Reforming experimental teaching all-roundly to cultivate students' innovative abilities courses. Res. Explor. Lab. 24(1), 4–6 (2005)
12. Ren, A., Sun, W., Shi, G.: Reform and practice of joint teaching mode for EDA experiment and digital circuit. Exp. Technol. Manag. 26(4), 200–208 (2009). (in Chinese)

Sub-pixel Edge Contour Detection Algorithm Based on Cubic B-Spline Interpolation

Jianzhao Cao, Ruwei Ma, Renning Pang, and Yuanwei Qi

Abstract It was proposed, in this paper, an algorithm based on machine vision to deal with the problems of low efficiency and large error in profile detection of industrial steel plates. The image filtering operation is used to remove the noise of the steel plate picture, and a method for judging the image filtering effect is proposed. And an image segmentation method combining OTSU and Canny algorithm is designed to achieve dynamic segmentation of steel plate images to obtain the best segmentation effect. In order to fit steel plates with head and tail deformation or camber, a sub-pixel edge fitting method based on cubic B-spline interpolation was proposed to obtain the sub-pixel coordinates of steel plate contour, which provides data basis for crop shearing. The experimental results show that this method has high detection speed and precision and can meet the actual production needs.

Keywords B-spline · Sub-pixel edge · Profile detection · Crop shearing

1 Introduction

In the modern steel plate industry, due to the production process restrictions and order requirements, steel plates are often cut. Steel plate cutting is an important process for the production of steel plates of contract dimensions and is also the main means to improve the success rate of steel plates [1]. Therefore, how to improve cutting efficiency and reduce the loss of the cutting process has become an urgent problem to be solved in increasing the production capacity of steel plates. In order to solve this problem, on the one hand, it is necessary to improve the shear accuracy of the steel plate, and on the other hand, it is necessary to establish an efficient shear model for precise guidance.

The accuracy of steel plate cutting is highly dependent on the precision of contour recognition. At present, most domestic factories use manual detection or contact steel plate contour detection [1, 2]. These two methods have low detection accuracy, low

J. Cao · R. Ma (✉) · R. Pang · Y. Qi
Faculty of Information and Control Engineering, Shenyang Jianzhu University, Shenyang, China
e-mail: 158564259@qq.com

© The Author(s), under exclusive license to Springer Nature Singapore Pte Ltd. 2021
Y. Li et al. (eds.), *Advances in Simulation and Process Modelling*,
Advances in Intelligent Systems and Computing 1305,
https://doi.org/10.1007/978-981-33-4575-1_4

efficiency, and high cost. In recent years, the contour detection algorithm based on machine vision has been widely used in industrial production. Its non-contact, low cost, and high degree of automation make it have great development potential in the steel industry. At present, machine vision inspection has been widely used in surface quality detection [3], material tracking [4, 5], billet recognition [6], and other fields.

The common contour detection algorithms use first-order and second-order differential operators for edge positioning [7]. The first-order differential operator is simple to calculate, but the anti-noise ability is weak, and it is easy to detect the false edge. The second-order differential operator emphasizes the abrupt change of image gray rather than the region with slow change of gray and has stronger edge positioning ability. The positioning accuracy of the above operators is all at the pixel level. In order to meet the requirements of high precision, we need to further subdivide the detected edge, so the image sub-pixel detection is introduced. There are three existing sub-pixel edge detection algorithms: moment method, interpolation method, and fitting method [8, 9]. The interpolation method has high computational efficiency. Although it is easily affected by noise, the image processing operation before detection can reduce noise well. Qu et al. [10] proposed an edge detection method combining Zernike moment operator and Sobel operator to extract sub-pixel coordinates. Breder et al. [11] proposed to use global B-spline model to estimate sub-pixel edges and remove Gaussian noise very well. Jin et al. [12] used b-spline curve three times to detect the lung tumor and fitted the tumor edge well. Chen et al. [13] solved the UAV path planning problem by using the curve B-spline curve.

Based on the above analysis, a sub-pixel edge contour detection algorithm based on cubic B-spline interpolation is proposed to obtain the sub-pixel edge position of steel plate.

2   Steel Plate Image Preprocessing

Due to the harsh environment in the steel production workshop, poor camera shooting conditions, on-site moisture, dust and CMOS imaging equipment, etc., will cause interference to image. If the contour calculation is directly performed, these interferences will also participate in the extraction of the required features. Resulting in excessive calculation errors, or inability to extract relevant features, so the image should be preprocessed.

2.1   Image Filtering

Image filtering can remove noise very well, and common filtering algorithms can be divided into spatial filtering and frequency domain filtering [14]. Spatial filtering is to directly process image data on pixel coordinates, while frequency domain filtering is to first transform images from spatial domain to frequency domain and then transform into the spatial domain after processing. Considering the execution efficiency and required resources, the spatial domain filtering algorithm is selected in this paper.

This paper uses three spatial filtering methods: mean filtering, median filtering, and Gaussian filtering to process the original image. In order to better compare the quality of the three filtering methods, choose the filtering algorithm that is more suitable for this paper. In this paper, a simple image filtering evaluation standard is established; that is, a horizontal line horizontal line is drawn at the same position of the steel plate, the pixel of the horizontal line is the x-axis, and the corresponding second-order grayscale derivative is the y-axis to establish the image grayscale derivative coordinate system. The comparison effect of the three filtering methods is shown in Fig. 1.

The part of the original image with large grayscale changes is likely to be the edge part of the steel plate image, so it should be retained. In the middle is the steel plate itself, and the fluctuation can be regarded as noise, so it should be eliminated. It can

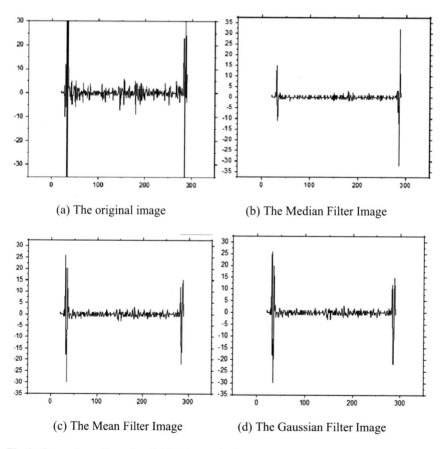

(a) The original image (b) The Median Filter Image

(c) The Mean Filter Image (d) The Gaussian Filter Image

Fig. 1 Comparison effect of spatial filtering method

be seen from the comparison that the effect of median filtering is relatively good, so this paper chooses median filtering to filter out noise.

2.2 Edge Algorithm Based on Canny Operator and Threshold Segmentation

The image threshold segmentation method [15] is to classify each pixel in the image according to the difference in grayscale characteristics between the target in the image and the surrounding background by selecting an appropriate threshold, thereby separating the target from the background area separate, extract, and convert the image from a complex grayscale image to a simple binary black and white image.

Due to the complexity of the actual production environment and the automatic requirement of steel plate processing, a single threshold value cannot be used for threshold segmentation. Instead, a dynamic threshold value should be used to enhance the adaptability of the algorithm to different environments. This paper chooses OSTU [16], which is $\sigma^2(T)$ binarization algorithm based on global adaptive threshold. According to the grayscale characteristics of the image, the algorithm proposes that when the segmentation effect is the best, the contrast between the foreground region and the background region should be the largest, and the standard used in the algorithm to measure the contrast between the two regions is $\sigma^2(T)$. The greater $\sigma^2(T)$ between the two regions, the greater the difference between the foreground region and the background region of the image calculated by this threshold. The calculation of $\sigma^2(T)$ is as follows:

$$\sigma^2(T) = \omega_A(\mu_A - \mu)^2 + \omega_B(\mu_B - \mu)^2 \tag{1}$$

where μ_A is the gray mean of the foreground image; ω_A the ratio of the number of pixels in the region to the total number of pixels in the image. μ_B the grayscale mean of the background image, and its ratio is ω_B. Calculate different $\sigma^2(T)$ by constantly adjusting the value of T.

In order to verify the image segmentation effect of OSTU, the evaluation method established in this paper is similar to the image filtering algorithm. A horizontal line is drawn at the same position of the steel plate with different brightness. And the pixels on the horizontal line are taken as the X-axis, and the corresponding gray value is taken as the Y-axis to establish the image gray-level coordinate system. The OSTU processing effect of different image brightness is shown in Fig. 2.

It can be seen from Fig. 2 that OSTU algorithm has better treatment effect on steel plates with different surface brightness.

The edge detection of the image is based on the discontinuity of image gray level change and quantifies the gray change rate of each pixel field based on the criterion that the first derivative of the image at the edge is the extreme value or the zero crossing of the second derivative. Its purpose is to find the point of grayscale mutation in the

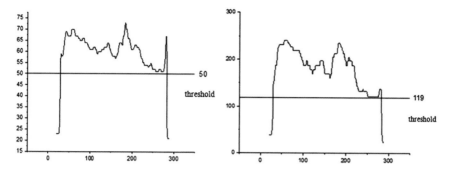

Fig. 2 OSTU algorithm processing results of different image

image. The Canny edge detection operator is still widely used because of its excellent detection effect and the optimal performance of step edge-shaped affected by white noise [17]. Traditional Canny requires manual input of high and low threshold. The method of determining high and low threshold in Ref. [18] in this paper takes the global threshold T calculated by OSTU algorithm as the high threshold in the double threshold of Canny operator, and half of T as the low threshold. The effect diagram of steel plate edge detection is shown in Fig. 3.

It can be seen from Fig. 3 that the edge detection algorithm based on Canny operator and threshold segmentation has a better effect on the contour detection of steel plate. The accuracy of Canny edge detection is pixel level. In order to meet the increasing requirements of precision in industry, it is necessary to adopt sub-pixel-level edge detection algorithms.

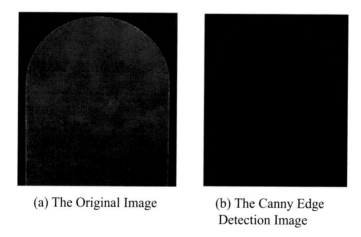

(a) The Original Image   (b) The Canny Edge
                                 Detection Image

Fig. 3 Renderings of Canny edge detection

3 Sub-pixel Edge Contour Detection Algorithm Based on Cubic B-Spline Interpolation

Due to the process, the steel plate produced may have head and tail deformation and sickle bending deformation. For deformed steel plates, straight lines cannot be used to directly fit the contour, and a more flexible third-order curve is needed to better fit the edges.

3.1 B-Spline Curve

The core idea of cubic B-spline is to replace higher-order polynomials with piecewise lower-order polynomials by continuous joining. A B-spline interpolation curve of order N is as follows:

$$P(u) = \sum_{i=0}^{n} p_i N_{i,k}(u) \tag{2}$$

where p_i is the coordinate of the control point, and the edge pixel obtained by Canny edge detection is used here. $N_{i,k}(u)$ is the basis function, i is the number of nodes, and k is the degree of the basis function.

The third-order B-spline is sufficient to describe the full shape of the steel plate and to fit the edge contour well, taking into account the complexity of the steel plate bending shape in the field environment. Each curve of a cubic B-spline requires four control points, and each curve is a Bessel curve. The basis function expression is as follows:

$$N_{0,3}(u) = \frac{1}{6}(1-u)^3$$

$$N_{1,3}(u) = \frac{1}{6}(3u^3 - 6u^2 + 4)$$

$$N_{2,3}(u) = \frac{1}{6}(-3u^3 + 3u^2 + 3u + 1)$$

$$N_{3,3}(u) = \frac{1}{6}u^3 \tag{3}$$

Substituting (3) into (2) shows that the expression of cubic B-spline is as follows:

$$P(u) = \sum_{i=0}^{n} p_i N_{i,3}(u) = p_i N_{0,3}(u) + p_{i+1} N_{1,3}(u) + p_{i+2} N_{2,3}(u) + p_{i+3} N_{3,3}(u),$$

$$i = 0, 1, 2, \ldots, n \tag{4}$$

3.2 Edge Detection Procedure and Experimental Results

The detailed steps of the Sub-pixel edge contour detection algorithm based on cubic B-spline interpolation are as follows:

(1) The gray value of pixel coordinates of the steel plate edge was detected by Canny edge. The gray value corresponding to some edge pixel points is shown in Table 1.

(2) The cubic B-spline interpolation algorithm was used to interpolate the data to obtain the continuous edge gray distribution.

(3) Each piecewise interpolation function is calculated according to cubic B-spline interpolation algorithm formula. The interpolation curve is shown in Fig. 4.

(4) Take the second derivative of the function, find the second derivative of both sides of the edge point, and get the sub-pixel coordinate value chart, as shown in Table 2. According to the coordinate value, the entire steel plate profile has been determined.

Table 1 Edge gray distribution

Abscissa of pixel	X_i	296	259	233	44	60	88
ordinate of pixel	Y_i	282	283	312	70	56	39
Gray value	Z_i	24	37	54	29	35	61

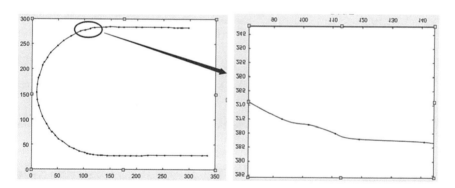

Fig. 4 Cubic B-spline interpolation

Table 2 Sub-pixel coordinates

Sub-pixel abscess	X_i	296.002	259.994	233.977	44.5457	60.0846	88.8285
Sub-pixel ordinate	Y_i	282.858	283.195	29.3844	70.5348	56.1174	39.4053

4 Conclusions

This paper proposes a sub-pixel edge contour detection algorithm based on cubic B-spline to detect steel plate contour. There are two innovations in this paper: First, a method for judging the effect of image filtering is proposed. The filtering effect is judged by drawing a horizontal line on the same position of the steel plate and selecting parameters to establish an image grayscale derivative coordinate system. Secondly, the contour fitting of the steel plate that may be deformed is carried out through the cubic B-spline to better fit the contour and calculate the corresponding sub-pixel coordinates to provide data basis for the shearing of the deformed part of the steel plate. This algorithm can well meet the actual environment's requirements for efficiency and accuracy and is of great significance for achieving rapid and high-precision steel plate contour detection.

Acknowledgements This research work was financially supported by the Youth Science Foundation of Liaoning Province, China.

References

1. Sun, W.Q., Yang, Q., Shao, J., He A.R., Li. M.X.: Research on high precision profile control technique of silicon steel for UCM tandem cold rolling mill. In: 2010 International Conference on Mechanic Automation and Control Engineering, pp. 5926–5929, Wuhan, China (2010)
2. Zhou, H., Zhang H., Yang, C.: Hybrid-model-based intelligent optimization of ironmaking process. IEEE Trans. Ind. Electroni. **67**(3), 2469–2479 (2020)
3. Martinez P., Ahmad R., Al-Hussein M.: A vision-based system for pre-inspection of steel frame manufacturing. Autom. Constr. **97**, 151–163 (2019)
4. Carruthers-Watt, B., Xue, N.Y., Morris, A.J.: A vision based system for strip tracking measurement in the finishing train of a hot strip mill. In: 2010 IEEE International Conference on Mechatronics and Automation, pp. 1115–1120. Xi'an, China (2010)
5. Schausberger, F., Steinboeck, A., Kugi, A., Jochum, M., Wild, D., Kiefer, T.: Vision-based material tracking in heavy-plate rolling. IFAC-PapersOnLine **49**(20),108–113 (2016)
6. Lee, S.J., Kim, S.W., Kwon, W., Koo, G., Yun, J.P.: Selective distillation of weakly annotated GTD for vision-based slab identification system. IEEE Access **7**, 23177–23186 (2019)
7. Li, Z., Zhang, J., Zhuang, T., Wang, Q.: Metal surface defect detection based on MATLAB. In: 2018 IEEE 3rd Advanced Information Technology, Electronic and Automation Control Conference (IAEAC), pp. 2365–2371, Chongqing, China (2018)
8. Yu, W., Liu, C., Yang, H., Wang, G.: A method for improving the detection accuracy of subpixel edge. In: 2019 Chinese Automation Congress (CAC), pp. 158–162, Hangzhou, China (2019)
9. Yu, T., Ni, J., Zhang, K.: Subpixel edge detection of inner hole for ceramic optical fiber ferrules. In: 2016 15th International Conference on Optical Communications and Networks (ICOCN), pp. 1–3, Hangzhou, China (2016)
10. Qu, Y.D., Cui, C., Chen, S.: A fast subpixel edge detection method using Sobel – Zernike moments operator. Image Vis. Comput. **23**(1), 11–17 (2005)
11. Breder, R.L.B., Estrela, V.V., de Assis, J.T.: Sub-pixel accuracy edge fitting by means of B-spline. 2009 IEEE International Workshop on Multimedia Signal Processing, pp. 1–5. Rio De Janeiro, Brazil (2009)

12. Jin, R., Liu, Y., Chen, M., et al.: Contour propagation for lung tumor delineation in 4D-CT using tensor-product surface of uniform and non-uniform closed cubic B-splines. Phys. Med. Biol. **63**(1), 015–017 (2017)
13. Chen, Y., Mei, Y., Yu, J., Su, X., Xu, N.: Three-dimensional unmanned aerial vehicle path planning using modified wolf pack search algorithm. Neurocomputing **266**, 445–457 (2017)
14. Siddique, M.A.B., Arif, R.B., Khan, M.M.R.: Digital image segmentation in Matlab: a brief study on OTSU's image thresholding. In: 2018 International Conference on Innovation in Engineering and Technology (ICIET), pp. 1–5. Dhaka, Bangladesh (2018)
15. Zhang, Y., Xia, Z.: Research on the image segmentation based on improved threshold extractions. In: 2018 IEEE 3rd International Conference on Cloud Computing and Internet of Things (CCIOT), pp. 386–389. Dalian, China (2018)
16. Lu, H., Yan, J.: Window frame obstacle edge detection based on improved Canny operator. In: 2019 3rd International Conference on Electronic Information Technology and Computer Engineering (EITCE), pp. 493–496. Xiamen, China (2019)
17. Ma, Q.D.Y., Ma, Z., Ji, C., Yin, K., Zhu, T., Bian, C.: Artificial object edge detection based on enhanced Canny algorithm for high-speed railway apparatus identification. In: 2017 10th International Congress on Image and Signal Processing, BioMedical Engineering and Informatics (CISP-BMEI), pp. 1–6. Shanghai, China (2017)
18. Wan, N., Xu, D., Ye, H.: Improved cubic B-spline curve method for path optimization of manipulator obstacle avoidance. In: 2018 Chinese Automation Congress (CAC), pp. 1471–1476. Xi'an, China (2018)

Spectrum Sensing Algorithm Based on Random Forest in Dynamic Fading Channel

Zhijun Gao and Xin Wang

Abstract Efficient utilization and energy saving of spectrum is a crucial technology in 5G, and spectrum sensing is an important foundation and core of efficient utilization of spectrum resources. At present, good results have been achieved for spectrum sensing in Gaussian channels, but in dynamic fading channels, due to fading and multi-path transmission factors, the spectrum sensing performance is poor. To solve this problem, this paper first proposes a perceptive system model under dynamic fading channel, and on the basis of this model, a spectral sensing algorithm based on the random forest is proposed. The algorithm extracts the fading amplitude gain, the energy value of the received signal, and the characteristic parameters of the signal cycle spectrum in the scene of the dynamic fading channel as the sample parameters to construct the random forest and then senses and classifies the constructed random forest signals to determine the state of the channel occupancy.

Keywords 5G · Spectrum sensing · Dynamic fading channel · Random forest

1 Introduction

5G is a new generation of the mobile communication system for the development needs after 2020, which has received extensive attention and research from enterprises, research institutes, and universities around the world. Considering that the current shortage of spectrum resources makes the development of wireless communication seriously restricted, how to efficiently use the limited spectrum resources is an urgent problem to be solved in 5G research [1]. Spectrum sensing provides an important basis for solving this problem and is a new technology that is expected to alleviate the problem of spectrum resource depletion and low utilization. Currently, spectrum sensing technology can be divided into three methods: Energy Detection (ED), Matched Filter Detection (MFD), and Cyclostationary Detection (CD) [2]. The energy detection method is simple and easy to operate, and the computational

Z. Gao · X. Wang (✉)
Faculty of Information and Control Engineering, Shenyang Jianzhu University, Shenyang, China
e-mail: wangx7988@sjzu.edu.cn

© The Author(s), under exclusive license to Springer Nature Singapore Pte Ltd. 2021   39
Y. Li et al. (eds.), *Advances in Simulation and Process Modelling*,
Advances in Intelligent Systems and Computing 1305,
https://doi.org/10.1007/978-981-33-4575-1_5

complexity is low, so it has been applied in many kinds of research. Unfortunately, the performance of energy detection is poor under low SNR [3]. Cyclic stationary feature detection can distinguish the non-coherent stationary noise signal energy from the received signal, and it has good robustness and noise resistance to the detection of unknown noise variables, so it can be applied to spectrum sensing problems in various noise environments. However, this algorithm has a high computational complexity [4]. The matched filter detection method can achieve higher processing gain and shorter detection time. However, this method needs to obtain the prior information of the primary user signal in advance. If the information is not accurate enough, then the performance of matching filter detection will be greatly affected. Besides, there must be an independent matching filter for each primary user receiver [5].

In recent years, many researchers have studied spectral sensing methods and achieved good results. In [6], Ahmed et al. proposed a GUESS algorithm with low complexity and communication overhead to realize efficient collaboration between sub-users and reduce the impact of network changes on perceived results. In [7], Ganesan et al. proposed a distributed collaboration spectrum sensing strategy based on the distance between secondary users and primary users and matched each secondary user based on the distance between secondary users and primary users, thus improving the overall perception performance. With the development of machine learning, scholars have begun to use machine learning methods to solve the problem of spectrum perception. In [8], a feature-based automatic modulation type classification algorithm is proposed by using the high-order cumulant as the recognition feature. These results improve the spectral sensing performance of Gaussian channels with high SNR to some extent. But in dynamic fading channels, the spectral sensing performance is poor due to fading and multi-path transmission factors.

In order to solve the above problems, this paper proposes a spectrum sensing algorithm based on random forest in dynamic fading channel. The contributions of this paper are as follows:

- A sensing system model in dynamic fading channel is proposed. For any channel, whether the channel state is idle or not can be reduced to a binary hypothesis testing model;
- In the dynamic fading channel, the fading amplitude gain a_k, the received signal energy value y_n, the mathematical expectation, and variance of the signal cyclic spectrum characteristics are extracted as the sample parameters to construct the random forest; and
- The constructed random forest signals are perceived and classified to determine the channel occupancy state.

The remainder of the paper is organized as follows. In Sect. 2, system model is described. The proposed algorithm is investigated in Sect. 3 and is well validated with simulation in Sect. 4. The concluding remarks are made in Sect. 5.

2 System Model

2.1 Primary User Status Model

For secondary users, their perception of whether the channel is occupied, that is, whether there are primary users in the channel, can be summarized as a binary hypothesis testing model.

$$\begin{cases} H_0 : y(t) = n(t) \\ H_1 : y(t) = s(t) + n(t) \end{cases} \tag{1}$$

Under this model, the main user energy and can be expressed by the following formula.

$$E_y = \begin{cases} \sum_{i=1}^{M} n_i^2, & H_0 \\ \sum_{i=1}^{M} (a_i x_i + n_i)^2, & H_1 \end{cases} \tag{2}$$

where M represents the number of sampling points in T_s; channel noise n is the Additive White Gaussian Noise (AWGN), which means that value is 0, and variance is σ^2. H_0 and H_1 correspond to the presence and absence of primary user signals in the detected frequency band, respectively.

The main goal of spectrum sensing is to judge hypothesis testing by receiving signals and finally determine whether there is a primary user signal in the current detection cycle. The traditional ED spectrum sensing method sets the threshold according to certain criteria and compares the received signal energy with the threshold value to obtain the decision result. However, in time-varying fading channels, the variation of channel gain with time will undoubtedly greatly increase the difficulty of threshold determination, thus significantly reducing the spectrum sensing performance in practical applications. In view of this, this paper not only takes the received signal energy as the sensing parameter, but also introduces other sensing parameters to obtain more accurate results.

2.2 Dynamic Fading Channel Awareness System Model

In this paper, it is assumed that time-varying fading channels are subject to Rayleigh time-slowing fading characteristics; that is, the probability distribution of random channel gain is Rayleigh distribution.

$$f(a) = \frac{a}{\sigma_R^2} \exp\left(-\frac{a}{2\sigma_R^2}\right), a \in [0, \infty) \tag{3}$$

where a represents channel gain, and σ^2 represents variance of Rayleigh distribution. The channel gain is divided into K discrete states $(K \geq 3)$. If $[v_K, v_{k+1})$ is used to represent the boundary value of the K-th discrete state channel gain, then the channel gain corresponding to that state can be defined as follows.

$$a_k = \frac{\int_{v_k}^{v_{k+1}} af(a)da}{\int_{v_k}^{v_{k+1}} f(a)da}, \quad k = 0, \ldots, K - 1 \tag{4}$$

According to the main user's working state model and the main user's energy and expression, the dynamic fading channel awareness system model proposed can be expressed as

$$\begin{cases} S_i = f(s_i) \\ a_k = h(a), & n = 0, \ldots, N - 1 \\ y_n = g(a, s_i, n_i) \end{cases} \tag{5}$$

where S_i represents the state of the primary user at time i, and the migration is carried out according to the specific state function $F(\cdot)$. s_i represents the signal transmitted by the primary user at time i, when the primary user signal does not exist, that is, when the authorized frequency band is free, $S_i = 0$; when there is a primary user signal, $S_i = 1$. a_k represents the amplitude gain of fading channel at time K, which is updated according to the specific state transfer function $H(\cdot)$. For cognitive users, their perceived signal y_n is the energy sum of the sampled signal in a specific observation time window, as shown in Eq. (2). When there is no primary user signal, the perception signal y_n obeys the center chi-square distribution with M degree of freedom, $y_n \sim \chi_M^2$; when there is the perception signal y_n obeys the non-center chi-square distribution with M degree of freedom, $y_n \sim \chi_M^2(K)$, its non-center parameter $K = M(a_k x_k)^2$.

3   The Proposed Algorithm

Based on the perceptive system model of dynamic fading channel in Sect. 2.2, a spectrum sensing algorithm based on random forest is proposed. The algorithm extracts fading amplitude gain, energy value of received signal, and characteristic parameters of signal cycle spectrum in the scene of dynamic fading channel as sample parameters for constructing random forest and perceiving and classifying the constructed random forest signals.

If the received signal has multiple cycle frequencies, then the cycle spectrum with the maximum energy is taken as $S(k)$. $S(k)$ obeys the Gaussian distribution.

$$\begin{cases} H_0 : S(k) \sim N\left(0, \sigma_0^2\right) \\ H_1 : S(k) \sim N\left(\mu, \sigma_0^2 + \sigma_s^2 + \sigma_{sn}^2\right) \end{cases} \tag{6}$$

Cyclic spectrum characteristics can be characterized by mean and variance of Gaussian distribution; that is, when there is no main user signal, the mean value of characteristic parameters corresponding to $S(k)$ is 0, and the variance is σ_0^2. When there is a primary user, the mean value of characteristic parameters corresponding to $S(k)$ is μ, and the variance is $\sigma_0^2 + \sigma_s^2 + \sigma_{sn}^2$.

According to the above system model and the analysis of cyclic spectrum parameters, the amplitude gain of fading channel a_k, the energy of perceptive signal y_n, and the mean and variance of Gaussian distribution that can reflect the characteristics of cyclic spectrum are obtained as sample parameters. When the primary user does not exist, the corresponding eigenvector is $x_0 = (a_{k0}, y_{n0}, 0, \sigma_0^2)$. Otherwise, the corresponding eigenvector is $x_1 = (a_{k1}, y_{n1}, \mu, \sigma_0^2 + \sigma_s^2 + \sigma_{sn}^2)$.

3.1 Random Forest Algorithm

Samples are generated according to the eigenvectors of the presence and non-presence of the main user, and the sample set G is formed. Using the samples in G, the random forest is constructed, and the details of the random forest algorithm are described as following.

Random forest by Leo Breiman (2001) proposed a classification algorithm, which uses bootstrap resampling technology to repeatedly randomly extract N samples from the original training sample set N to generate a new training sample set training decision tree, and then generate M decision trees to form a random forest according to the above steps. The classification ability of a single tree may be very small, but after a large number of decision trees are randomly generated, a test sample can select the most likely classification through the classification results of each tree. The process of random forest is as follows:

1) N samples were selected from random samples;

2) K features are randomly selected from all the features, and a decision tree is built based on these features for the selected samples;

3) Repeat the above two steps m times to generate m decision trees and form a random forest;

4) For new data, after each tree decision, the final vote to confirm which category.

3.2 Dynamic Fading Channel Sensing Algorithm Based on Random Forest

According to the construction process of the random forest in algorithm 1, the constructed random forest is used to realize the perception of the occupying state of the primary user signal to the channel.

Dynamic fading channel sensing algorithm based on random forest called DFCS-RF is proposed in this section

The fading amplitude gain, energy value of received signal and characteristic parameters of signal cycle spectrum are extracted in the H_0 and H_1 respectively. According to these parameters, positive and negative samples are generated, in which positive samples correspond to the channel state in H_1 and negative samples correspond to the channel state in H_0. Then, according to the process of the random forest, the random forest is constructed and used to classify and detect the samples.

4   Simulation Results

In this section, simulation results of detection performance of spectrum sensing method in Rayleigh fading channel are given. We first conduct simulations to compare the performance of detection probability (P_d) under our DFCS-RF algorithm with two classic algorithms, sensing algorithm based on Support Vector Machine (SVM) and Artificial Neural Network (ANN). BPSK and OFDM are used as primary user signals. ANN adopts BP neural network with two input nodes, ten hidden layer nodes, and two output nodes. SVM classifier adopts cross-validation method, $C = 50, \sigma = 0.875$. The number of random forest decision trees is $K = 100$.

It is shown, in Fig. 1, that the comparison of detection rates of the proposed algorithm, ANN algorithm and SVM algorithm for BPSK signal under different SNR environments. As can be seen from the figure, when the SNR is -15 dB, the detection rate of the proposed DFCS-RF algorithm is 0.87, and that of the ANN and SVM algorithms is 0.58 and 0.69, respectively. The detection rate of DFCS-RF algorithm is 29% higher than that of ANN algorithm and 18% higher than that of SVM algorithm. When SNR is -20 dB, the detection rate of DFCS-RF algorithm, ANN algorithm, and SVM algorithm is 0.73, 0.31 and 0.43, respectively. Compared with ANN algorithm and SVM algorithm, the detection rate of DFCS-RF algorithm is improved by 42% and 30%, respectively.

Figure 2 shows that when OFDM signals appear, the detection rate of each algorithm is analyzed, and the detection rate of the proposed DFCS-RF algorithm decreases from 0.96 to 0.73 with the decrease of SNR. The detection rate of ANN algorithm decreased from 0.85 to 0.36. The detection rate of SVM algorithm decreased from 0.89 to 0.50. The detection rate of the proposed algorithm is significantly higher than that of the comparison algorithm.

The above results show that the proposed DFCS—RF algorithm has good detection performance, the algorithm can effectively extract the characterization of primary user signal characteristics of four parameters as sample parameter, and by using the random forest factors such as noise has the very good tolerance, not easily seen fitting phenomenon, etc., and thus effectively to achieve the dynamic spectrum sensing under the fading channel.

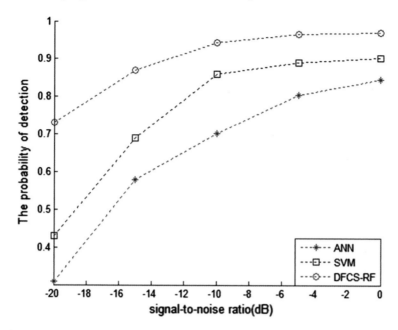

Fig. 1 Probability of detection of the proposed algorithm versus ANN, SVM, and DFCS-RF algorithms for BPSK

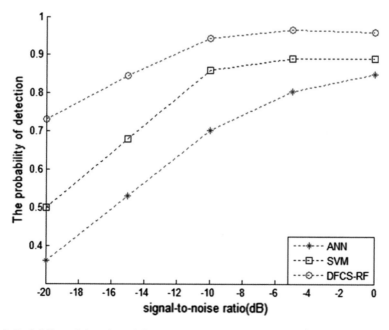

Fig. 2 Probability of detection of the proposed algorithm versus ANN, SVM, and DFCS-RF algorithms for OFDM

5 Conclusions

Spectrum energy saving is a crucial issue in 5G development. Based on spectrum energy saving, this paper proposes a random forest-based spectrum sensing algorithm for spectrum sensing under the scenario of dynamic fading channel. Based on the sensing system model of the dynamic fading channel, fading amplitude gain, energy value of received signals, and characteristic parameters of signal cycle spectrum in dynamic fading channel scenes are extracted as sample parameters for constructing random forest in the algorithm and perceived and classified by the constructed random forest. Simulation results show that the proposed DFCS-RF algorithm in the dynamic fading channel has good perceptive performance.

Acknowledgements This work was supported by the National key R and D plan tasks under contract 2018YFF0300304-04, National Natural Science Foundation of China under contract 61903357, Scientific research project of Liaoning Provincial Department of Education under contract LNJC201912, Liaoning Provincial Natural Science Foundation of China under contract 2020-MS-032 and China Postdoctoral Science Foundation under contract 2020M672600.

References

1. Gupta, A., Jha, R.K.: A survey of 5G network: architecture and emerging technologies. IEEE Access **3**, 1206–1232 (2015)
2. Choo, K.R., Gritzalis, S., Park, J.H.: Cryptographic solutions for industrial Internet-of-Things: research challenges and opportunities. IEEE Trans. Ind. Inform **14**(8), 3567–3569 (2018)
3. Liu, X., Jia, M., Zhang, X., Lu, W.: A novel multi-channel Internet of Things based on dynamic spectrum sharing in 5G communication. IEEE Internet Things J. **6**(4), 5962–5970 (2019)
4. Sahoo, P.K., Mohapatra, S., Sheu, J.: Dynamic spectrum allocation algorithms for industrial cognitive radio networks. IEEE Trans. Ind. Inform **14**(7), 3031–3043 (2018)
5. Zhang, K., Ni, J., Yang, K., Liang, X., Ren, J., Shen, X.S.: Security and privacy in smart city applications: challenges and solutions. IEEE Commun. Mag. **55**(1), 122–129 (2017)
6. Ahmed, N., Hadaller, D., Keshav, S.: GUESS: gossiping updates for efficient spectrum sensing. In: ACM2006, pp. 12–17. International Workshop on Decentralized Resource Sharing in Mobile Computing and Networking. USA, Los Angeles, USA (2006)
7. Ganesan, G. and Li, Y.: Cooperative spectrum sensing in cognitive radio networks. In: Proceedings of IEEE-DySPAN, 2005, pp. 137–143. Baltimore, USA (2005)
8. Liao, Y., Wang, T., Song, L., Han, Z.: Listen-and-talk: protocol design and analysis for full-duplex cognitive radio networks. IEEE Trans. Veh. Technol. **66**(1), 656–667 (2017)

A Study on Escape Path Planning of Multi-Source/Multi-Sink for Public Buildings

Yi Zhang, Tianqi Liu, Chi Wang, and Chenlei Xie

Abstract With the development of urbanization and the rise of commercial center, the complex internal structure of public commercial buildings and the high density of people make the evacuation of people more aimless, and accurate path planning plays an important role in escape and evacuation. Dijkstra algorithm can only satisfy the single point to single point shortest path planning situation and cannot actually solve the shortest path planning problem with many source points and sink points. Thus, this paper studies the shortest path of escape with multi-source/multi-sink, and proposes an improved Dijkstra algorithm, which can solve the shortest path planning task of escape from multi-source point to multi-sink point and has fast and efficient characteristics of the original Dijkstra algorithm, thus improving the escape efficiency of personnel. In the end, this paper simulates the improved algorithm using the data of different distribution density of different people. The experimental results show that the proposed algorithm is more practical for the shortest path planning from multi-source to multi-sink under the condition of relatively dispersed personnel.

Keywords Dijkstra algorithm · Multi-source/multi-sink (MSMS) · Algorithm optimization · Path planning

1 Introduction

With the development of economy, the volume and function of buildings are increasing, and the structure is becoming more and more complex. Nowadays, large buildings often carry a large number of people who engage in various kinds of work for a long time. It is worrying that the complexity of the structure of the building makes the personnel in the interior often unable to move to a safe place in time

Y. Zhang · C. Xie (✉)
Anhui Province Key Laboratory of Intelligent Building and Building Energy Saving, Anhui Jianzhu University, Hefei, China
e-mail: 1033904749wc@gmail.com

T. Liu · C. Wang
School of Electronic and Information Engineering, Anhui Jianzhu University, Hefei, China

© The Author(s), under exclusive license to Springer Nature Singapore Pte Ltd. 2021   47
Y. Li et al. (eds.), *Advances in Simulation and Process Modelling*,
Advances in Intelligent Systems and Computing 1305,
https://doi.org/10.1007/978-981-33-4575-1_6

when they encounter emergency disasters such as fire, thus causing certain casualties. Therefore, in order to reduce or even avoid such a thing, we need to design a safe evacuation algorithm strategy, so that in the event of an emergency, the relevant institutions can conduct timely and effective path guidance to a large number of people in the building according to the algorithm strategy, so that people can evacuate in the shortest time through a safe route.

For the research about the path selection strategy, the Dijkstra algorithm [1] refers to the idea of greedy algorithm, which the source node traverses each node in the search graph until it reaches the target node, so that the shortest path from the source node to each node including the target node in the graph is obtained. And because Dijkstra algorithm is more usable and accessible, many scholars, based on the original algorithm, have proposed their own ideas about how to use or improve the Dijkstra algorithm. Gbadamosi and Aremu [2] noticed that sometimes finding the shortest path could cost huge energy, so they modified the Dijkstra algorithm by entailing a specially designed text file to get the alternate routes. And by comparing with the classical algorithm, the outcome of the new method was more effective. Deng, et al. [3] discussed how to properly use the Dijkstra algorithm under an uncertain environment, especially when the parameters are all fuzzy. And by adopting the graded mean integration representation of fuzzy numbers, they successfully improved the classical Dijkstra algorithm, and got a good outcome. And Fusic, et al. [4] also tried to implement Dijkstra algorithm in the mobile robot to let it find the most efficient path under the complicated environment. They modified the parameters of Dijkstra algorithm and adopted V-REP simulation software, the results were inspiring, and this proposed method could be used for path planning of robots. In this paper, we will discuss a new method to deal the multi-source point to multi-sink point problem by improving the Dijkstra algorithm.

2 Dijkstra Algorithms and Building Models

2.1 Dijkstra Algorithm

Dijkstra algorithm is proposed by Edsger Wybe Dijkstra in 1959 to find the shortest path of single point to single point [5], which is mainly used to solve the problem of finding the shortest path between vertices in directed weight graph. The algorithm has been used to build the intelligent fire evacuation system [6]. In addition, the algorithm adopts greedy algorithm strategy, which means that each time it will traverse to the node, which has not been visited, with the least distance from the source node, then one round of traversal is terminated. This will be repeated until searching to the sink point. According to the weight of the edge connected between vertices, the value of the distance between each vertex, which has been traversed, and corresponding source node is stored or updated, which can be used to update the "backtracking" of the weights.

The basic idea of Dijkstra algorithm is that first supposing a formula, named $G = (V, E)$, where G is used to represent a weight graph, which there are some connected edges, vertexes all with positive weights in it, and V E are used to represent the vertex set and the edge set, respectively. Then, dividing V into two groups, one group is named as S, where S is the vertex set that storing the node whose shortest path to the source node has been found, and the source node is represented as V_0. At the beginning, there is only V_0 in S, which can uses the formula, $S = \{V_0\}$, to represent. Then, a new vertex, whose shortest path to V_0 has been obtained, is added to S after each traversal. This loop will be continued until all vertexes in G are added to S, and if there are n nodes, the algorithm will have the n rounds. The other group is named as U, where U is the vertex set that used to store the node whose shortest path to V_0 has not been found, which can uses the formula, $U = \{V_1, V_2, V_3, \ldots, V_n\}$, to represent, which also means that all vertexes except V_0 are in U at the beginning. During each traversal, the length of the distance from each vertex in U to V_0 is arranged in an increasing order, and the vertex which has the smallest length will be selected to join S. In addition, each vertex corresponds to one particular length weights, which uses V_k to represent. And for one vertex in U, if it is found that the weights obtained from the previous rounds are greater than those obtained from the current traversal, then the number of the weights will be updated. The algorithm also stipulates that the length corresponding to each vertex in S set is the shortest distance from the source node to each of them. And the length corresponding to each vertex in U is the current shortest distance from the source node to each of them, which is determined only by the current conditions.

2.2 Building Mathematical Model Construction

Building a reasonable mathematical model is the prerequisite for the effective evacuation simulation. The core of the algorithm is how to choose the optimal path to reduce the evacuation time, and it should also limit the number of people. However, only the length weight of the edge is defined in the original Dijkstra algorithm, the edge, or saying corridor in fact, usually has the limited ability of carrying people. For that reason, based on the original length weight, we define the capacity weight to form the binary weight of the edge. In addition, we also adopt the method of two-dimensional model to find the path in the form of topological map. This map consists of nodes and edges, where nodes represent rooms, exits, etc., and edges represent corridors, passage, etc. To better clarify things in the map, we define some parameters. First, N_{p_i} is the position of a vertex, named i, which can also represent its coordinate, (x_i, y_i). And C_{ij} is used to represent the capacity of path, which means the maximum number of simultaneous evacuees allowed in the path from i to another vertex, named j. W and fp are used to represent the adjacency matrix of all the path and the position of fire points, respectively. n_k means the number of people now still in the fire point, named k. T is the total time consumed by the completion of the evacuation. And v is the average speed of evacuees. And at the beginning of the algorithm, we will initialize all these parameters, and the process is shown in Fig. 1.

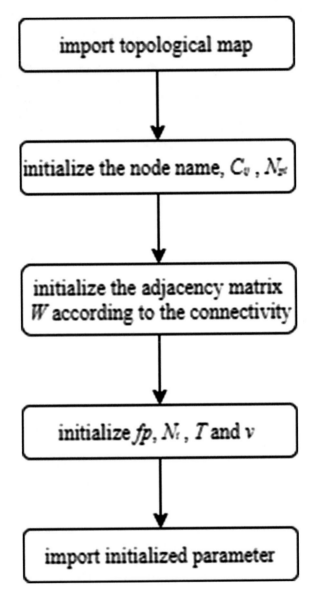

Fig. 1 Flowchart of parameters' initialization

3   Improved Dijkstra Evacuation Path Algorithm Design

3.1   Multi-Source/Multi-Sink Design

Typically, buildings have multiple emergency exits. And whether these buildings are used for office operations or commercial catering, people in them will not just gather in one place, but often disperse in many areas [7]. Therefore, when there is an emergency happening in the building, people in many internal areas all will need to be evacuated. In that case, both exits of building and internal areas with many people are more than one, so we need to consider a situation named multi-source/multi-sink (MSMS) [8]. And this paper designs an evacuation algorithm for MSMS building emergency based on Dijkstra algorithm, using Dijkstra algorithm to plan the shortest path. Dijkstra algorithm uses traversal as the way of searching, the shortest distance from the source point to the target node in the topology which can be found after one round. So at the beginning of our algorithm, the shortest path from each source point to each sink point is found by multiple calls of the Dijkstra algorithm, then these paths are sorted to find the shortest path out. By constantly looking for the shortest path, the personnel can be quickly allocated at the same time in order to make full use of the escape path, so as to solve the MSMS evacuation problem.

3.2   Algorithm Designed

Considering that the original algorithm used alone can only deal with the path searching problem of single point to single point, which cannot really meet the needs of evacuation in real life. So we take "using more escape paths in the shortest time" as the main idea of our algorithm. And this algorithm is implemented by two modules, which are first round of allocation and second to N round of allocation.

(1)   First round of allocation
    In the actual process, if the path allocation is used properly, there will be a higher probability that the all people can be allocated at once.
    First we need to judge whether there is still at least one path that is accessible to exits and whether there is still a person who needs to escape at the fire point. And if so, we will call Dijkstra algorithm to traverse the path between each fire point and exit to find the shortest path. Then the minimum capacity of the whole path, which uses C_{min} to represent, will be allocated, and the C_{ij} and W will be updated. Finally, until all paths are occupied or all personnel have been evacuated, the first round is over. The specific process is shown in Fig. 2:

(2)   Second to N round of allocation
    Because there might be too many people in the building, the allocation cannot be down just in the first round. Therefore, we need to design second to N round to evacuate people. And first, we need to release the path capacity consumed by

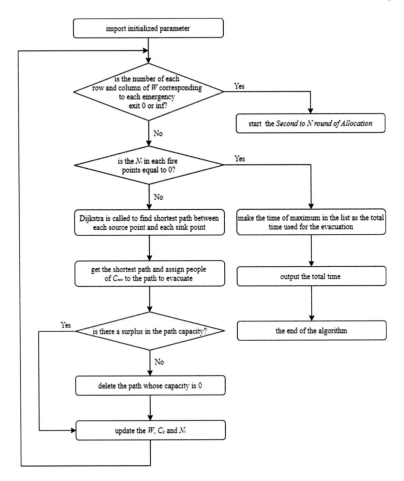

Fig. 2 First round allocation flowchart

the path which takes up the least escape time among all the paths allocated in the first round, and we use $P_{t_{min}}$ to name it, and add the evacuation time to the time list. Then the W, path capacity and remaining number of fire points will be updated, and the same process as in the first round of the evacuation will also be continued. At last, either iteration will end when the exits are all not accessible, and then, the algorithm will enter the next round or the algorithm is over and outputs the evacuation time when there are no remaining personnel at the fire points. The specific process is shown in Fig. 3.

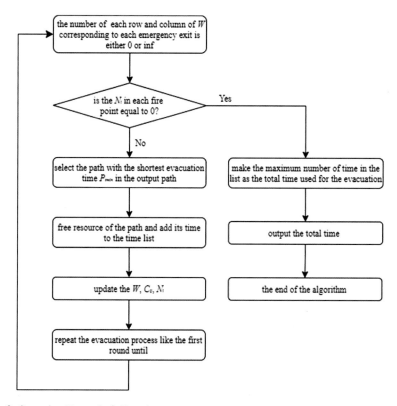

Fig. 3 Second to N round of allocation

4 Simulation Study and Performance Analysis

4.1 Simulation Scenario Description

In order to test the practicability of the algorithm, this simulation selected a commercial market as an example.[1] The mall has six safety exits and several internal areas. As the layout is complex, like that all kinds of stores, counters, corridors are mixing together, which makes the recognition very difficult, so in the topology map construction, we selected 17 internal typical areas, representing as nodes, and 38 paths to carry out the simulation of the algorithm [9]. The specific topology is modeled in Fig. 4: (circular nodes are internal regions and square nodes are emergency exits).

Under the condition that the algorithm can basically realize the evacuation of people in buildings, in order to further verify the performance of the algorithm, we have carried out the tests on different population density distribution with different

---

[1]The simulation study of this research work has been conducted with approval obtained from the owner of one market in Fuyang, China.

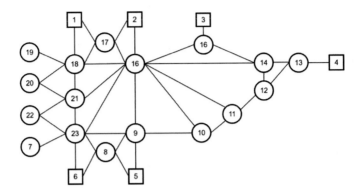

Fig. 4 Simulation building topology

number of people. And we also adopted the basic Dijkstra algorithm, which uses the way of single source to single sink, to compare with the improved algorithm under the same distribution. And the main focus of these tests are comparing the performance of above two algorithms under the different conditions and judging the influence of the difference in population density distribution on the simulation results. In addition, this paper designs two situations of crowd concentration and crowd dispersion, and simulates the evacuation process under the condition that the total number of evacuees is 180, 300, and 450, respectively.

In addition, in order to make the simulation of the evacuation algorithm be more realistic, this algorithm is preset in the following conditions:

(1) The location of the fire point and the distribution of the escape personnel are always initialized in a disorderly and random manner to ensure that the evacuation results are not affected by the regular distribution of the personnel;

(2) This algorithm is defined as a simulated evacuation drill. In the case of simulation, **it is easier for people to calm down, thus avoiding the possibility of making the wrong choice.** In addition, the simulation drill makes people more likely to follow the route planning made by the algorithm, so as to better test algorithm performance;

(3) In the evacuation simulation, the relevant personnel need to be evacuated in the form of trot to reduce the impact of the travel speed on the overall evacuation completion time. Taking into account the different physical fitness of individuals, so the speed of people in this simulation will take a fairly average value of 2 m/s, and evacuees, on the same round, in the same path will act together at the same time. After the path has been determined, the time required for the evacuation is obtained by subtracting 2 m/s from the path length.

4.2 Simulation Results

When the final result is output, each path determined in each round is shown by outputting a topology path selection diagram. The total output of the three simulations is 76 graphs, because of the limited space, here only random displaying of the distribution under the number and density of various people. The simulation results are shown in Figs. 5 and 6.

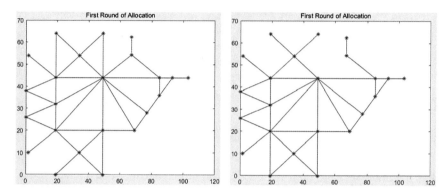

Fig. 5 Second time of allocation in the first round with 180 people in the state of dispersion and the fifth time of allocation in the first round of 180 people in the state of concentration

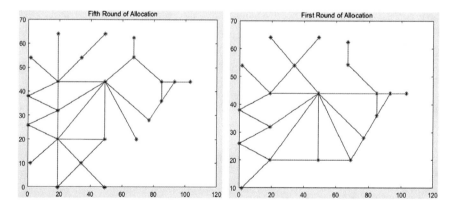

Fig. 6 Second time of allocation in fifth round with 300 people in the state of concentration and the tenth time of allocation in the first round with 450 people in the state of concentration

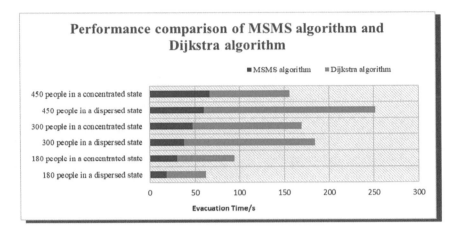

Fig. 7 Performance comparison of MSMS algorithm and Dijkstra algorithm

4.3 Simulation Performance Analysis

Both the improved algorithm and the basic algorithm conducted simulation tests in the MATLAB. And the results are shown in the form of a statistic chart as in Fig. 7 (the improved algorithm is named as MSMS algorithm).

From the chart, we can clearly see that the MSMS algorithm performs much better than the basic Dijkstra algorithm under different conditions. And by computing all the records, which records of MSMS algorithm are, respectively, 66.184 s, 60.000 s, 47.000 s, 37.621 s, 30.000 s and 18.028 s, and the records of Dijkstra algorithm are, respectively, 90.000 s, 191.962 s, 122.127 s, 146.446 s, 64.185 s, and 43.762 s, we concluded that **the performance of MSMS algorithm is improved by 154% than the basic Dijkstra algorithm.**

And by comparing the total evacuation time of the tests, which use the MSMS algorithm, we find that the performance of the MSMS algorithm designed in this paper is more outstanding and excellent in the state of dispersion, so it is more suitable for the evacuation in buildings with people in the state of dispersion.

5   Conclusion and Prospect

This paper improves the Dijkstra algorithm from some aspects. First, the shortest path from each source point to each sink point can be obtained by calling multiple times, and then, the shortest path in the global sense can be obtained by sorting. Second, the paths in the graph all have the corresponding carrying capacity. Therefore, except the basic weight of the edge length, the algorithm also defines the weight value of path capacity according to the actual situation and adopts the method of minimum capacity. Finally, the suitable paths are selected by two modules named the first round of allocation and the second to N round of allocation, so as to solve the multi-source point to multi-sink point personnel escape route planning problem. At the same time, this paper selects a shopping mall as the modeling object and simulates the actual path selection and output of the algorithm under different personnel distribution, and finally draws the conclusion that the improved algorithm designed in this paper has better performance under the condition of disordered dispersion of personnel distribution.

The current research in this paper focuses on the algorithm improvement of the theoretical model, and fails to add the unstable factors caused by the subjective blindness of the crowd and the environmental interference factors of the integrated buildings of business and residence buildings. Therefore, it is necessary to consider the prediction of crowd behavior model [10] and the simulation of complex commercial and residential environment in the future research to make the algorithm more in line with the needs of real life.

Acknowledgements This work is supported by the following two funds as National Key Research and Development Project of China No. 2017YFC0704100 (entitled New Generation Intelligent Building Platform Techniques) and Foundation of Anhui Jianzhu University, Hefei, China, No. JZ192012.

References

1. Dijkstra, E.W.: A Note on two problems in connection with graphs. Numerische Math. **1**, 269–271 (1959)
2. Gbadamosi, O.A., Aremu, D.R.: Design of a modified Dijkstra's algorithm for finding alternate routes for shortest-path problems with huge costs. In: International Conference in Mathematics, Computer Engineering and Computer Science (ICMCECS), pp. 1–6, IEEE, Ayobo, Ipaja, Lagos, Nigeria (2020)
3. Deng, Y., Chen, Y., Zhang, Y., et al.: Fuzzy Dijkstra algorithm for shortest path problem under uncertain environment. Appl. Soft Comput. **12**(3), 1231–1237 (2012)
4. Fusic S.J., Ramkumar, P., Hariharan, K.: Path planning of robot using modified dijkstra Algorithm. In: National Power Engineering Conference (NPEC), pp. 1–5, IEEE, Madurai (2018)
5. Sedeño-Noda, A., Colebrook, M.: A biobjective Dijkstra algorithm. Eur. J. Oper. Res. **276**(1), 106–118 (2019)
6. Xu, Y., Wang, Z., Zheng, Q., Han, Z.: The application of Dijkstra's algorithm in the intelligent fire evacuation system. In: 4th International Conference on Intelligent Human-Machine Systems and Cybernetics, pp. 3–6, IEEE, Nanchang, Jiangxi (2012)
7. Zhang, F., Xu, B., Shao, Z.: Study on optimization of emergency evacuation mode of "Multi Source and Multi Sink" in high-rise apartments. IOP Conf. Ser. Earth Environ. Sci. **252**(5), 052151(2019)
8. Lin, P., Lo, S.M., Huang, H.C., et al.: On the use of multi-stage time-varying quickest time approach for optimization of evacuation planning. Fire Saf. J. **43**(4), 282–290 (2008)
9. Wolfgang, M., Feldmann, A., Maennel, O., et al.: Building an AS-topology model that captures route diversity. ACM SIGCOMM Comput. Commun. Rev. **36**(4), 195–206 (2006)
10. Rozo, R.K., Arellana, J., Santander-Mercado, A., et al.: Modelling building emergency evacuation plans considering the dynamic behaviour of pedestrians using agent-based simulation. Saf. Sci. **113**, 276–284 (2019)

Design of 3D Building Model Based on Airborne LiDAR Point Cloud

Maohua Liu, Manwen Li, and Jiahua Liang

Abstract Airborne LiDAR is a new technology that can quickly obtain high-precision 3D information on the ground and ground objects. The accurate 3D spatial coordinate information and DSM data it obtains provide a new and effective observation method for the establishment of a smart city 3D model. How to obtain useful terrain information and feature information, and to study the application of point clouds in the construction of smart cities is a problem we need to solve. According to these problems, we propose a method for building extraction and 3D reconstruction using LiDAR point clouds. First, we process the airborne laser point cloud and then fit the roof plane by clustering the roof plane point cloud, and finally determine the outer boundary of the roof and the boundary of each plane, and use the boundary to obtain the three-dimensional coordinates of each corner of the roof to reconstruct three-dimensional model of the building. In addition, it explores the design methods and operations of different building 3D models and compares the effects of the two modeling methods.

Keywords Airborne LiDAR · 3D modeling · Point cloud

1 Introduction

Cities are the areas with the most frequent production and living activities. With the acceleration of the construction of smart cities in China, the changes and development of cities are also changing with each passing day [1]. Buildings as urban infrastructure are the most critical landmark information for cities, and they are also the basic spatial data information required for the construction of smart cities and other projects [2]. Traditional two-dimensional plane information can no longer meet people's

M. Liu · M. Li
School of Transportation Engineering, Shenyang Jianzhu University, Shenyang, China

J. Liang (✉)
Shenyang Geotechnical Investigation and Surveying Research Institute Co., Ltd., Shenyang, China
e-mail: hyukcici@163.com

needs, and city three-dimensional data, especially three-dimensional building data, is becoming more and more important [3]. How to obtain the high-precision digital elevation model data and building model of the required target area quickly and in large quantities is the focus of current research.

The LiDAR system has a wide range of applications. It can directly obtain the 3D coordinate information of the surface points of the target object through the observation data such as position, angle, and distance to realize surface extraction and 3D scene reconstruction [4]. It can be used in urban construction, cultural relics restoration, forestry planning, and other fields [5]. The obtained LiDAR point cloud contains rich natural and artificial feature information, mainly including vegetation, buildings, roads, green spaces, and other features [6]. How to extract the building information we need from the numerous laser point clouds has been a hot spot of research. Vosselman and others at Delft University in the Netherlands used 3D Hough transform to obtain a plan image of a building to achieve the purpose of three-dimensional reconstruction of the building [7]. Li Shukai and others used DSM combined with images to extract the building borders, which were implemented in three steps, followed by laser ranging point analysis, shadow analysis, and building boundary reconstruction [8]. Liang Xinlian uses a split-based minimum mean square error line segment approximation method to extract building contours from discrete LIDAR data. This experiment will study the method of building 3D modeling [9].

3D building modeling is divided into four sub-modules: building point cloud extraction, roof patch segmentation, building model, and automated modeling. First, the point cloud data is spliced, cropped, and denoised. Then, the building point cloud extraction, roof patch segmentation and other steps are performed, and finally the building is modeled. Another method is to import the processed point cloud into Revit software after initial processing of the data. Use window family and door family to manually construct 3D models of buildings, and import external family for detailed building modeling.

2 Airborne LiDAR Point Cloud Processing

In order to construct a three-dimensional model of a building, a series of processing must be performed on the acquired point cloud data, such as point cloud denoising, filtering, and building point cloud extraction [10].

2.1 Point Cloud Denoising

In the process of acquiring point cloud data, due to the defects of the scanner itself, environmental interference and other factors, the point cloud often contains many noise points. The process of removing these noise points is called point cloud denoising [11]. The effective value and noise of the point cloud have similar characteristics. The effective value reflects the unique characteristics of the target surface, but the noise is useless interference information. When dealing with actual problems, it is difficult to accurately identify the effective value and noise. Therefore, in the process of processing point cloud data, it is very important to maintain the unique characteristics of the target information. We need to identify the noise in time and then reduce it. In the experiment, we use a denoising algorithm based on the statistics of the local spatial distribution of the point cloud and mark the points with large differences between the local point density and the overall point density as noise points and remove them.

2.2 Point Cloud Filtering

Filtering is an important part of LiDAR point cloud preprocessing. Its purpose is to separate ground point cloud and non-ground point cloud, so as to provide data sources for data post-processing such as obtaining DEM and digital elevation models [12]. The main idea of morphological filtering is to use the corrosion and expansion operations in mathematical morphology to remove the higher point cloud in the point cloud and retain the lower point cloud to achieve the purpose of extracting ground points. After filtering, the results are ground point cloud and non-ground point cloud are shown in Figs. 1 and 2.

3 3D Modeling of Buildings

Three-dimensional building modeling includes four sub-modules: building point cloud extraction, roof patch segmentation, building model construction, and automated modeling. First, we perform initial processing such as splicing, cropping, and

Fig. 1 Ground points after processing

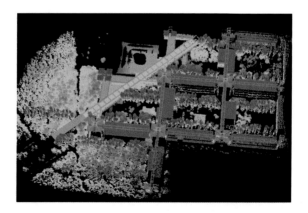

Fig. 2 Non-ground points
after processing

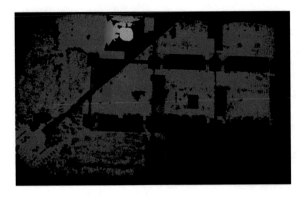

point cloud denoising on the point cloud data. Then we perform point cloud filtering, building point cloud extraction, roof patch segmentation and other steps, and finally model the building.

3.1 Building Point Cloud Extraction

Use normal vector to detect non-ground point data for building wall point cloud data. Vegetation points are removed by the intensity of the point cloud and the angle between the normal vector and the Z-axis, and then non-ground point data clustering and ground point data interpolation [13]. Finally, the separation of the building roof point cloud is by separating the building area and height information, and the results are shown in Figs. 3 and 4.

Fig. 3 Ground point clouds
and non-ground point clouds

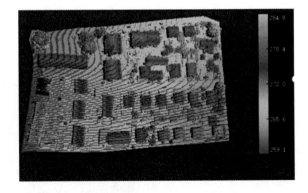

Fig. 4 Building point clouds

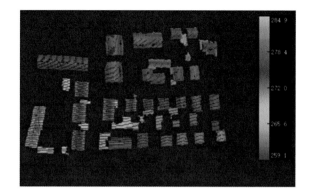

3.2 Roof Patch Split

There are two methods of roof patch segmentation for building point cloud, the normal vector method and the RANSAC method. Generally, the normal vector distribution of the roof patch point cloud is consistent [14]. The normal vector method uses this feature to segment the roof patch. The parameter setting options and specific thresholds can be set in combination with data and prior knowledge. RANSAC is an effective and robust estimation algorithm, which has a certain inhibitory effect on noise points, and can better segment the roof patches in the building point cloud data. This method is to find the best-fitting plane through repeated iterations. After clustering, the segmented roof patches are obtained. Segmentation results are shown in Figs. 5 and 6.

Fig. 5 Normal vector segmentation result

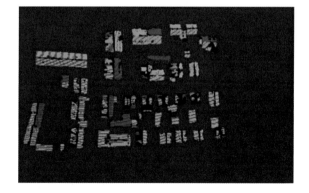

Fig. 6 RANSAC
segmentation result

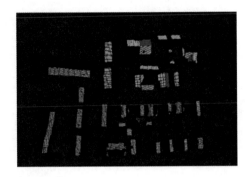

3.3 3D Modeling

To process the results of the above process, first extract the boundary point clouds of the two layers, respectively, then extract the boundary line and the ridge line, and regularize, then construct the roof polygon, and finally combine the ground elevation to obtain the building model; the result is shown in Fig. 7.

4 3D Modeling with Revit

Using the Revit2019 modeling method is to perform the initial processing of the data, and build the 3D model of the building manually by manual methods, and perform fine modeling of the building by using the window family, door family, and importing external family of the building.

Fig. 7 Building model result

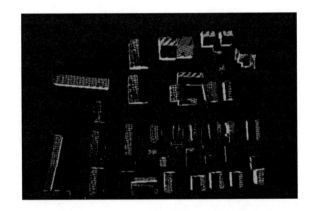

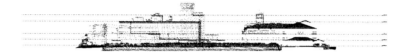

Fig. 8 Elevation results

Fig. 9 Result of establishing coordinate system

4.1 Grid and Elevation Establishment

Construct grids and elevations, as much as possible to establish an elevation for each floor plan, which is convenient for subsequent construction of the model, and can be fine-tuned to make the elevation line coincide with the floor plan. The result is shown in Fig. 8.

By moving and rotating functions, the corner of the building is placed at the intersection of the grid, one side is attached to the grid, and the point cloud of the building edge is placed parallel to the grid. The final result is shown in Fig. 9.

4.2 Construction of Building Exterior Walls, Floors, and Roofs

Select the "wall" function key to draw a line along the outer contour of the building point cloud to draw a closed geometry. Each side can be fine-tuned to make the outer wall line coincide with the building point cloud contour line and set the wall height to the previously established. At all levels of elevation, the walls of the building are completely drawn. Select the function key "floor" to start building the floor, pick up the building exterior wall line to automatically generate the floor.

Fig. 10 3D view of exterior
wall results

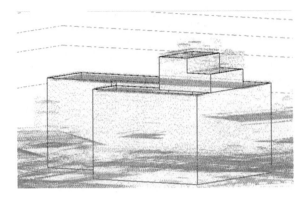

Fig. 11 3D view of roof
results

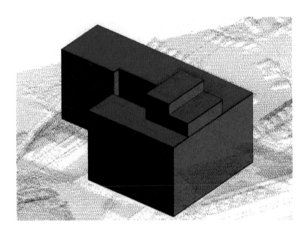

Select the function key "roof" to start building the roof, pick up the exterior wall line of the building, and control the bottom elevation, the bottom offset from the elevation, and the roof slope through the constraint function. After the adjustment is correct, click OK to automatically generate the roof. The results of wall and roof are shown in Figs. 10 and 11.

4.3   Construction of Building Doors and Windows

By loading the family, selecting the appropriate doors and windows, and selecting the function key "window" to start building doors and windows, and by changing the type attributes to change the doors and windows to the appropriate size, the result is shown in Fig. 12.

Fig. 12 Door and window construction results

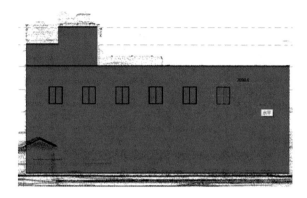

4.4 Building 3D Model Results

In the experiment, the method used in the previous experiment was used to process the denoising, filtering, and classification of the point cloud data, and the required point cloud data of the target building has been extracted. Taking the building point cloud as the research object, the Revit software is used to construct the 3D model of the building based on manual operation. The model results are shown in Figs. 13 and 14.

However, the roofs of complex and irregular buildings are mostly polygonal or curved, and the relationship between the edge of the roof of the building and the direction of the main axis of the building is not fixed. It may be neither parallel nor perpendicular to each other. In the existing building 3D model construction methods, some methods are defaulting that each edge of the roof of the building is perpendicular or parallel to the direction of the main axis of the building. For a building that is completely asymmetrical, it cannot be determined. Direction, and for most irregular and complex roofs, the edges are not necessarily parallel or perpendicular to the main direction.

The structural structure of buildings in real life is generally cumbersome. The basic building 3D model construction method can never meet the needs of 3D modeling of most buildings in life. Considering that there are many irregular and complicated

Fig. 13 Right view of the 3D model

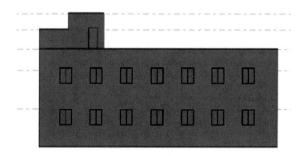

Fig. 14 Stereo view of 3D model

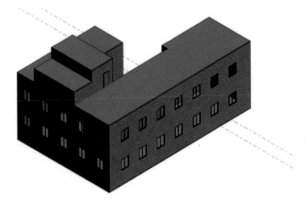

building roof shapes in real life, based on the basic building 3D model construction method, a method that can construct a complex 3D building model is redesigned.

5 Conclusions

In the 3D building model construction experiment, the airborne LiDAR point cloud is the main research object. It analyzed the characteristics of the data obtained by the airborne LiDAR system and used a series of LiDAR point cloud data filtering and classification methods for processing. Finally, it carried out the research on building extraction and 3D modeling based on LiDAR point cloud. The main conclusions are as follows:

1. Taking the extracted building roof point cloud as the research object, a 3D building reconstruction method based on LiDAR point cloud is proposed. By clustering the roof plane points, fitting the roof plane, determining the outer boundary of the roof, and the boundary of each plane, so as to obtain the three-dimensional coordinates of each corner of the roof to reconstruct the three-dimensional model of the building. Experimental results show that this method not only reconstructs simple and regular buildings better, but also reconstructs buildings with complex roof planes and irregular structures.

2. Comparing the two experiments, the three-dimensional model of the building constructed by fitting the roof and then modeling method mentioned in the paper has better roof effect, lower side accuracy and can be automatically modeled, but it depends more on computer hardware configuration and point cloud data. No results when running. The 3D model of the building in Revit2019 is fine modeling, and the accuracy of the roof side is not much different. Manual modeling is not required. The model can be adjusted according to the needs, which is more suitable for the mass production of simple models.

References

1. Wu, B., Yu, B., Wu, Q., Yao, S., Zhao, F., Mao, W., Wu, J.: A graph-based approach for 3D building model reconstruction from airborne LiDAR point clouds. Remote Sens. **9**(1), 92 (2017)
2. Li, P.C., Xing, S., Xu, Q., Zhou,Y., Liu, Z.Q., Zhang, Y., Geng, X.: Automatic reconstruction of complex building models based on key point detection. J. Remote Sens.**18**(06), 1237–1246 (2018). (in Chinese)
3. Schwalbe, E., Maas, H.G., Seidel, F.: 3D building model generation from airborne laser scanner data using 2D GIS data and orthogonal point cloud projections. In: Proceedings of ISPRS WG III/3, III/4, vol. 3, pp. 12–14 (2005)
4. Sun, S., Salvaggio, C.: Aerial 3D building detection and modeling from airborne LiDAR point clouds. IEEE J. Sel. Top. Appl. Earth Obs. Remote Sens. **6**(03), 1440–1449 (2013)
5. Haala, N., Peter, M., Kremer, J., Hunter, G.: Mobile LiDAR mapping for 3D point cloud collection in urban areas—a performance test.Int. Arch. Photogram. Remote Sens. Spat. Inf. Sci. **37**, 1119–1127 (2008)
6. Du, J.L., Chen, D., Zhang, Z.X., Zhang, L.Q.: Research progress on the reconstruction method of architectural point cloud geometric model. J. Remote Sens. **23**(03), 374–391 (2019). (in Chinese)
7. Vosselman, G.: Fusion of laser scanning data, maps, and aerial photographs for building reconstruction.In: IEEE International Geoscience and Remote Sensing Symposium, vol. 1,pp. 85–88 (2002)
8. Li, Q., Xiao, C.L., Chen J., Yang, D.C.: Method of constructing DSM based on airborne LiDAR point cloud and building outline.Remote Sens. Land Resour. **25**(02), 95–100 (2013). (in Chinese)
9. Jochem, A., Höfle, B., Wichmann, V., Rutzinger, M., Zipf, A.: Area-wide roof plane segmentation in airborne LiDAR point clouds.Comput. Environ. Urban Syst. **36**(1), 54–64 (2012)
10. Tao, J.H., Su, L., Li, S.K.: Method of extracting building model from LiDAR point cloud. Infrared Laser Eng. **38**(2), 340–345 (2009) (in Chinese)
11. Cheng, L., Gong, J., Li, M., Liu, Y.: 3D building model reconstruction from multi-view aerial imagery and lidar data. Photogram. Eng. Remote Sens. **77**(2), 125–139 (2011)
12. Sha, J., Li, Z.X., Zhang, W.Y.: Progress in large-scale 3D city modeling. J. Surv. Mapp. **48**(12), 1523–1541 (2019) (in Chinese)
13. Niemeyer, J., Rottensteiner, F., Soergel, U.: Conditional random fields for LiDAR point cloud classification in complex urban areas. ISPRS Ann. Photogram. Remote Sens. Spat. Inf. Sci. **3**, 263–268 (2012)
14. Sun, Y., Zhang, X.C., Luo, G.W.: An improved method for extracting active contour model of building roof boundary from airborne LiDAR point cloud. J. Surv. Mapp.**43**(06), 620–626+636 (2014) (in Chinese)

Research on the Connection Performance of SMA Pipe Coupling for Sports Equipment

Xudong Yang, Fan Gu, Yuyu Zhang, and Yuhan Jiang

Abstract Based on the constitutive relation of shape memory alloy (SMA) proposed by Auricchio and Taylor, a numerical simulation study was carried out on the connection performance and the functional failure mechanism of SMA pipe coupling, and the effect of interference magnitude, the wall thickness of SMA connector on the stress distribution and tensile strength was investigated. The simulation result shows that with the increase of interference magnitude, the Mises stress of the SMA pipe coupling increases accordingly, as well as the ultimate pullout load increases. With the increase of the wall thickness of SMA connector, the Mises stress of the SMA connector decreases, whereas the Mises stress of connected steel pipe increases. Furthermore, it is suggested that the gradual internal chamfer should be machined at both ends of SMA connector to alleviate the stress concentration. To improve the connection performance and the tensile strength of SMA pipe coupling, it should take into consideration concurrently to the increase of the wall thickness and the length of the SMA connector to avoid the yield failure of steel pipe.

Keywords Shape memory alloy · Pipe coupling · Connection performance · Constitutive relation · Numerical simulation

1 Introduction

In the early 1940s, Olander discovered the shape memory effect in Au-Cd alloy for the first time, after that scholars successively discovered the shape memory effect in Cu-Sn alloy and Cu–Zn alloy. At present, with the advantaged characteristics of shape memory effect, super-elasticity, fatigue resistance, corrosion resistance, high damping, and high resistance, the shape memory alloy (SMA) has been widely

X. Yang
Sports Department, Shenyang Jianzhu University, Shenyang, China

F. Gu (✉) · Y. Zhang · Y. Jiang
School of Civil Engineering, Shenyang Jianzhu University, Shenyang, China
e-mail: guzhaozheng@yeah.net

© The Author(s), under exclusive license to Springer Nature Singapore Pte Ltd. 2021
Y. Li et al. (eds.), *Advances in Simulation and Process Modelling*,
Advances in Intelligent Systems and Computing 1305,
https://doi.org/10.1007/978-981-33-4575-1_8

Fig. 1 Application of titanium alloy in sports products

applied in the fields of aerospace, civil engineering, machinery, medicine, bioengineering, sports equipment, and so on [1–5]. According to statistics, 16–20% of the total titanium alloy is used in sports and leisure products, becoming the third largest application market of titanium alloy, such as golf clubs, tennis rackets, badminton bats, fencing masks, sprinting shoe nails, climbing tools, diving suits, fishing tackle, tent poles, racing parts, and so on, as shown in Fig. 1.

Some sports products make use of titanium alloy on its characteristics of high strength, lightweight, and super-elastic properties, such as titanium alloy golf clubs with lightweight can improve golfer's hitting ratio and hitting distance significantly, the titanium alloy handle applied on tennis rackets and badminton bats can increase the instantaneous inertia force and striking force effectively, and racing bike parts produced by titanium alloy can reduce the weight and wind resistance concurrently. On the other hand, some sports products make use of titanium alloy on the shape memory effect. For example, the mechanical vibration of automobile gearbox occurs at elevated temperature due to the different thermal expansion coefficient of each gear, SMA gasket on the driving shaft can provide elastic recovery force constantly with the increase of temperature, which can improve the fastening force and reduce running noise. For another example, the traditional automotive oil pipe joint usually adopts welding technology, and the corrosion protection layer on the inner surface near the welding seam is easy to fail at elevated temperature. As an alternative product, SMA automotive pipe coupling can effectively avoid corrosion leakage, and the shape memory effect guarantees it good connection performance at elevated temperature.

At present, scholars mainly concentrate on the perspective of metallography and microstructure and have carried out the studies of SMA on the evolution of microstructure, phase transformation, magnetization behavior, shape memory behavior, and its influence factors such as annealing time and transformation temperatures. [6–11]. With the characteristic of shape memory effect and recovery driving force, Ni–Ti SMA and Fe–Mn–Si SMA are beginning to be used in pipe coupling of sports automobile and airplane [12]. Comparatively, from the point of macroscopic mechanical, the researches on the connection performance of SMA pipe coupling is insufficient. Zhang et al. [13, 14] and Gu et al. [15] carried out experimental and numerical researches on the connection performance of SMA pipe coupling and proposed the influence of wall thickness and inner diameter on the radial compressive stress of SMA pipe coupling. Jin et al. [16] studied the influence of predeformation temperature on the recovery performance of Ni–Ti–Nb SMA pipe joint. Chen et al. [17] conducted numerical research on the relationship of Ni–Ti–Nb SMA pipe

Table 1　Geometric parameters of SMA pipe coupling model

SMA pipe connector	Outer diameter/mm	Inner diameter /mm	Wall thickness/mm	Inner diameter before expanding/mm	Prestrain/%
	41.2	37.1	2.0	1.5	1.5
Connected steel pipe	Outer diameter/mm	Inner diameter /mm	Wall thickness/mm	Fit clearance/mm	Interference magnitude /mm
	36.78	34.78	1.0	0.160	0.115

coupling between radial stress, tension force, and temperature. Due to the complexity of the constitutive relation of SMA, the research on the connection performance of SMA pipe coupling was mainly conducted by experiment. In this paper, based on the constitutive relation of SMA proposed by Auricchio and Taylor [18], the numerical research was investigated on the stress distribution and failure mechanism of SMA pipe coupling, in order to provide the theoretical reference to the optimization design of SMA connector.

2　Numerical Model of SMA Pipe Coupling

According to the geometric parameters in Table 1, the numerical model of SMA pipe coupling was developed, and the axisymmetric solid element CAX4R was adopted for representing SMA connector and connected steel pipe. Due to the axisymmetric characteristics of the SMA pipe coupling, the generatrix model was adopted to establish. According to the constitutive relation of SMA proposed by Auricchio and Taylor, the SMA material parameters were evaluated as shown in Table 2 and Fig. 2, where E_M and E_A are the elastic modulus of martensite and austenite, respectively. At the reference temperature, σ_{TL}^S and σ_{TL}^E are the starting critical stress and the ending critical stress in the process of tensile loading, respectively, σ_{TU}^S and σ_{TU}^E are the starting critical stress and the ending critical stress in the process of unloading, respectively, and σ_{CL}^S is the starting critical stress in compressive loading process.

3　Connection Performance Analysis

3.1　Stress Distribution Analysis

When two steel pipes and SMA connector were assembled together, the SMA connector was heated to the predefined temperature of 600 °C. With the process of reverse martensitic transformation, SMA connector deforms to its original dimension

Table 2 Material parameters of SMA connector

Poisson's ratio	Elasticity modulus (GPa)	Critical stress of phase transformation (MPa)	Constant of phase transformation (MPa °C^{-1})	Strain of phase transformation	Reference temperature (°C)
0.33	$E_M = 55$ $E_A = 170$	$\sigma_{TL}^S = 520$ $\sigma_{TL}^E = 750$ $\sigma_{TU}^S = 550$ $\sigma_{TU}^E = 200$ $\sigma_{CL}^S = 520$	$\left(\frac{\delta_\sigma}{\delta_t}\right)_L = 8.0$ $\left(\frac{\delta_\sigma}{\delta_t}\right)_U = 13.8$	0.067	600

	Martensite's Young's Modulus	Martensite's Poisson's Ratio	Transformation Strain	Start of Transformation (Loading)	End of Transformation (Loading)	Start of Transformation (Unloading)	End of Transformation (Unloading)	Start of Transformation in Compression (Loading)	Reference Temperature	Loading	Unloading	Value of field variable that triggers shape setting
1	55000000000	0.33	0.067	520000000	750000000	550000000	200000000	520000000	600	8000000	13800000	0

Fig. 2 Setting in ABAQUS of the material parameters and hyper-elastic parameters of SMA

and makes two steel pipes connect together as a whole. The Mises stress distribution nephogram of SMA pipe coupling can be obtained, as shown in Fig. 3.

According to the symmetry of SMA pipe coupling model in axial direction, the semi-structure in axial direction was selected as analysis object, and the radial stress nephogram and the circumferential stress nephogram of SMA pipe coupling are shown in Fig. 4. It can be seen that the radial stress of SMA connector is distributed evenly along the axial direction and decreases gradually along the radial direction from inner surface to outside surface, and the radial stress of connected steel pipe is distributed evenly along the axial direction and increases gradually along the radial direction from inner surface to outside surface. The circumferential stress distributions of SMA connector and connected steel pipe are evenly both in axial direction and radial direction, and their absolute values vary within the range of [134.7, 146.5 MPa] and [275.8, 302.3 MPa], respectively. In addition, the stress concentration phenomenon occurs in the contact area near the ends of SMA connector, where the radial stress of interaction is about −25.2 MPa, and the absolute values of circumferential stress of SMA connector and connected steel pipe reach 146.5 and 302.3 MPa, respectively. As is known that SMA connector and connected steel pipe

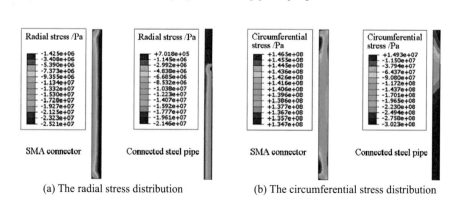

Fig. 3 Mises stress distribution nephogram of SMA pipe coupling

Radial stress /Pa	Radial stress /Pa	Circumferential stress /Pa	Circumferential stress /Pa
-1.425e+06 -3.408e+06 -5.390e+06 -7.373e+06 -9.355e+06 -1.134e+07 -1.332e+07 -1.530e+07 -1.728e+07 -1.927e+07 -2.125e+07 -2.323e+07 -2.521e+07	+7.018e+05 -1.145e+06 -2.992e+06 -4.838e+06 -6.685e+06 -8.532e+06 -1.038e+07 -1.223e+07 -1.407e+07 -1.592e+07 -1.777e+07 -1.961e+07 -2.146e+07	+1.465e+08 +1.455e+08 +1.445e+08 +1.436e+08 +1.426e+08 +1.416e+08 +1.406e+08 +1.396e+08 +1.386e+08 +1.377e+08 +1.367e+08 +1.357e+08 +1.347e+08	+1.493e+07 -1.150e+07 -3.794e+07 -6.437e+07 -9.080e+07 -1.172e+08 -1.437e+08 -1.701e+08 -1.965e+08 -2.230e+08 -2.494e+08 -2.758e+08 -3.023e+08
SMA connector	Connected steel pipe	SMA connector	Connected steel pipe

(a) The radial stress distribution   (b) The circumferential stress distribution

Fig. 4 Stress nephogram of the semi-structure of SMA pipe coupling

are thin-walled circular tube structure, the absolute value of its circumferential stress is much greater than that of radial stress according to Lamè formula [19].

3.2 Connection Performance and Influence Factors

Pullout load was exerted on the end of steel pipes, the relation between load and displacement is as shown in Fig. 5, and the radial stress nephogram of SMA connector is as shown in Fig. 6. It can be seen that the axial displacement of SMA pipe coupling is proportional to the pullout load in the initial stage of axial tension, and there is no sliding occurrence between SMA connector and steel pipes, which means that the axial displacement of SMA pipe coupling is completely caused by elastic deformation. When the pullout load reaches the ultimate value, sliding between SMA connector and steel pipe occurs leading the axial displacement increases rapidly. With the increase of axial displacement, the friction area between SMA connector and steel pipes decreases gradually, leading to the decrease of pullout load.

Fig. 5 Relation between load and displacement

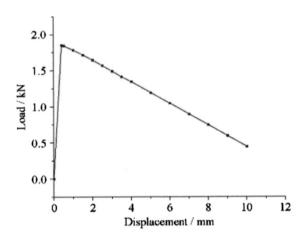

Fig. 6 Radial stress nephogram of connector

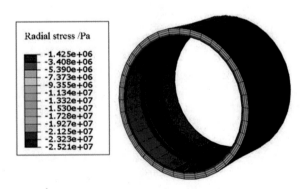

Five numerical models of SMA pipe coupling were developed with difference interference magnitude of 0.065, 0.115, 0.165, 0.215, and 0.265 mm, respectively, and other geometric parameters being the same as those in Table 1. With the difference interference magnitude, the stress nephogram of SMA pipe coupling could be gotten as shown in Fig. 7. It can be seen that the stress concentration occurs in the contact area near the ends of SMA connector. The absolute value of the radial stress at this location was extracted, and the relation between the maximum absolute value of radial stress and the interference magnitude could be obtained, as shown in Fig. 8. In addition, the relation between the ultimate pullout load and the interference magnitude could be investigated as shown in Fig. 9. It can be seen that with the increase of the interference magnitude, the maximum absolute value of radial stress and the ultimate pullout load increase accordingly.

Similarly, five SMA pipe coupling numerical models were established with difference wall thickness of SMA connector of 1, 2, 3, 4, and 5 mm, respectively, and the relation between the maximum absolute value of radial stress and the wall thickness of SMA connector, the relation between the ultimate pullout load and the wall thickness of SMA connector could be obtained as shown in Figs. 10 and 11. It can be seen that with the increase of the wall thickness of SMA connector, the maximum absolute value of radial stress and the ultimate pullout load increase accordingly, as soon as the circumferential stress of steel pipe increases whereas the circumferential stress of SMA connector decreases simultaneously.

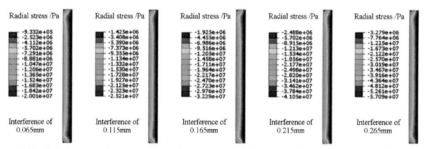

(a) Radial stress nephogram of SMA connector with different interference magnitude.

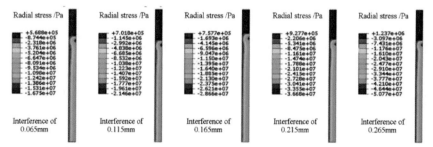

(b) Radial stress nephogram of steel pipe with different interference magnitude.

Fig. 7 Radial stress nephogram of SMA pipe coupling with different interference magnitude

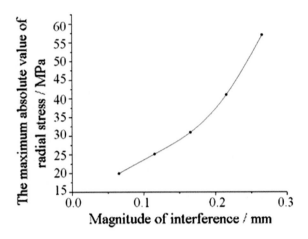

Fig. 8 Relation between the maximum value of radial stress and the interference magnitude

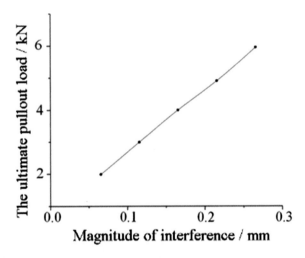

Fig. 9 Relation between the ultimate pullout load and the interference magnitude

4 Conclusions

Based on the constitutive relation of SMA proposed by Auricchio and Taylor, the numerical simulation research on the connection performance of SMA pipe coupling was carried out, and numerical simulation result shows the conclusions as follows.

1. The stress concentration phenomenon occurs in the contact area near the ends of SMA connector, and the gradual internal chamfer should be machined at both ends of SMA connector to alleviate the stress concentration.

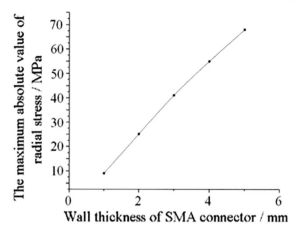

Fig. 10 Relation between the max. value of radial stress and the wall thickness of connector

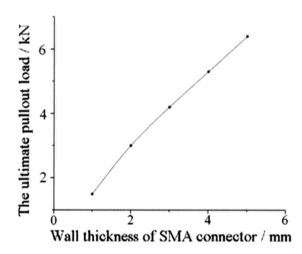

Fig. 11 Relation between the ultimate pullout load and the wall thickness of connector

2. Under the action of pullout load, in the initial stage, the axial displacement of SMA pipe coupling caused by elastic deformation is proportional to the pullout load.
3. When the pullout load reaches the ultimate value, the axial displacement of SMA pipe coupling increase rapidly caused by frictional sliding between SMA connector and steel pipes. With the increase of the axial displacement of SMA pipe coupling, the friction area decreases gradually, leading to the decrease of pullout load accordingly.
4. The increase of the wall thickness of SMA connector is equivalent to the increase of the relative stiffness of SMA connector, which will lead to the decrease of the

Mises stress of SMA connector and the increase of the Mises stress of steel pipe. Therefore, in the process of the design of SMA pipe coupling, to improve the tensile strength of SMA pipe coupling, it should take into consideration concurrently to the increase of the wall thickness and the length of SMA connector to avoid the yield failure of steel pipe.

Acknowledgements This research work is supported by Scientific Research Project of Liaoning Science and Technology Department (Project Codes: 2019-ZD-0297) and Scientific Research Project of Shenyang Jianzhu University (Project Codes: 2017007).

References

1. Khan, M.I., Pequegnat, A., Zhou, Y.N.: Multiple memory shape memory alloys. Adv. Eng. Mater. **15**(5), 386–393 (2013)
2. Lee, S.H., Kim, S.W.: Improved position control of shape memory alloy actuator using the self-sensing model. Sens. Actuators Phys. **297**(1), 64–72 (2019)
3. Ren, H.M., Peng, G.S.: Research on application of SMA in structure strengthening and rehabilitation. Struct. Eng. **34**(6), 175–180 (2018) (in Chinese)
4. Niskanen, A.J., Laitinen, I.: Design and simulation of a magnetic shape memory (MSM) alloy energy harvester. Adv. Sci. Technol. **78**(1), 58–62 (2013)
5. Ren, Y.S., Tian, J.S., Liu, Y.L., Du, C.G.: Nonlinear deformation, thermal buckling and vibration of SMA fiber composite beams. J. Shandong Univ. Sci. Technol. (Nat. Sci.) **38**(1), 99–110 (2019) (in Chinese)
6. He, Z.R., Liu, M.Q.: Effects of annealing and deforming temperature on microstructure and deformation characteristics of Ti-Ni-V shape memory alloy. Mater. Sci. Eng., B **177**(12), 986–991 (2012)
7. Velmurugan, C., Senthikumar, V., Kamala, P.S.: Microstructure and corrosion behavior of NiTi shape memory alloys sintered in the SPS process. Int. J. Miner. Metall. Mater. **26**(10), 1311–1321 (2019)
8. Peng, H.B., Yong, L.Q., Wang, S.L., Wen, Y.H.: Role of annealing in improving shape memory effect of as-cast Fe-Mn-Si-Cr-Ni shape memory alloys. Metall. Mater. Trans. A **50**(1), 3070–3079 (2019)
9. Zhang, J., Ma, Y.H., Wu, R.L., Wang, J.M.: Shape memory effect of dual-phase NiMnGaTb ferromagnetic shape memory alloys. J. Iron. Steel Res. Int. **26**(1), 321–328 (2019)
10. Haidar, M.A., Saud, S.N., Hamzah, E.: Microstructure, mechanical properties, and shape memory effect of annealed Cu-Al-Ni-xCo shape memory alloys. Metall. Microstruct. Anal. **7**(1), 57–64 (2018)
11. Shuwadi, M.F., Saud, S.N., Hamzah, E.: Deformation influences on microstructure, mechanical properties, and shape memory behavior of Cu-Al-Ni-xTi shape memory alloys. Metall. Microstruct. Anal. **8**(1), 406–414 (2019)
12. Gu, F., Zhang, L., Wang, W., Zhang, Y.Y.: Review of shape memory alloy pipeline coupling. Constr. Budg. **260**(12), 30–37 (2017) (in Chinese)
13. Zhang, H.B., Wang, J., Jin, W., Yang, R.: Coupling force simulation of TiNiFe shape memory alloy pipe-coupling. Chin. J. Nonferrous Meta. **20**(Special 1), 510–513 (2010) (in Chinese)
14. Zhang, H.B., Jin, W., Yang, R.: 3D finite element simulation of pull-out force of TiNiFe shape memory pipe coupling with inner convex. ACTA Metall. Sinca. **48**(12), 1520–1524 (2012) (in Chinese)
15. Gu, F., Song, J.R, Hou, Y.X., Zhang, L.: Research on connection performance of shape-memory-alloy pipe joint. J. Shenyang Jianzhu Univ. (Nat. Sci.) **36**(2), 307–313 (2020) (in Chinese)

16. Jin, L.L., Lu, S.Q., Li, G.F., Wang, K.R., Liu, J.W.: Effect of pre-deformed temperature on recovery property of $Ni_{47}Ti_{44}Nb_9$ alloy Φ8mm pipe joint. J. Nanchang HangKong Univ. (Nat. Sci.) **26**(4), 71–77 (2012) (in Chinese)
17. Chen, Q., Wang, K.L., Lu, S.Q., Li, G.F.: Numerical simulation analysis of Φ10mm NiTiNb shape memory alloy pipe-coupling. Hot Working Technol. **46**(2), 74–77 (2017) (in Chinese)
18. Auricchio, F., Taylor, R.L.: Shape-memory alloys: modelling and numerical simulations of the finite-strain superelastic behavior. Comput. Methods Appl. Mech. Eng. **143**(1), 175–194 (1997)
19. Xu, Z.L.: Elasticity. High Education Press, Beijing (2016)

A New Shingling Similar Text Detection Algorithm

Peng Li, Tianling Qiao, Yongxing Guang, and Lan Zhang

Abstract Since the traditional Shingling algorithm removes duplicate text in Chinese web pages, the amount of data that needs to be processed is particularly large, the efficiency of the algorithm is low. In this paper, a new text duplicate checking algorithm is constructed by changing the traditional Shingling algorithm. The meaningless dummy words are deleted in the text. Then the text is segmented according to semantics. Finally, the text similarity calculation formula is used to calculate the text similarity. The above work focuses on removing meaningless words from the text. In this way, the calculation rate of text similarity can be improved, and the accuracy of the text query and the complete query rate can be improved. The simulation results show that the algorithm is simple and feasible and has good text similarity calculation effect and certain advantages.

Keywords Space vector model · Text detection · Shingling algorithm · Word segmentation

1 Introduction

With the popularization of computers and the rapid development of the network, the number of Internets and various electronic documents has increased at an unprecedented rate, and the way people acquire knowledge has undergone profound changes. In the face of such a huge ocean of knowledge, it's very important to quickly find relevant information. If there is no effective organization and extraction method, the average user may take a longer time to find the information he wants than the information itself. Text similarity means that two or more texts a measurable parameter of the degree of matching between the two, if the similarity is large, indicating that the text-similarity is high, and conversely the text similarity is low. For text clustering, information retrieval, web page deduplication, text classification and many other

P. Li (✉) · T. Qiao · Y. Guang · L. Zhang
School of Science, Shenyang Jianzhu University, Shenyang, China
e-mail: andylee@sjzu.edu.cn

© The Author(s), under exclusive license to Springer Nature Singapore Pte Ltd. 2021   83
Y. Li et al. (eds.), *Advances in Simulation and Process Modelling*,
Advances in Intelligent Systems and Computing 1305,
https://doi.org/10.1007/978-981-33-4575-1_9

fields, the effective calculation problem of text similarity is the key to information processing.

Shingling algorithm is used to calculate the similarity of two documents [1–3]. The core idea of this algorithm is to transform the file similarity problem into the set similarity problem. When the document contains a large amount of data, the system overhead is high if similarity processing is performed on all shingle. In addition, due to the complex semantic and grammatical structure of Chinese text, when the common text similarity algorithm is used to classify specific Chinese text, the accuracy rate often fails to meet the requirements of relevant applications. To solve these problems, an improved Shingling (New Shingling) algorithm is proposed in this paper.

2 Text Similarity Detection Algorithm

2.1 Space Vector Model

In the late 1960s, the space vector model (VSM) was proposed by Salton et al. It is a kind of algebraic model and a widely used model [4]. The basic idea is: Assume that words are not related to words, the text is represented by a vector, each dimension corresponds to a separate word, then $(w_1, w_2, w_3, \ldots w_n)$ each entry a corresponding weight w_i, where the documentation d_k available vector $(w_1, w_2, w_3, \ldots w_n)$ representation. Document similarity meter in vector space model:

$$\text{sim}\,(d_k \times d_p) = \frac{\sum\limits_{i=1}^{n} w_{ki} \times w_{pi}}{\sqrt{\sum_{i=1}^{n} w_{ki}^2} \times \sqrt{\sum_{i=1}^{n} w_{pi}^2}} \tag{1}$$

where w_{ki} and w_{pi} are words t_i in d_k with d_p. The weight of n is the dimension of the vector. The premise of the vector space model is to assume that the words are not related to each other, but this assumption is unrealistic because there is often a semantic correlation of the words.

However, in terms of many research methods based on VSM words are used as feature items. In the actual context, the relationship between words often appears "skew", which is difficult to meet the orthogonal assumption. The use of word granularity will lead to low accuracy of text classification, which will affect the reliability of calculation results [5].

2.2 Shingling Algorithm

Shingling is mainly to find roughly the same documents, that is, the same content, except for changes in format, minor modifications, signatures, and logos. It can also find that one document is "substantially contained" in another. First, the concept of mathematics is used to strictly define what is called "approximately the same": The similarity between A and B of two documents is a number between 0 and 1, so if this number is close to 1, then the two documents are "approximately the same." The definition of inclusion is the same as this. The similarity and inclusion between two documents is calculated, and only a few hundred bytes of sketch are reserved for the two documents. Sketch's calculation efficiency is relatively high, and it has a linear relationship with the size of the document in time, and given two sketches, the calculation of the similarity and inclusion of the corresponding documents is linearly related to the size of the two sketches in time [6–9].

The algorithm treats the document as a sequence of words. First, it uses word to analyze it as a series of tokens, ignoring small details such as format, HTML commands, and case. Then it consists of document D and the substring of the token. The set $s(D, W)$ is connected [10].

The adjacent substrings in D are called shingles. Given a document D, define its w-shingling $s(D, W)$ as all the different shingles of size W in D. For example, (a, rose, is, a, rose, is, a, rose) the 4-shingling of is the set: {(a, rose, is, a), (rose, is, a, rose), (is, a, rose, is)}.

Given the size of the shingle, the similarity r between the two documents A and B is defined as

$$r(A, B) = \frac{s(A) \cap S(B)}{s(A) \cup S(B)} \tag{2}$$

Therefore, the similarity is a value between 0 and 1, and $r(A, A) = 1$, i.e., a document is 100% similar to itself [11–13].

Given a shingle size W, U is the set of all shingles of size W. Without loss of generality, U can be regarded as a set of values. Now set a parameter, for a set $W \subseteq U$, definition $\text{Min}_s(W)$ for

$$\text{Min}_s(W) = \begin{cases} \text{A set of the smallest elements in } W : |W| > s \\ \text{Others} \end{cases} \tag{3}$$

"Minimum" in formula (3) refers to the numerical order of the elements in U and defines $\text{Mod}_m(W)$. A set of all elements in set W divisible by m.

Theorem: let $\pi: U \to U$ U is a random arrangement selected uniformly by U, $F(A) = \text{MIN}_s(\pi(S(A)))$, $V(A) = \text{MOD}_m(\pi(S(A)))$, $F(B)$, $V(B)$ are also defined, then

$$\frac{|\text{MIN}_s(F(A) \cup F(B)) \cap F(A) \cap F(B)|}{\text{MIN}_s(F(A) \cup F(B))} \tag{4}$$

This is an unbiased estimate of the similarity of A and B

$$\frac{|V(A) \cap V(B)|}{|V(A) \cup V(B)|} \tag{5}$$

From above (3) and (4), we can select a random arrangement, and then keep a sketch for each document, which consists only of $F(D)$ and $V(D)$. Only through these sketches, we can estimate the similarity or inclusion of any pair of documents without the original file. In the system of text, the sketch method is generated as follows:

1. Remove the HTML format of the document and convert all text to lowercase;
2. The size of shingle is 10;
3. Use the improved 40-bit fingerprint function based on robin fingerprints to arrange randomly; and
4. Select the shingle by taking the remainder of the modulus, and choose the value of the dividend of the modulus as 25;

Specific steps to apply this algorithm to the entire network include:

1. Obtain documents on the Internet;
2. Calculate the sketch of each document;
3. Compare the sketch of each pair of documents to see if they exceed a certain similarity threshold; and
4. Cluster similar documents.

Shingle refers to a group of adjacent ordered words in the document. The shingling-based algorithm requires that a series of shingles are selected from the document, and then the shingles are mapped to the hash table, one shingling corresponds to a hash value, and finally, the same shingling in the hash table is counted. Amount or ratio is used as a basis for judging text similarity. Many references are commonly used single-based algorithms, such as reference. To achieve large-scale document detection, researchers have adopted different sampling strategies to reduce the number of shingled taken to compare.

2.3   Analysis of Shingling Algorithm

The Shingling algorithm is very simple, but it is impractical to be implemented simply. The size of the 30,000,0000 HTML and text document collections grabbed from the Internet is to be tested. $O(10^{15})$ This pairwise comparison is obviously not feasible. The amount of input data has great limitations on the design and algorithm of data structure. Each document in the data structure requires 4 M. Each document requires an 800-byte sketch. The calculation of each document in microseconds takes a total of 8 h. If the algorithm has random hard disk access or page activity, it is completely infeasible.

In the algorithm design in the text, a simple method is used to deal with the problem of large data volume: segmentation, calculation, and merging. The data is separated into blocks, each block is calculated separately, and then the results are merged. The size of the block is selected, which can be done in memory. The result of the merge is simple but very time-consuming, because it involves I/O operations. A single merge is a linear level, but all operations require $\log(n/m)$, so the complexity of the whole process is $O(\log(n/m))$.

Although the Shingling algorithm has a good mathematical basis and can give strict and accurate proofs, it is difficult to achieve because the intersection cost of calculating documents is too large. As a result, the original algorithm was improved by supershingle and sketching documents.

Heintze [14] selected Shinglings with the smallest Hash value and removes frequently occurring shingles. Bharat [15] selected shingles with a hash value of 25, and selects up to 400 shingles per document. Broder [16] combined multiple shingles to form a supershingle and calculates the similarity of documents by comparing the hash value of supershingle. Although the supershingle algorithm has a smaller calculation amount, Broder found that it is not suitable for the detection of short documents. Fetterly regarded five consecutive occurring words as one shingle, each document samples 84 shingles, and then combined these shingles into six supershingles; documents with two identical supershingles are considered as documents with similar content.

3   An Improved Shingling Text Similarity Detection Algorithm (New Shingling)

As in Chinese, real words in text often recognize the nature of text, such as nouns, verbs, adjectives, etc., while some functional words in text, such as prepositions, prepositions, conjunctions, etc., often have no effect on the category characteristics of the recognized text, that is, words that have no meaning in determining the category of text. If these functional words that are not meaningful to the text category are used as text feature words, they will bring a lot of noise, directly reducing the efficiency and accuracy of text classification. Therefore, when extracting text features, consider removing these function words that are not useful for text classification. Nominally, nouns and verbs are the most expressive features of the text category, so we can only extract nouns and verbs from the text as the first character words in the text.

On the other hand, for the problem of taking one piece of n Chinese characters in the text, in principle, the selection of word pieces is not divided in principle; it seems that it is possible to take several Chinese characters as one piece. But in fact, the appropriateness of the selection of slices directly affects the calculation of text similarity. There are too many Chinese characters in the phrase, although the amount of calculation is reduced, the accuracy may be reduced, such as "I love you Beijing." And "Beijing I love you." The text means the same. But if you use four words as a segment, I love you Beijing {I love you north, love you Beijing}; Beijing I love you {Beijing I love, Beijing I love you}; in fact, the meaning is totally different. Thus, the choice of word fragments is very important. Conversely, if the word fragments are too small, it will also cause the problem of inaccurate calculation.

New Shingling similar text detection algorithm implementation steps are as follows:

1. What can identify the textuality is often the real words in the text, such as nouns, verbs, adjectives, etc. Some function words in the text, such as interjections, prepositions, conjunctions, etc., do not contribute to the category characteristics of the identified text. These words have no meaning in determining text categories. Therefore, the function words in the text must be removed. Function words such as also, all, from, to, same, follow …

2. According to the Chinese language lexicon, the sentence is divided into one word, two words, or three words. According to the divided words, the words are divided into blocks. According to the meaning of the words, the block is divided into more than the Shingling algorithm. It is more feasible for each N character to take a word. According to the division of the meaning of words, the text can be divided into more practical, more accurate and meaningless words, so as to improve the similarity of the text, improve the degree of accuracy. According to the properties of common sub-sequences, the solution of the longest common sub-sequences can be obtained recursively. The formula is as follows:

$$
\text{LCS}(X_i, Y_j) = \begin{cases} \varnothing & \text{if } i = 0 \text{ or } j = 0 \\ \text{LCS}(X_{i-1}, Y_{j-1}) \cap x_i & \text{if } x_i = y_j \\ \text{longest}(\text{LCS}(X_i, Y_{j-1}), \text{LCS}(X_{i-1}, Y_j)) & \text{if } x_i \neq y_j \end{cases}
$$

$$(6)$$

whereas sequences $X = (x_1, x_2, \ldots, x_n)$, $Y = (y_1, y_2, \ldots, y_n)$, X_i, Y_j is the words blocks.

3. According to the calculation, the number of completely identical shingle:

   (a) Calculate the total shingle number of the two documents minus the consistent shingle number. This step uses Eq. (2).
   (b) Calculate the result of the number of consistent singles divided by the result of the number of inconsistent singles. This step uses Eq. (4).

4. According to the similarity between texts, follow-up work such as specific text classification.

$$
r(A, B) = \frac{2 * \text{length}(\text{SWAL}(A, B))}{\text{length}(A) + \text{length}(B)}
$$

$$(7)$$

where r is the text similarity, its value is $0 \leq r \leq 1$, A and B are two texts.

Table 1 Text set class

Computer software design	Medicine and health	Astronomy and geography	Computer image processing	Artificial intelligence	Entertainment magazine	Military world
18	8	13	10	13	11	7

4 Test Results and Analysis

4.1 Data Acquisition

Based on the theory of this article, the author made relevant experiments and selected 80 text collections. These texts are roughly divided into seven categories, including computer software design, medicine and health care, astronomy and geography, computer image processing, artificial intelligence, military world, entertainment magazines. Their distribution is shown in Table 1.

4.2 Simulation Results

First, we need to pre-process text messages, which are segmented of text and removed from function words. Second, calculate the weight of words, select entries with higher weights, and calculate the relevance of other terms to them to find potential feature words that are highly correlated. Finally, according to the New Shingling algorithm introduced in the article, it is rationally optimized to establish the feature vector of the text.

By using the Shining algorithm and the New Shingling algorithm to calculate the text similarity, the experimental results are as follows: Figs. 1, 2, 3, 4, 5, 6, and 7.

It can be seen from the above simulation results, the New Shingling algorithm text similarity algorithm is superior to the traditional Shingling algorithm.

Fig. 1 Computer software design text result

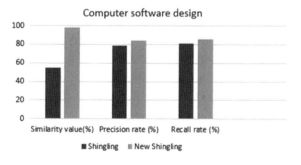

Fig. 2 Medicine and health text result

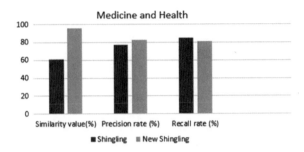

Fig. 3 Astronomy and geography text result

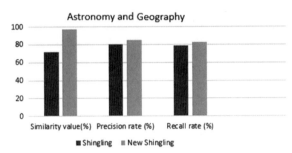

Fig. 4 Computer image-processing text result

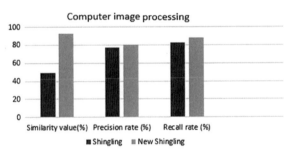

Fig. 5 Artificial intelligence text result

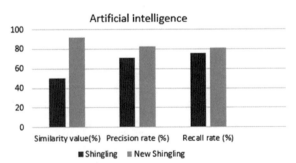

Fig. 6 Entertainment
magazine text result

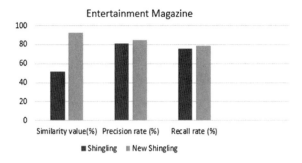

Fig. 7 Military world text
result

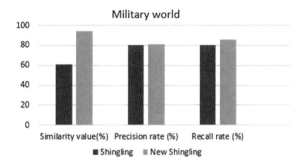

The purpose of the experiment was to compare the similarity of multiple texts to find duplicate text. The steps involved include text segments, high weight word selection, potential feature word selection, feature vector model establishment, and text similarity calculation. The text set used is relatively small, so the difference between the two methods is not obvious. If the text set is large, the method is considered to have more significant advantages in similarity accuracy.

The dash-dotted line is the fitting deviation curve of the OLS algorithm; the solid line is the fitting deviation curve of the weed optimization algorithm.

5 Conclusions

A new method of text similarity calculation proposed by the author, compared with the traditional text similarity calculation algorithm, uses a word meaning segmentation to express the characteristics of the text. It can not only simplify the calculation method, but also improve the text similarity calculation accuracy. Experiments show that the method is better than this method. In the future, according to the above shortcomings and the existing experimental results, the algorithm of this paper should be further improved and applied to the text similarity search system, to improve the efficiency of text retrieval.

Acknowledgements This research work has received some support from the basic scientific research projects of universities in Liaoning Province (Project Codes: LJZ2017029).

References

1. Gurmeet, S.M., Arvind, J., Anish, D.S.: Detecting near duplicates for web crawling. In: International World Wide Web Conference Committee (IW3C2), pp. 141–149 (2007)
2. Wu, P.B., Chen, Q.X., Ma, L.: The study on large scale duplicated web pages of Chinese fast deletion algorithm based on string of feature code. J. Chin. Inf. Process. **17**(2), 28–35 (2003) (in Chinese)
3. Mohammad, A.K., Vijaypal, S.D., Sanjeev, K.S.: Web crawler based on mobile agent and java aglets. Int. J. Inf. Technol. Comput. Sci. (IJITCS) **5**(10), 85–91 (2013)
4. Manish, K., Ankit, B., Robin, G., Rajesh, B.: Keyword query based focused web crawler. Procedia Comput. Sci. **125**, 584–590 (2018)
5. Armand, J., Edouard, G., Piotr B., Tomas, M.: Bag of tricks for efficient text classification. In: Proceedings of the 15th Conference of the European Chapter of the Association for Computational Linguistics, vol. 2, no. 1, pp. 427–431 (2016)
6. Jacob, D., Chang, M.W., Lee, K., Toutanova, K.: BERT: Pre-training of deep bidirectional transformers for language understanding. arXiv preprint arXiv:1810.04805 (2018)
7. Li, M.G., Ma, Z.X.: Chinese text similarity algorithm based on part-of-speech tagging and word vector model. In: 11th International Conference on Computer Science and Information Technology, vol. 5, no. 1, pp. 98–106 (2018)
8. Zhao, Q., Jing, Q., Li, A.P.: A short text similarity calculation method based on semantic and syntactic structure. Comput. Eng. Sci. **40**(7), 1287–1294 (2018) (in Chinese)
9. Arora, S.,Liang, Y.,Ma, T.: A simple but tough-to-beat baseline forsentence embeddings.In: Proceedings of the 5th International Conferenceon Learning Representations.Toulon. ICLR, vo. 1, no. 3, pp. 1–16 (2017)
10. Bojanowski, P., Grave, E., Joulin, A.: Enriching word vectors with subword information. Trans. Assoc. Comput. Linguist. **5**(1), 135–146 (2017)
11. Wang, Z., Mi, H.A.: Sentence similarity learning by lexical decomposition and composition. In: COLING 2016, 26th International Conference on Computational Linguistics, Proceedings of the Conference: Technical Papers, 11–16 December 2016, Osaka, Japan, pp. 1340–1349 (2016)
12. Grover, A., Leskovec, J.: node2vec: scalable feature learning for networks. In: Proceedings of the 22nd ACM SIGKDD International Conference on Knowledge Discovery and Data Mining, pp. 855–864. ACM (2016)
13. Li, F., Hou, J.Y., Zeng, R.R.: Research on multicharacter sentence similarity calculation method of fusion word vector. J. Comput. Sci. Technol. **11**(4), 608–618 (2017) (in Chinese)
14. Heintze, N.: Scalable document fingerprinting. In: Proceedings of the Second USENIX Electronic Commerce Workshop (Oakland), vol. 2, no. 5, pp. 191–200 (1996)
15. Bharat, K., Broder, A.Z., Dean, J.: A comparison of techniques to find mirrored hosts on the WWW. J. Am. Soc. Inform. Sci. (JASIS) **5**(2), 1114–1122 (2000)
16. Broder, A., Glassman, S., Manasse, S.: Syntactic clustering of the web. In: Proceedings of the Sixth International World Wide Web Conference (WWW), pp. 391–404 (1997)

Kalman Consensus Filtering Algorithm Based on Update Scheduling Scheme for Estimating the Concentrations of Pollutants

Rui Wang and Yahui Li

Abstract In this paper, a Kalman consensus filtering algorithm is proposed based on the status updates scheduling scheme (US-KCF) to estimate the concentrations of pollutants in the cabin. By introducing the concept of age of information (AoI), the freshness of status updates and the average AoI of wireless sensor network (WSN) are measured. Under the condition of network energy constraint, this paper designed a status updates scheduling scheme to minimize the average AoI of the network by selecting the status updates that need to be transmitted on the cluster head which can improve the convergence speed and energy saving performance of KCF algorithm. Simulation results show that compared with other consensus algorithms, this algorithm can get the estimation value of the target state more quickly and capable of reducing the network energy consumption effectively.

Keywords Average AoI · Schedule model · US-KCF modeling · Distributed WSN · System simulation

1 Introduction

Distributed estimation algorithms in WSN are focused on the network system with interferences. Distributed KCF algorithm is widely concerned because of its fast convergence speed, high fusion precision, and strong robustness. In the case of packet loss and path loss, Kalman consensus filtering algorithm is used to monitor the state of the target system with a better estimation effect [1]. In practical WSN applications, the energy, bandwidth, information computing, storage, and other resources of sensor nodes are limited. Therefore, how to reduce the system energy consumption becomes the key to extend the service life of the network. In [2], it introduced the event triggering mechanism into Kalman consensus filtering calculation and achieved the goal of reducing the energy consumption of the network by reducing the number of

R. Wang · Y. Li (✉)
College of Information Engineering and Automation, Civil Aviation University of China, Haihang Building, South Campus, Tianjin 300300, China
e-mail: 1805225032@qq.com

© The Author(s), under exclusive license to Springer Nature Singapore Pte Ltd. 2021 93
Y. Li et al. (eds.), *Advances in Simulation and Process Modelling*,
Advances in Intelligent Systems and Computing 1305,
https://doi.org/10.1007/978-981-33-4575-1_10

consensus calculations during samplings. Although existing studies have solved the problem of network bandwidth limitation to a certain extent, in the actual monitoring of the target system (such as monitoring the concentrations of pollutants in the cabin), the promptness of the estimated results is of great significance. If the current estimated value cannot reflect the concentrations of pollutants in real time, it may cause hazards such as delayed alarm. In terms of improving the real-time performance of the monitoring system, many scholars have carried out the following researches: In the embedded system, Zhou et al. [3] propose a status update scheduling scheme to minimize the average AoI of the system, so that the system has the better real-time performance. Tang et al. [4] built an asymptotic optimal truncated policy that can satisfy the hard bandwidth constraint under the power limit. This policy realizes the real-time update of user information by minimizing the average AoI of the system. Talak et al. [5] study the relationship between the average AoI of the system and the real-time performance, and simulations show that the smaller the average AoI of the system, the better the real-time performance. Based on the above research, in order to ensure that the distributed WSN can timely estimate the concentrations of pollutants in the cabin under the energy restriction, a Kalman consensus filtering algorithm based on the status update scheduling scheme is designed by minimizing the average AoI of the system.

The main contribution of this paper is to design a status update scheduling scheme to minimize the average AoI of the system under the constraint of network energy. Kalman consensus filtering algorithm based on this scheme can timely estimate the concentrations of pollutants in the cabin and the algorithm is superior to other consensus filtering algorithms in terms of consensus estimation and energy saving.

2 System Model

When using a distributed WSN to monitor the cabin pollutants concentrations, the cluster head receives status updates from sensor nodes and fuse measurement values of pollutants concentrations in a cabin with its neighbor nodes based on Kalman consensus filter to obtain the consensus-estimated values. Then, the consensus-estimated values are transmitted to the data center to judge whether pollutants concentrations exceed bid. The topology diagram of WSN is defined as $G = (V, E, A)$, where $V = \{v_1, v_2, \cdots, v_n\}$ is the set of sensor nodes within a cluster, and $E = V \times V$ is the set of edges between nodes. Define N_i represents the set of neighbor nodes for the node i, i.e., $N_i = \{v_j \in V : (v_i, v_j) \in E\}$. The state model of the target system and the observation model of the sensor are [1]:

$$\begin{cases} x_{i,k+1} = A_k x_{i,k} + B_k w_{i,k} \\ z_{i,k} = H_{i,k} x_{i,k} + F_{i,k} v_{i,k} \end{cases} \tag{1}$$

where $x_{i,k}$ and $w_{i,k}$ represent the state vector and the process noise vector of the target system, respectively. $z_{i,k}$ is the observation vector, and $v_{i,k}$ is the observed noise vector of the sensors. A_k and B_k are the system matrices with appropriate dimensions. $Hi_{,k}$ and $F_{i,k}$ are the measurement matrix and the fault matrix that are assumed to be invertible.

3 KCF Algorithm Based on Scheduling Scheme

The main content of this section has two parts: (1) Design a status update scheduling scheme by determining which status updates (generated by the nodes within the cluster) will be transmitted in the cluster head with the goal of minimizing the average AoI of distributed WSN. (2) Let the status updates obtained under the above update scheduling scheme participate in KCF algorithm to obtain the consensus-estimated value of the concentrations of pollutants in the cabin.

Sensors: The sensors monitor the real-time status of the cabin and it can be assumed that the sensor sends status updates to the cluster head based on a Poisson process with rate λ. This assumption satisfies the M/M/1 queue mode [6].

Energy harvesting (EH): It can be assumed that the energy supply process obeys the Poisson distribution with the parameter of η [7].

Transmit process: It can be assumed that the transmission service rate is μ following the M/M/1 queueing system [8]. Therefore, the service (i.e., transmission service) times follow an exponential distribution with μ, and $1/\mu$ is the mean service time.

3.1 Status Updates Scheduling Scheme

Aiming at minimizing the average AoI of the network, a status update scheduling scheme is designed to make the cluster head select or discard status updates to obtain the status update sequence, $X_{i,k} = \{\ldots, x_{i,k}, \ldots\}$ that needs to be transmitted.

According to [7], the average AoI ($\overline{\Delta}$) of network is determined by

$$\overline{\Delta} = \eta \left(\frac{1}{\mu} E[Y] + \frac{1}{2} E[Y^2] \right) \tag{2}$$

where Y is the time interval of two sequential transmission of updates (since the energy arrival rate is fixed to η and the total transmission interval of updates should be equal to the total arrival interval of energy units, the sum of Y_i would be fixed to n / η where n is the number of harvested energy units).

According to the Cauchy inequality [9], the minimum value of the average AoI can be obtained when Y_i is close to the mean value of $\sum_{i=1}^{n} Y_i$ i.e. $1/\eta$. At the same time, the optimal updates x_i can also be obtained by minimizing $|Y_i - 1/\eta|$.

Algorithm 1 Updates scheduling scheme

Input: $\{\cdots \; T(x_{i,k}) \; \cdots\}$, $\{E_1, E_2, \; \cdots, E_n\}$, $e, E, \; \eta$

Output: The update which will be transmitted

1: Init $i_{tx} = 0$;

2: **while** True **do**

3:      **if** Harvested energy arrives **then**

4:          $e=e+1$;

5:      **end if**

6:      **if** Update arrives **then**

7:          **if** $e{=}{=}E \; \vee \; (e > 0 \; \wedge \; abs(i_{tx} - 1/\eta) < abs[(i_{tx} + T(x_{next})) - 1/\eta])$

then

8:              Transmit the current update, $i_{tx} = 0$ and $e=e-1$;

9:          **else**

10:              Discard the current update; update i_{tx}.

11:          **end if**

12:      **end if**

13: **end while**

Algorithm 1 presents the updates scheduling scheme. The input includes the time intervals $\{\ldots \; T(x_i) \; \ldots\}$ of all updates $\{\ldots, \; x_{i,k}, \; \ldots\}$ the time intervals $\{E_1, E_2, \ldots, E_n\}$ of all the harvested energy units, the energy buffer capacity E and e is used to indicate the number of harvested energy units in the energy buffer, η is the harvested energy rate. The output returns the update which will be transmitted. For the arrived update, it will be transmitted for two cases: (1) when the energy buffer is full; (2) when energy is available and the current update minimizes $|Y_i - 1/\eta|$ based on the time interval of the next update (marked as $T(x_{next})$). Besides, if 7 is true, i_{tx} and e are updated. For all the other cases, this update is discarded and i_{tx} is updated. Thus, this scheduling scheme can effectively improve the convergence speed of the KCF algorithm and reduce the bandwidth pressure.

3.2 KCF Algorithm Based on the Updates Scheduling Scheme (US-KCF)

The cluster head brings the state update $\{x_1, x_2, \ldots, x_m\}$ $(m < N)$, which can minimize the average AoI of the network into the KCF algorithm introduced in [1], which not only can improve the convergence speed of filtering estimation and ensure the real-time monitoring of pollutants concentrations in the cabin but reduce the energy consumption of the network.

Algorithm 2 Kalman consensus filter algorithm based on scheduling policy

Input: The updates sequence $\{x_1, x_2, \cdots, x_m\}$ which can minimize the average AoI of system

Output: The Kalman Consensus state estimate value of the target detection system

1: Init $P_i=P_0$, $x_i=x_0$ at time $k=0$ and given messages $m_j = \{u_j, U_j, x_j\}, j \in N_i$;

2: Obtain measurement z_i with covariance R_i.

3: Compute information vector and matrix of cluster head node i

$$\begin{cases} u_{i,k} = (H_{i,k})^T (R_{i,k})^{-1} \alpha_{i,k} \cdot z_{i,k} \\ U_{i,k} = (H_{i,k})^T (R_{i,k})^{-1} \alpha_{i,k} \cdot H_{i,k} \end{cases}$$

4: Broadcast message $m_i = \{u_i, U_i, x_i\}$ to neighbor cluster head nodes.

5: Receive messages from all neighbors.

6: Fuse information matrices and vectors.

$$y_{i,k} = \sum_{j \in N_i} \beta_{i,k} \cdot u_{j,k} + u_{i,k}, \quad S_{i,k} = \sum_{j \in N_i} \beta_{i,k} \cdot U_{j,k} + U_{j,k}, \quad \forall j \in N_i$$

7: Compute the Kalman Consensus state estimate

$$\hat{x}_{i,k+1} = A_k \hat{x}_{i,k} + \varpi_1 \cdot A_k K_{i,k} (Z_{i,k} - H_{i,k} \cdot \hat{x}_{i,k}) + \varpi_2 \cdot A_k C_{i,k} \sum_{j \in N_i} [(1 - \theta_{ij}) \hat{x}_{j,k}^o - \hat{x}_{i,k})]$$

8: Update the gain matrix and error covariance matrix.

$$K_{i,k} = P_{i,k} H_{i,k}^T (H_{i,k} P_{i,k} H_{i,k}^T + R_{i,k})^{-1}$$

$$P_{i,k+1} = (A_k - A_k K_{i,k} H_{i,k}) P_{i,k} (A_k - A_k K_{i,k} H_{i,k})^T$$
$$+ A_k K_{i,k} R_{i,k} K_{i,k}^T A_k^T + B_k Q_{i,k} B_k^T$$

In addition, this paper will consider the packet drop phenomenon. Binary variables $\alpha_{i,k}$ and $\beta_{i,k}$ are defined to describe the packet arrival process on cluster head node i at time k. $\alpha_{i,k} = 1$ indicates the observed packet successfully received. Similarly, $\beta_{i,k} = 1$ indicates that the communication packet was received successfully. Furthermore, $P\{\alpha_{i,k} = 1\} = \varpi_1$, $P\{\beta_{i,k} = 1\} = \varpi_2$. When considering the path loss of wireless signal transmission, assume that θ_{ij} is the path loss rate between cluster head node i and cluster head node j. Algorithm 2 introduces the calculation process of US-KCF.

Input is status updates $\{x_i, x_2, \cdots, x_m\}$ and output is the estimation values of the pollutants concentrations.

4 Performance Analysis of the US-KCF Algorithm

The Monte Carlo method is used in the simulation process to carry out a large number of independently repeated experiments. Using the statistical mean value of each time, the error response of monitoring network is analyzed. By adopting the following performance indexes given in Wang et al. [10].

Mean estimation error (MEE) and mean consistency error (MCE) are

$$\text{MEE}_k = \sqrt{\frac{\sum_{i=1}^{m} (e_{i,k}^T e_{i,k})}{m}} \quad e_{i,k} = \hat{x}_{i,k} - x_{i,k}, \text{MCE}_k = \sqrt{\frac{\sum_{i=1}^{m} (\delta_{i,k}^T \delta_{i,k})}{m}} \quad \delta_{i,k} = \hat{x}_{i,k} - \frac{\sum_{i=1}^{m} \hat{x}_{i,k}}{m}$$

where k is the instantaneous time, and m is the number of updates.

In order to compare to the other method, ET-KCF algorithm in Wang et al. [2], the paper selects the same parameters for US-KCF algorithm proposed in this paper.

The initial value is set as $x_0 = (10, 8)^T$ and $P_0 = 8I_2$. Process noise and observed noise are independent Gaussian white noise with covariance of $10i$ and $100i$, respectively, where i is the update index. The initial energy of each node is 8 J. The path loss rate θ_{ij} between node i and j is 0.3, and the observed packet loss rate ϖ_1 and communication packet loss rate ϖ_2 are 0.4. $\mu = 0.5$, $\eta = 0.4$, $\lambda = 0.8$ ($\eta < \lambda$, s.t. energy is not enough).

As shown in Fig. 1, when packet loss and path loss exist, the ET-KCF algorithm is proposed in Wang et al. [2], and the US-KCF algorithm in this paper can converge stably. Besides, the US-KCF algorithm can converge after about 100 steps of sampling while the ET-KCF algorithm needs around 300 steps which can be seen the superiority of the proposed algorithm in convergence speed.

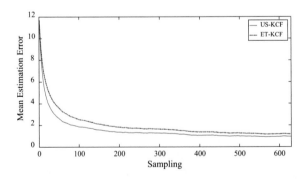

Fig. 1 Comparison of average estimation errors for different filters under packet loss and path loss

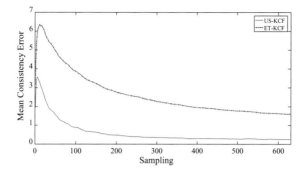

Fig. 2 Comparison of average consensus error for different filters under packet loss and path loss

According to Fig. 2, the average consensus error performance of US-KCF is superior to the ET-KCF. At the same time, the US-KCF algorithm can achieve consensus after about 200 steps of sampling while the ET-KCF algorithm needs around 400 steps which proves the superiority of the proposed algorithm in convergence speed also.

Figure 3 is a comparison of energy consumptions between US-KCF algorithm and ET-KCF algorithm. US-KCF consumes network energy after about 750 steps of sampling, while ET-KCF consumes energy after about 200 steps. The simulation results demonstrate that the US-KCF algorithm can reduce the bandwidth pressure while ensuring the promptness and accuracy of the estimation algorithm.

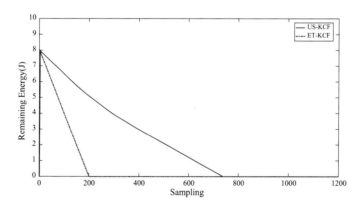

Fig. 3 Comparison of residual energy between two filtering algorithms

5 Conclusions

In this paper, a new Kalman consensus filter algorithm based on an update scheduling scheme is proposed to monitor the concentrations of pollutants in the cabin. Update scheduling scheme is designed to minimize the average AoI of the system to enable the US-KCF algorithm can obtain the consensus estimation value timely and accurately with the influence of observed packet loss, communication packet loss as well as the path loss in distributed WSN. Besides, the US-KCF algorithm can converge quickly and reduce energy consumption to a certain extent because of the scheduling scheme. Simulation results show that the algorithm has advantages in convergence speed, consensus estimation performance and energy saving.

References

1. Olfati-Saber, R.: Kalman-consensus filter: optimality, stability and performance. In: Proceedings of the 48th IEEE Conference on Decision and Control (CDC) held jointly with 2009 28th Chinese Control Conference, Shanghai, China, pp. 7032–7046 (2009)
2. Wang, R., Wang, X.Y., Sun, H., Chen, Z.Q.: Analysis of estimator and energy consumption with multiple faults over the distributed integrated WSN. Int. J. Model. Ident. Control **32**(2), 154–168 (2019)
3. Zhou, Z.M., Fu, C.C., Xue, C.J., Han, S.: Transmit or discard: optimizing data freshness in networked embedded systems with energy harvesting sources. In: 2019 56th ACM/IEEE Design Automation Conference (DAC), Las Vegas, NV, USA, pp. 1–6 (2019)
4. Tang, H.Y., Wang, J.T., Song L.Q., Song, J.: Scheduling to minimize age of information in multi-state time-varying networks with power constraints. In: 2019 57th Annual Allerton Conference on Communication, Control, and Computing (Allerton), Monticello, IL, USA, pp. 1198–1205 (2019)
5. Talak, R., Karaman, S., Modiano, E.: Distributed scheduling algorithms for optimizing information freshness in wireless networks. In: 2018 IEEE 19th International Workshop on Signal Processing Advances in Wireless Communications (SPAWC), Kalamata, pp. 1–5 (2018)
6. Kaul, S., Yates, R., Gruteser, M.: Real-time status: how often should one update? In: 2012 Proceedings IEEE INFOCOM, Orlando, FL, pp. 2731–2735 (2012)
7. Zhou, Z.M., Fu, C.C., Xue, C.J., Han, S.: Energy-constrained data freshness optimization in self-powered networked embedded systems. IEEE Trans. Comput. Aided Des. Integr. Circ. Syst. (2019, to be published)
8. Farazi, S., Klein, A.G., Brown, D.R.: Age of information in energy harvesting status update systems: when to preempt in service? In: 2018 IEEE International Symposium on Information Theory (ISIT), Vail, CO, pp. 2436–2440 (2018)
9. Steele, J.M.: The Cauchy-Schwarz Master Class: An Introduction to the Art of Mathematical Inequalities. Cambridge University Press (2004)
10. Wang, R., Li, Y.X., Sun, H., Chen, Z.Q.: Analyses of integrated aircraft cabin contaminant monitoring network based on Kalman consensus filter. ISA Trans. **71**(Pt 1), 112–120 (2017)

Modeling and Simulation
of Manufacturing and Production
Processes

Numerical Simulation of Welding Quality of Reinforcement Framework Under Different Welding Sequence

Shuwen Ren, Shizhong Chen, Zijin Liu, Zhongxian Xia, Yonghua Wang, and Songhua Li

Abstract The reinforcement framework is welded by the closed stirrup and the main reinforcement. The welding sequence of different welding points on the same stirrup has an important impact on the welding quality of the reinforcement framework. In order to optimize the welding sequence of reinforcement framework, the three-dimensional finite element model of reinforcement framework of six main reinforcement is established to simulate the maximum stress of reinforcement skeleton under three different welding schemes (Scheme 1: welding the upper and lower four welding points first and then welding the middle two welding points; Scheme 2: welding the middle two welding points first and then welding the upper and lower four welding points; Scheme 3: simultaneous welding of six welding points) Residual stress and strain are analyzed and compared to select the best welding scheme. The results show that the residual stress mainly occurs at the solder joint after welding, and the maximum stress value of the solder joint closest to the clamping end is 216.20 MPa. The thermal deformation caused by welding has little effect on the overall size. Compared with Scheme 1 and Scheme 3, the reinforcement framework welded under Scheme 2 has better welding quality.

Keywords Reinforcement framework · Welding sequence · History maximum stress · Residual stress · Strain

1 Introduction

The automatic weld-forming technology of reinforcement framework is a relatively new technology in the forming and processing of reinforcement framework at this

S. Ren · S. Chen · Z. Xia · Y. Wang · S. Li (✉)
School of Mechanical Engineering, Shenyang Jianzhu University, Hunnan District, 25 Hunnan Middle Road, Shenyang, Liaoning 110168, China
e-mail: rick_li2000@163.com

Z. Liu
China Academy of Building Sciences Limited Construction Mechanization Research Branch, 61 Jinguang Road, Guangyang District, Langfang, Hebei 065000, China

© The Author(s), under exclusive license to Springer Nature Singapore Pte Ltd. 2021   103
Y. Li et al. (eds.), *Advances in Simulation and Process Modelling*,
Advances in Intelligent Systems and Computing 1305,
https://doi.org/10.1007/978-981-33-4575-1_11

stage. Compared with the traditional framework forming and binding technology, it has many advantages, such as lower cost and higher work efficiency, and has gradually become the main processing method of skeleton forming [1, 2]. However, from the point of view of welding technology, the two welded parts to be connected are cylindrical surface with unsmooth surface, and the welding parts are in cross lap state, and the contact area is small, so the solder joint is small, which belongs to special spot welding. During welding, the welding heat effect caused by welding heat source will produce welding thermal stress and thermal deformation, which will affect the welding quality of skeleton to a certain extent. The process maximum stress and welding residual stress will directly affect the tensile and shear properties of the whole skeleton, and the welding thermal strain caused by welding heat will affect the dimensional accuracy of the skeleton to a certain extent [3].

With the continuous development of computer technology, the method of using numerical simulation to predict the residual stress in weldment has been widely used. Xu et al. [4] conducted numerical simulation on the residual stress distribution of CT70 continuous steel pipe under different welding parameters and found that the residual stress of extrusion amount and welding power is inversely proportional, while the welding speed is proportional to the residual stress. Qiao and Han [5] carried out simulation research on different shapes of plates and found that processing the plates into U-shape can significantly reduce the residual stress and strain caused by welding. The relationship between creep failure and residual stress of heat-resistant steel was studied in [6, 7]. It showed that excessive residual stress leads to direct failure of welded parts. However, the research on the stress and strain of steel skeleton in the welding process is still lacking.

Based on the above research, aiming at the improvement requirements of the existing steel framework welding equipment, the equipment transformation is carried out for the situation that the interference phenomenon of each welding joint will appear in the space when six welding points are welded at the same time, and the scheme of batch welding of six welding points on the same stirrup is proposed, and the finite element analysis is carried out. Two improved welding schemes are put forward: For Scheme 1, the upper and lower four solder joints are welded first, and then the middle two solder joints are welded; for Scheme 2, the middle two solder joints are welded first, and then the upper and lower four solder joints are welded; and the original six points simultaneous welding is set as Scheme 3. Through numerical simulation, the overall residual stress and deformation of the reinforced framework after welding under the three schemes can be compared and analyzed, and the optimized processing method is obtained.

Fig. 1 Finite element model
of reinforcement joint
framework welding design

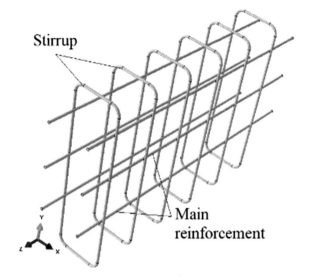

2   Modeling and Simulation Process Design

2.1   *Establishment of Finite Element Model*

The main components of reinforcement framework are main reinforcement and stirrup. As shown in Fig. 1, in order to adhere to the actual reinforcement framework specification, reduce the simulation calculation process and ensure the reliability of the simulation results, and the framework model is designed to be composed of six main reinforcements with length of 1 m and six stirrups (but only the first five stirrups are welded, and the sixth stirrup is regarded as the observation object), and the stirrup spacing is set as 150 mm. According to the characteristics of the whole structure of the reinforcement framework, the model is divided into 69,333 elements by tetrahedral meshing technology, and the grid independence is verified. Whether the number of elements continues to increase or decrease, the final calculation results are not affected. Therefore, considering the cost, time and accuracy of calculation results, it is appropriate to divide the model into 69,333 units. The meshing of solder joint is shown in Fig. 2.

2.2   *Establishment of Analysis Step*

According to the actual welding situation, the sequential coupling analysis method is adopted, and the subroutine is used to move the input of body heat source. Since the welding method of reinforcement framework is carbon dioxide protection welding,

Fig. 2 Grid division of solder

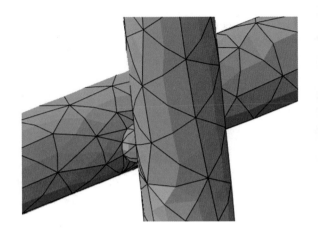

the moving speed of heat source in subprogram should be set as 0 m/s. Due to the limitation of heat source applied by subroutine, only single coordinate point can be input with heat source each time. Therefore, the subprogram is used for the design simulation process to apply body heat source to single solder joint. From the second simulation, the results of the previous simulation are used as the pre-defined temperature field. For the whole welding process, each welding may be 2, 4, or 6 points at the same time. Therefore, when using the last simulation results as the pre-defined temperature field in the same welding step, only one analysis step needs to be set, and the last simulation results are applied to this analysis step as the pre-defined temperature field. If two adjacent simulations of different welding steps are performed, two heat transfer analysis steps need to be set in the first simulation of the next welding step. The results of the previous simulation are applied to the first analysis step as the pre-defined temperature field of the first analysis step, and the setting is not transferred to the second analysis step, and the second analysis step is set as the welding step to be simulated this time. Therefore, for the whole model, a total of 30 times of temperature field simulation analysis are needed, and finally, the temperature field simulation results on the whole reinforcement framework after welding are obtained. After that, a static simulation is carried out. According to the results of temperature field after welding, a static analysis step is set, and full constraints are applied at one end of the six main reinforcements to simulate the clamping of the equipment on one end of the skeleton during welding. The constraint position is shown in Fig. 3. Finally, a static analysis step with a duration of 1 s is set to simulate the state after clamping removal, and the constraints set in the previous step are canceled in the analysis unit [8].

Fig. 3 Clamping force constraint

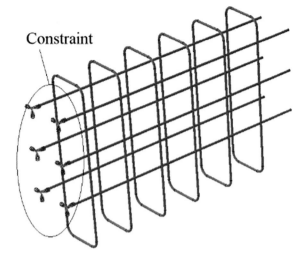

Constraint

2.3 Heat Input

According to the different input methods of heat source, the heat source can be generally divided into concentrated heat source, plane distribution heat source and volume distribution heat source. CO_2 gas shielded welding is a kind of high-temperature melting electrode produced by the contact between the weldment and the welding wire. After the welding rod melts, it forms molten droplets and enters into the molten pool under the influence of its own gravity and electromagnetic force, so as to fill the welding seam. According to the characteristics of CO_2 gas shielded welding, the fixed-point surface heat source is input to the weldment, so the plane Gaussian distribution heat source is selected as the heat source input form. Heat input is carried out for each solder joint on the model, and the coordinates of different solder joints are set in the subroutine to apply to the corresponding simulation [9]. According to the available data, the heat flux density input formula of moving plane Gaussian heat source is as follows:

$$q(r, x, y, z) = \frac{3\eta U I}{2\pi r^2} \exp\left(-\frac{3(z - vt)^2}{r^2}\right) \tag{1}$$

Because the heat source will not move in the actual welding process, the velocity $v = 0$ m/s; 'r' is the action radius of the heat source, according to the actual measurement, the solder joint radius $r = 6$ mm; '$q(r, x, y, z)$' is the instantaneous heat flux density of any spot; the thermal efficiency can be selected as $\eta = 0.6$; the welding current is $I = 300$ A, the welding voltage is $U = 25$ V; t is the welding time.

2.4 Relationship Between Stress and Strain Caused by Welding Heat

The heating model of cantilever welding is analyzed. The initial welding temperature is T_0 and the length is L_0. When it is heated to raise its temperature to T_1, its length will change to L_1, then the free deformation is ΔL_T:

$$\Delta L_T = L_1 - L_0 = \alpha L_0(T_1 - T_0) \tag{2}$$

where α is the coefficient of thermal expansion of the material. The free deformation rate ε_T is as follows:

$$\varepsilon_T = \Delta L_T / L_0 = \alpha(T_1 - T_0) \tag{3}$$

However, when the weldment is blocked and cannot be completely deformed, the deformation can only be partially realized, and the part that cannot be shown is internal deformation. According to Hooke's law, the stress and strain in the elastic range meet the linear relationship:

$$\sigma = E\varepsilon = E(\varepsilon_e - \varepsilon_T) \tag{4}$$

2.5 Parameter Setting of Skeleton Material

In the process of finite element analysis, the setting of material parameters is a crucial part, which directly affects the authenticity of simulation results. Especially in the simulation of welding process, the thermal effect of welding process makes the properties of materials change nonlinearly and transiently with the change of temperature. The HRB400 steel bar used in this simulation belongs to low carbon structural steel, and its physical property and mechanical properties can be obtained as shown in Table 1. In addition to setting the material parameters of the skeleton, there are some necessary parameters to be set, such as setting the ambient temperature to 20 °C, the thermal emissivity does not change with the temperature change and is set to 0.7 [10].

Table 1 Material parameters

Temperature/°C	Thermal conductivity/10^3 W m^{-1} °C^{-1}	Density 10^{-3} Kg m^{-3}	Specific heat capacity J kg^{-1} °C^{-1}	Poisson ratio/μ	Expansivity/10^{-5} °C^{-1}	Elastic modulus/10^5 MPa	Yield stress/MPa
20	0.050	7.80	460	0.28	1.10	2.05	400.0
250	0.047	7.70	480	0.29	1.22	1.87	375.0
750	0.027	7.55	675	0.35	1.48	0.70	280.0
1000	0.030	7.49	670	0.40	1.34	0.20	200.0
1500	0.035	7.35	660	0.45	1.33	0.19	150.0
1700	0.140	7.30	780	0.48	1.32	0.18	120.0

3   Analysis of Simulation Results

3.1   Comparison of Maximum Stress and Residual Stress During Welding Process Under Different Welding Schemes

According to the structural characteristics of the whole skeleton, the maximum stress and residual stress at each point of the skeleton are studied and compared. As shown in Fig. 4, the overall skeleton is ZOY plane symmetric, so only the stress values of points on one side are collected in X direction. After welding, the main stress concentration part of the skeleton should be at each welding point (in the actual welding, the connection failure between stirrup and main reinforcement will occur due to the process maximum stress or excessive residual stress). According to the simulation results, it can be seen that the stress is mainly concentrated at the solder joint, and there is a small stress distribution in the stirrup and other parts of the main reinforcement, as shown in Fig. 5. The solder joint is divided into several grid elements, and the largest maximum history variable unit is collected as the reference point. According to the requirements, collect the solder joints of three main reinforcement and stirrup on one side, a total of 15 points, marking the three main reinforcement as 'a', 'b', and 'c' in sequence, and mark the solder joints on each main reinforcement from left to right as number 1–5 solder joints.

From Fig. 6, it can be concluded that the maximum stress of the three main reinforcements on one side is kept at 40–60 MPa except for number 1 solder joint. The maximum stress is concentrated at number 1 solder joint, and the maximum value is 216.20 MPa. By observing and comparing the characteristics of the values of number 1 solder joint under the three schemes, the historical maximum stress values of number 1 solder joint on main reinforcement 'a' and main reinforcement

Fig. 4 Location and sequence number of acquisition points

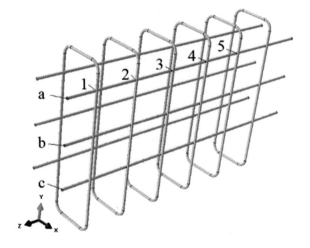

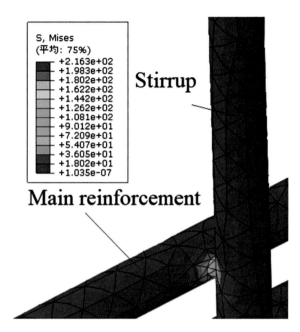

Fig. 5 Residual stress distribution at solder joint

'c' of Scheme 1 and Scheme 3 reach 100 MPa+ and 200 MPa+, respectively, even to the extent that the historical maximum stress values e of number 1 solder joint of main reinforcement 'c' in Scheme 1 is 216.20 MPa, while that of Scheme 2 is only 202.00 MPa on main reinforcement 'b' the historical maximum stress values is the minimum. Therefore, according to the collected historical maximum stress values of each characteristic solder joint, Scheme 2 has the characteristics of optimizing the historical maximum stress values at the solder joint.

The residual stress values of each characteristic solder joint were directly collected 1 s after the constraint was removed, and the following broken line diagram was made.

Observe and analyze the residual stress values of each characteristic solder joint recorded in Fig. 7. Similarly, several solder joints with maximum residual stress are studied and analyzed. In Scheme 1 and Scheme 3, the main reinforcement 'a' and the main reinforcement 'c' have the maximum residual stress at the number 1 solder joint, while, however, in Scheme 2, the maximum residual stress only occurs at the number 1 solder joint of main reinforcement 'b', and its value is smaller than the maximum residual stress 210.2 MPa in Scheme 1 and 201.6 MPa in Scheme 3. Except for a few maximum solder joints, the residual stress of other solder joints in each scheme has little difference, which is less than 30 MPa. Moreover, the number of maximum values in Scheme 2 is less, and the ratio of value to Scheme 1 and Scheme 3 is also lower. According to the analysis of the residual stress at each welding spot, it is shown that the residual stress of the whole skeleton after welding can be significantly improved by adopting Scheme 2.

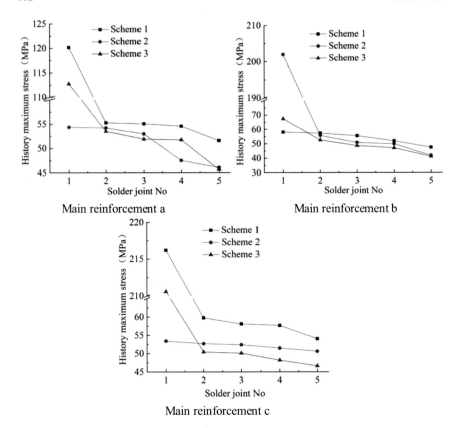

Fig. 6 Comparison of maximum stress of solder joint on main reinforcement under three schemes

3.2 Comparison of Different Strain Under Different Welding Schemes

To study the strain characteristics of the whole skeleton under different schemes, the strain at the maximum strain point of each point on the skeleton is mainly collected. According to the observation of the strain on a single stirrup, 4 characteristic points are collected on one stirrup as the observation points of strain value, as shown in Fig. 8, with a total of 20 recording points. They are numbered according to the sequence from left to right and from top to bottom.

In order to observe the strain of the skeleton at each characteristic point under each scheme, the collected data are made into a broken line chart, as shown in Fig. 9. Compared with the values in Fig. 9, the strain of Scheme 1 is smaller than that of Scheme 2 at each point, especially on the fifth stirrup, the value has a large deviation. It can be seen that the strain of each point in Scheme 2 is the largest. According to the above analysis, a smaller deformation can be obtained by adopting Scheme 1. Therefore, according to the deformation obtained from the simulation, Scheme 1 is

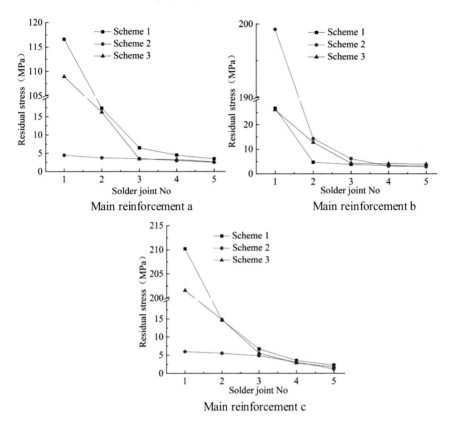

Fig. 7 Comparison of residual stress of solder joint on main reinforcement under three schemes

the better one. This is in contradiction with the results of maximum stress analysis and residual stress analysis.

3.3 Scheme Analysis and Selection

According to the analysis of the maximum stress and residual stress in the course, the quality of the steel frame formed by welding is better under Scheme 2, the welding dangerous points are less, and the residual stress at the welding joint is generally small. However, according to the strain, the overall strain of the welded reinforcement framework is smaller under Scheme 1. However, the quality after welding is comprehensively compared from two aspects of stress and strain, in which the contradiction of large residual stress, small strain and small residual stress and large strain appears. Therefore, it is necessary to continue to analyze the influence of strain characteristics and stress on the quality of the overall skeleton, determine the primary and

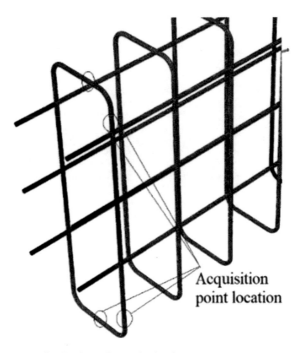

Fig. 8 Location mark of collection point on single stirrup

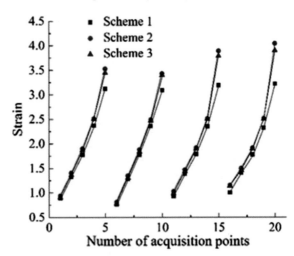

Fig. 9 Strain of each acquisition point

Table 2 Strain extremum under each scheme

Extremum program	$U1_{\text{Max,Min}}$/(mm)	$U2_{\text{Max,Min}}$/(mm)	$U3_{\text{Max,Min}}$/(mm)
Scheme 1	0.21, −0.17	0.95, −0.10	0.60, −3.50
Scheme 2	0.19, −0.15	1.12, −0.10	0.30, −3.90
Scheme 3	0.29, −0.25	1.01, −0.10	0.64, −3.80

secondary relationship, and then select the scheme. The strain characteristics were further analyzed, and the extreme values of skeleton strain components under each scheme were collected, as shown in Table 2.

From the analysis of the data in the table, the size of each strain component in each scheme shows $\Delta X < \Delta Y < \Delta Z$, and the strain in X direction and Y direction determines the change of cross-sectional area of skeleton. The change in X direction shows that the deformation of Scheme 1 and Scheme 2 is smaller than that of Scheme 3, and Scheme 2 is the smallest; the change in Y direction shows that the deformation of Scheme 1 is smaller than that of Scheme 2; the main strain occurs in Z direction, and the strain difference of three schemes is very small compared with that of Scheme 3, which can be ignored.

In Y direction, the difference of strain between Scheme 1 and Scheme 2 is 0.17 mm, and the overall length of the skeleton is 1 m. Therefore, for the skeleton of general length, the difference is the minimum, which has little impact on the overall quality of the skeleton compared with the residual stress of 100 MPa. Therefore, Scheme 2 is selected as the optimal scheme for welding.

3.4 Analysis and Discussion

According to the analysis results, there are many large residual stresses at the number 1 solder joint in each scheme. Therefore, the number 1 solder joint on the main reinforcement 'a' in Scheme 1 is selected as the research point, and the historical stress curve here is extracted. As shown in Fig. 10, the stress value of all the welding joints on the stirrup where number 1 solder joint is welded is up to 76.21 MPa. The value of the later stage is still increasing, which indicates that the heat input of the later welding also has an effect on the stress value of the solder joint. Moreover, the stress value of number 1 solder joint is much larger than that of the later solder joint due to its very close distance to the clamping point, which indicates that the near-point clamping results in greater stress. According to this phenomenon, a scheme can be put forward. In the welding of the overall main reinforcement framework, the first stirrup welded is designed as the preset stirrup, and it is removed from the overall skeleton after welding. In this way, there will be no extremely dangerous parts in the reinforcement skeleton, so as to ensure the quality of the whole skeleton.

Fig. 10 Stress extraction value at number 1 solder joint

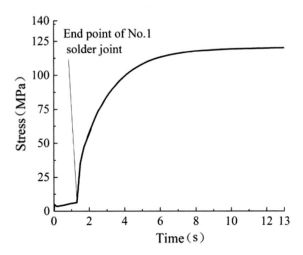

4 Conclusion

In this paper, the welding quality of reinforcement framework under three different welding sequences are numerical analyzed and discussed, and the following conclusions are drawn:

1. In the welding process, the welding spot and nearby heat affected zone appear larger stress, and the stress of base metal is smaller;

2. In the welding process of reinforcement framework, the welding spot on the first stirrup formed by welding has great welding residual stress. Therefore, the first stirrup in the reinforcement skeleton belongs to extremely dangerous type, the maximum residual stress can reach more than 200 MPa after welding, and the mechanical properties of steel bars in other parts are very poor;

3. In the welding process of reinforcement framework, the main direction of skeleton deformation occurs in its length direction (i.e., Z direction of this simulation). There is also a small amount of strain on the cross-section, which mainly occurs in the parallel direction of the side with more stirrups (i.e., the Y direction in this simulation process);

4. Through the comparison of the maximum stress and residual stress of the three schemes, it is found that the residual stress of the welded joints of the reinforcement skeleton is smaller than that of the first scheme, but the difference is smaller than that of the overall skeleton size, and the influence can be ignored. Therefore, in the comparison with the three schemes, Scheme 2 is selected as the optimal welding scheme.

Acknowledgements The authors appreciate the financial support provided by the National Key R&D Program of China (2017YFC0704002).

References

1. Chen, B.: The Automatic Control Technology of Forming and Welding for Reinforcing Cage Welding Machine. Changchun University of Technology (2018).
2. Liu, H., Sydora, C., Altaf, M.S., et al.: Towards sustainable construction: BIM-enabled design and planning of roof sheathing installation for prefabricated buildings. J. Cleaner Prod. **235**, 2289–1201 (2019)
3. Knoedel, P., Gkatzogiannis, S., Ummenhofer, T.: Practical aspects of welding residual stress simulation. J. Constr. Steel Res. **132**, 83–96 (2017)
4. Xu, Z., You, Q., Wang, Y., et al.: Numerical simulation of residual stress distribution of CT70 coiled tubing welded by high frequency resistance welding under different welding parameters. J. Netshape Forming Eng. **12**(1), 66–74 (2020)
5. Qiao, L., Han, T.: Effect of geometric shape of plate on residual stress and deformation distribution for butt-weld joint. China Welding **27**(3), 20–26 (2018)
6. Ni, Y., Xu, H., Chang, Y., et al.: Research on elastic-plastic creep damage of notched P92 steel specimens. Mater. High Temp. **35**(4), 335–342 (2018)
7. Zhang, J., Zhang, G., Guo, J.: Numerical simulation on interfacial creep failure of dissimilar metal welded joint between HR3C and T91 heat-resistant steel. J. Wuhan Univ. Technol.-Mate **31**(5), 1068–1074 (2016)
8. Mondal, A.K., Biswas, P., Bag, S.: Experimental and FE analysis of submerged arc weld induced residual stress and angular deformation of single and double sided fillet welded joint. Int. J. Steel Struct. **17**(1), 9–18 (2020)
9. Wang, P., Liu, Y., Chang, H., et al.: Brief analyses of thermo-mechanical coupling issue on welding structures. Trans. China Welding Inst. **40**(7), 006–011 (2019)
10. Jiang, W., Xu, X.P., Gong, J.M., et al.: Influence of repair length on residual stress in the repair weld of a clad plate. Nucl. Eng. Des. **246**, 211–219 (2012)

Theoretical Study and Experimental Test on Solenoid Actuator of Active Control Mount

Fang-Hua Yao, Rang-Lin Fan, and Song-Qiang Qi

Abstract Solenoid active control mount is excellent in vibration isolation, but the complex nonlinear relationship between electromagnetic force and exciting voltage/current makes it difficult to be controlled. The nonlinearity in solenoid actuator was studied by combining methods of theory, simulation, and experiment. Two theoretical equations for electromagnetic suction on voltage/current were obtained, then the equation with voltage was modified through electromagnetic simulation, and the test data were used to get the fitted parameters in equation with current. The results of simulation and theoretical calculation are consistent, and the fitted curve is close to the test, indicating the two equations are reliable. This reveals the nonlinear relationship between solenoid actuator and excitation and promotes follow-up control research.

Keywords Active control mount · Nonlinearity · Electromagnetic suction · Modeling · FEM · Automotive

1 Introduction

One of the main sources of vehicle vibration is the simple harmonic excitation of different harmonic orders and amplitudes generated by automobile engine [1]. The mounting between the powertrain and the frame can greatly reduce the vibration transmitted from the engine to the frame. Due to immature production technology and high price, active mount only appears in a few high-end cars. With the continuous improvement of new material technology and control level, active mount will be more

F.-H. Yao · R.-L. Fan (✉) · S.-Q. Qi
School of Mechanical Engineering, University of Science and Technology Beijing, Beijing 100083, P.R. China
e-mail: fanrl@ustb.edu.cn

R.-L. Fan
Shunde Graduate School of University of Science and Technology Beijing, Shunde 528309, P.R. China

© The Author(s), under exclusive license to Springer Nature Singapore Pte Ltd. 2021   119
Y. Li et al. (eds.), *Advances in Simulation and Process Modelling*,
Advances in Intelligent Systems and Computing 1305,
https://doi.org/10.1007/978-981-33-4575-1_12

and more popular [2, 3], so the research on active mount has practical engineering value.

The solenoid-type active mount is composed of a solenoid actuator on the basis of the traditional hydraulic engine mount (HEM). The actuator converts the electric energy into mechanical energy by using the unidirectional alternating suction produced by the electromagnet when alternating voltage/current being applied to the coil. These advantages of no permanent magnet, compact structure, wide frequency band, and large displacement make solenoid actuator widely used in active vibration control [4]. The principle of solenoid actuator is similar to electromagnet. Xiang et al. [5] using finite element method to obtain accurate magnetic field calculation results. The static and dynamic performance test of force and displacement type proportional electromagnet can be completed automatically by computer, and the corresponding performance index can be obtained [6]. In addition, according to the typical structure of the toroidal electromagnet, Mei et al. [7] have derived the calculation equation of the electromagnetic attraction of the circular ring electromagnet, which lays a foundation for the research of solenoid actuator. This paper focus on the nonlinearity of solenoid actuator in active force and input voltage/current.

2   The Structure of Solenoid Actuator

The structure of solenoid actuator is shown in Fig. 1. The actuator is formed by the coil, armature (mover), upper yoke, lower yoke, coil cover, and conical air gap, which can reduce the magnetic resistance, so that the magnetic flux generated by the energized coil is less leaked, the suction force is enhanced, and the power efficiency is improved.

The solenoid actuator is installed in the mount bottom case and incorporated with the decoupling membrane. When the coil is electrified, a magnetic field is generated, which produces a unidirectional suction on the mover; when the suction decreases, the mover moves upward by the elastic restoring force of the decoupling membrane.

When engine is running, unbalanced reciprocating inertia force and unbalanced overturning torque are transmitted to the frame end through the primary channel of the active mount [8, 9]. The working principle of solenoid actuator is that the coil generates alternating magnetic field through alternating current, which produces

Fig. 1 Solenoid actuator 3D model: 1. Decoupling membrane, 2. Ejector rod, 3. Ejector rod nut, 4. Coil, 5. Mount bottom case, 6. Positioning spring, 7. Upper yoke, 8. Sleeve, 9. Armature (mover), 10. Lower yoke

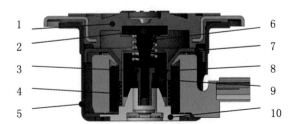

suction to the mover fixed on the decoupling membrane. The reaction force generated by the movement of the actuator counteracts the force transmitted by the engine to the frame end to reduce the vibration of the frame. Meanwhile, the dynamic characteristics of the HEM are also changed by the electromagnetic force. The active attenuating mechanism could be comprehended from the perspective of directly reducing the force transmitted from the engine to the frame end by the actuator, or from the perspective of changing the dynamic characteristics of HEM to harvest good vibration isolation.

3 Modeling of Solenoid Actuator

According to the structure of the actuator in Fig. 1, the mechanical model is shown in Fig. 2. The mathematical model of solenoid actuator is as follows:

$$m_3\ddot{y}_3 + c_4(\dot{y}_3 - \dot{y}_4) + k_4(y_3 - y_4) + c_3(\dot{y}_3 - \dot{y}_5) + k_3(y_3 - y_5) = 0 \tag{1}$$

$$m_4\ddot{y}_4 + c_4(\dot{y}_4 - \dot{y}_3) + k_4(y_4 - y_3) - f_a = 0 \tag{2}$$

$$m_5\ddot{y}_5 + c_5\dot{y}_5 + k_5y_5 + c_3(\dot{y}_5 - \dot{y}_3) + k_3(y_5 - y_3) + f_a = 0 \tag{3}$$

When the actuator works, in addition to the reaction force of electromagnetic force, the elastic force and damping force of decoupling membrane are transferred to the stator. Therefore, the net output force (positive upward) harvested on the stator is as follows:

Fig. 2 Solenoid actuator physical model

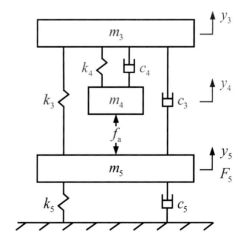

$$F_5 = -f_a + c_3(\dot{y}_3 - \dot{y}_5) + k_3(y_3 - y_5) = -(m_3\ddot{y}_3 + m_4\ddot{y}_4) \tag{4}$$

The equation shows that the force on the body end caused by the action of the actuator is essentially the sum of the inertia force of the mass m_4 of the mover, and the mass m_3 of the ejector bar, the decoupling membrane, and the attached liquid. The sum of the inertia forces is used to counteract the force transferred from the primary channel to the body end, so as to achieve the effect of vibration and noise reduction.

According to the mathematical model, the frequency response function (FRF) of net output force F_5 to electromagnetic force f_a can be obtained, and the FRF curve is obtained after the actual parameters of actuator in Table 1 are brought in, as shown in Fig. 3.

Table 1 Solenoid actuator parameters

Parameter	Name	Value
m_3	Mass of decoupling membrane/kg	0.266
m_4	Mass of mover/kg	7.100×10^{-2}
m_5	Mass of stator/kg	0.4
k_3	Dynamic stiffness of decoupling membrane/N·m^{-1}	8.641×10^4
k_4	Stiffness of mover/N·m^{-1}	1.547×10^6
k_5	Stiffness of stator/N·m^{-1}	2.855×10^6
c_3	Damping of decoupling membrane/N·s·m^{-1}	19.5
c_4	Damping of the mover/N·s·m^{-1}	16
c_5	Damping of the stator/N·m·s^{-1}	5.000×10^2

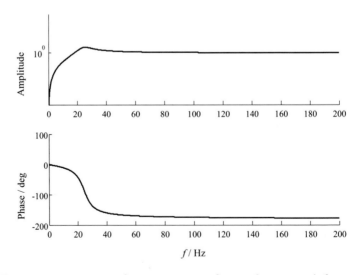

Fig. 3 Frequency response curve of actuator net output force to electromagnetic force

Fig. 4 Actuator mode curve

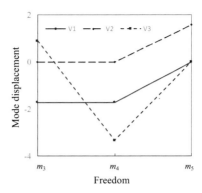

$$G(s) = \frac{F_5}{f_a} = \frac{-M_3 M_4 M_5 + M_3 M_4 C_3 + M_5 C_3 C_4 + M_5 C_4^2 - C_3^2 C_4 - C_3 C_4^2}{M_3 M_4 M_5 - M_4 C_3^2 - M_5 C_4^2}$$

(5)

where $M_3 = m_3 s^2 + (c_3 + c_4)s + (k_3 + k_4); \quad M_4 = m_4 s^2 + c_4 s + k_4$

$M_5 = m_5 s^2 + (c_3 + c_5)s + (k_3 + k_5); \quad C_3 = c_3 s + k_3; \quad C_4 = c_4 s + k_4$

It can be seen from Fig. 3 that the ratio of net output force to electromagnetic force is 1 in the frequency band of 25–200 Hz, that is, the net output force and electromagnetic force of the actuator are equal. Therefore, electromagnetic force can be used to replace the active force. Thus, the required electromagnetic force can be directly obtained by controlling the voltage or current to eliminate the vibration of the engine. The frequency band meets the requirements of active engine mount.

In fact, the modal analysis was shown in Fig. 4, which shows that the low-frequency peak value of the FRF in Fig. 3 is induced by the modal of the ejector rod, the decoupling membrane and its attached liquid mass m_3, and the mass m_4 of the mover on the elastic k_3 of the decoupling membrane. The modal frequency of this order is low, m_3 and m_4 vibrate in the same amplitude simultaneously, which was shown in the first mode shape in Fig. 4; the other two orders, the modal of the stator mass m_5 on the foundation elasticity k_5 and the modal of the mover mass m_4 on the ejector elasticity k_4, are higher than 400 Hz, which is far away from the working frequency band of the actuator.

4 Theoretical and Experimental Study on Solenoid Actuator

The results of the previous sections show that, in the working frequency band of the actuator, the active electromagnetic force f_a acts on the frame end with equal reverse force F_5, so as to attenuate the force transmitted from the passive channel to the frame end, achieving the purpose of vibration reduction. The following contents will study the relationship between the exciting current/voltage of solenoid actuator and active electromagnetic force.

4.1 Study on the Relationship Between Electromagnetic Attraction and Voltage/Current

The simplified structure of solenoid actuator is shown in Fig. 5. For the electromagnet with conical surface [10], the air gap magnetoresistance is:

$$R_\delta = \frac{1}{\mu_0 \times 10^{-2}\left(\frac{100\pi d_c^2}{4\delta\sin^2\alpha} - \frac{15.7d_c}{\sin^2\alpha} + 75d_c\right)} = \frac{10^8}{\left(\frac{0.139}{\delta}+1.452\right)} \tag{6}$$

where α is cone angle, 45°; d_c is outer diameter of mover, 26.5 mm; A_0 is cross-sectional area of air gap, 379.5 mm^2; μ_0 is air permeability, $4\pi \times 10^{-7}$ H/m; R_δ is air gap magnetoresistance, H^{-1}; δ is air gap length, m.

When the coil is energized, a certain amount of magnetic flux is generated on the magnetic circuit formed by the mover, upper yoke, lower yoke, and actuator shell. According to the theory of electromagnetism, electromagnet suction is $f = 10^7 B_0^2 A_0 / (8\pi)$, the premise that the core is not saturated, the magnetic field in the AC electromagnet is also AC. if the AC magnetic field $B_0 = B_m\sin\omega t$, the instantaneous value of electromagnetic attraction f, the maximum value F_m and the average value F_0 are, respectively:

$$f = \frac{1}{2}F_m - \frac{1}{2}F_m \cos 2\omega t \tag{7}$$

$$F_m = \frac{10^7}{8\pi} B_m^2 A_0 \tag{8}$$

$$F_0 = \frac{1}{2}F_m \tag{9}$$

Fig. 5 Simplified model of solenoid actuator: 1. Armature (mover), 2. Coil, 3. Outer shell, 4. Lower yoke

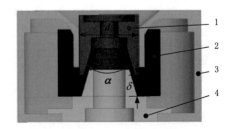

Based on the relationship $\Phi_m = B_m A_0$, and Ohm's law of magnetic circuit $\Phi_m R_\delta = N I_m$, the relationship between the maximum suction force F_m and the maximum current I_m is:

$$F_m = \frac{10^7}{8\pi} \frac{N^2}{A_0} \left(\frac{1}{R_\delta}\right)^2 I_m^2 = \left(\frac{0.00639}{\delta} + 0.06697\right)^2 I_m^2 \tag{10}$$

where N is the number of coil turns, with a value of 150.

For AC iron core-coil circuit, the relationship between the effective value of voltage U and the maximum magnetic flux Φ_m is $U = 4.44 f N \Phi_m$, combined with $\Phi_m = B_m A_0$, the relationship between the maximum suction force F_m and the effective voltage U is as follows:

$$F_m = \frac{10^7}{8\pi} \frac{1}{A_0} \left(\frac{1}{4.44N}\right)^2 \left(\frac{U}{f}\right)^2 = 2132.1 \left(\frac{U}{f}\right)^2 \tag{11}$$

Accordingly, the relationship between the amplitude of electromagnetic suction F_a and the maximum current I_m and the effective voltage U is as follows:

$$F_a = \frac{1}{2} F_m = \left(\frac{0.001598}{\delta} + 0.01674\right)^2 I_m^2 \tag{12}$$

$$F_a = \frac{1}{2} F_m = 1066.05 \left(\frac{U}{f}\right)^2 \tag{13}$$

4.2 Simulation Analysis Based on Ansoft Maxwell

This paper mainly simulates the transient magnetic field of solenoid actuator. Based on the actual structure and working principle of solenoid actuator, the simulation model is established and analyzed by Ansoft Maxwell. The finite element model (FEM) is shown in Fig. 6.

The relative permeability of the silicon steel sheet is 8000–10,000, the mover is pure iron, and the coil turns are 150. The size of the air gap is adjusted by the position of the actuator. When the air gap is 0, 2.5 and 5 mm, a voltage with effective value of 5 V and frequency of 50 and 100 Hz, respectively, are applied to simulate the magnetic induction strength, the magnetic flux generated at the air gap, the current in the coil, and the amplitude suction force of the actuator are extracted.

The simulation results and theoretical calculation results show that the current, magnetic flux, magnetic resistance, and electromagnetic force obtained by simulation have large errors compared with the theoretical calculation results. The causes of large error are analyzed as follows:

Fig. 6 Actuator FEM

(1) When the magnetic resistance equation in electromagnet design manual is applied to solenoid actuator, the actual structure between them is different, so the theoretical calculated reluctance is smaller than the simulated reluctance, and the simulation current is greater than the theoretical calculation current.

(2) There is magnetic flux leakage in the simulation process, which leads to the simulated magnetic flux less than the theoretical calculation, so the simulated electromagnetic force is less than the theoretical calculation.

In view of the fact that the electromagnetic force of theoretical calculation is increased by 30% compared with the electromagnetic force of FEM simulation, to sum up the error analysis and simulation results, Eq. (13) is modified as follows:

$$F_{a,m} = \frac{F_a}{1.3} = 820.04\left(\frac{U}{f}\right)^2 \tag{14}$$

The comparison between the calculation results by using the modified Eq. (14) and the simulation results is given in Table 2. The modified results agree well with the simulation ones, indicating that the calculation Eq. (14) of electromagnetic force on voltage is reliable.

4.3 Experimental Test on Solenoid Actuator

The experimental test is carried out. When the actuator being activated, the force sensor picks up the force transmitted to the frame end, and the current sensor picks up the loading current.

The first group of tests: the current is used as the excitation, the effective current remains unchanged, and the excitation frequency varies from 10 to 70 Hz. The test

Table 2 Comparison of simulation and theoretical calculation after voltage equation correction

	U/V	f/Hz	Φ_m/Wb	F_a/N
0 mm				
Theoretical results	5	50	1.502×10^{-4}	8.204
Simulation result	5	50	1.529×10^{-4}	8.195
Theoretical results	5	100	7.510×10^{-5}	2.050
Simulation result	5	100	7.430×10^{-5}	2.045
2.5 mm				
Theoretical results	5	50	1.502×10^{-4}	8.491
Simulation result	5	50	1.490×10^{-4}	8.861
Theoretical results	5	100	7.510×10^{-5}	2.123
Simulation result	5	100	7.780×10^{-5}	2.212
5.0 mm				
Theoretical results	5	50	1.502×10^{-4}	8.323
Simulation result	5	50	1.589×10^{-4}	8.841
Theoretical results	5	100	7.510×10^{-5}	2.081
Simulation result	5	100	7.350×10^{-5}	2.429

results of the maximum force and frequency under different current are shown in Fig. 7. The maximum force is basically not affected by the frequency, which is consistent with the theoretical result without frequency variable in Eq. (12), and is also consistent with the simulation results of the frequency response curve in Fig. 3 reaching horizontal straight line after 25 Hz.

Considering that the theoretical Eq. (12) of electromagnetic suction on current contains the air gap value δ between conical surfaces which is inconvenient to accurately measure. It can be seen from Table 2 when the air gap changes, the force almost

Fig. 7 Test curve of maximum force and frequency under different currents

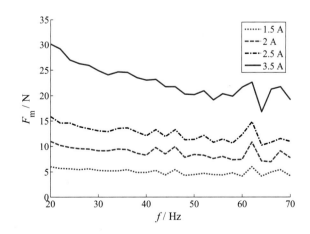

Fig. 8 Comparing
calculation results with test
results after fitting

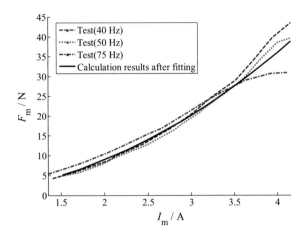

remains unchanged at the same frequency. Therefore, the second group of tests can be designed to fit the parameter δ of Eq. (12).

The second group of tests: the current is used as the excitation to which frequency is fixed but the effective value is gradually increased from 0 to 5 A; the maximum attraction force at multiple frequencies of 40 Hz, 50 Hz, 75 Hz, etc., were measured, respectively.

Fitting the test data at multiple frequencies to obtain $\delta = 0.00460$, it can be substituted into Eq. (12) for calculation, and the maximum electromagnetic attraction at 40 Hz, 50 Hz and 75 Hz is compared with the test results as shown in Fig. 8. The test curve is basically consistent with the equation calculation curve obtained after fitting the parameters, indicating that the parameter fitting of δ is successful. Therefore, the fitting value of δ can be directly used in Eq. (12), and the accurate electromagnetic suction equation for current can be obtained as follows:

$$F_0 = \frac{1}{2} F_m = \frac{1}{2} \left(\frac{0.00639}{0.00460} + 0.06697 \right)^2 I_m^2 = \frac{1}{2} \times 2.120 I_m^2 = 1.060 I_m^2 \qquad (15)$$

After obtaining Eq. (15), the problem that the required vibration isolation force cannot be calculated even when the air gap value could not be measured when the electromagnetic force is controlled by the current can be solved.

5   Conclusion

In this paper, the solenoid actuator is studied deeply, the frequency response curve of the net output of the actuator to the electromagnetic force is obtained through its mathematical and physical model. It is clear that the amplitude of electromagnetic suction and net output force are equal in the working range of the actuator, so the vibration transmitted from the engine to the body end can be offset by controlling electromagnetic suction.

Based on theoretical derivation, the electromagnetic attraction is studied. the calculation equations of electromagnetic suction with respect to voltage / current are obtained; electromagnetic simulation is carried out for the equation with voltage, and according to the error analysis between the theoretical calculation results and the simulation results, the equation is modified to make it close to the simulation results; In addition, the parameters in the electromagnetic suction equation about the excitation current are fitted through the test data. Finally, the simulation results are consistent with the modified theoretical calculation results, and the fitting calculation curve is basically consistent with the test curve, which shows that the calculation equation of electromagnetic suction of actuator about voltage / current obtained in this paper is accurate.

In conclusion, the research method of combining theory, simulation, and experimental test on solenoid actuator is effective and reveals the nonlinear relationship between solenoid actuator electromagnetic suction and excitation voltage / current, which lays a solid foundation for next research on the control strategy of solenoid active control mount.

Acknowledgements We are grateful to Anhui Eastar Active Vibration Control Technology Co. Ltd. and Anhui Eastar Auto Parts Co. Ltd. for providing us with the test site and related technical support. We are also grateful to the National Natural Science Foundation of China (No. 51175034) and the Scientific and Technological Innovation Foundation of Shunde Graduate School, USTB (No. BK19CE002) for supporting this work.

References

1. Sun, G.C.: Study on the Key Technology in Active Vibration Control of Automobile Powertrain. Jilin University, Changchun (2007).(in Chinese)
2. Min, H.T., Shi, W.K., Li, Yi., Cheng, M.: Dynamic characteristic simulation and experimental study on a semi-active powertrain mount. Autom. Technol. (07), 34–38 (in Chinese) (2007)
3. Lee, Y.W., Lee, C.W.: Dynamic analysis and control of an active engine mount system. Proc. Inst. Mech. Eng. Part D: J. Autom. Eng. **216**(11), 921–931 (2002)
4. Liang, T.Y., Shi, W.K., Tang, M.X.: The summary of study in engine mount. Noise Vib. Control **27**(1), 6–10 (2007). (in Chinese)
5. Xiang, H.G., Chen, D.G., Li, X.W., Wu, R., Liu, H.W., Geng, Y.S.: Construction of equivalent magnetic circuit for electromagnet based on 3-D magnetic field. J. Xi'an Jiaotong Univ. **37**(8), 808–811 (2003). (in Chinese)
6. Ma, L.L., Wang, J.Z., Zhao, J.B., Zhou, W.K., Zhu, D.H.: The study and application of the test system for static and dynamic character of propotional solenoid. Chin. Hydraul. Pneumatics **07**, 34–36 (2004). (in Chinese)
7. Mei, L., Liu, J.L., Fu, Z.Y.: Calculation of electromagnet attractive force and simulation analysis. Micromotors **45**(6), 6–9 (2012). (in Chinese)
8. Fan, R.L., Huang, Y.Y., Lu, Z.P.: The simplification of automotive power-plant mounting system and its illuminations. Mach. Des. Manuf. **20**(4), 8–10 (2010). (in Chinese)

9. Fan, R.L., Fei, Z.N., Qu, S.J., Shao, J.Y., Song, P.J.: Experimental study about the effect of powertrain mounting system on vehicle interior noise. Eng. Mech. **36**(9), 205–212 (2019). (in Chinese)
10. Wang, Z.W., Ren, Z.B., Chang, Z., Cui, P.F.: Simulation of the solenoid force based on ANSYS and AMESim. Missiles Space Vehicles **45**(6), 93–97 (2017). (in Chinese)

Research on the Problems and Countermeasures of Sustainable Development of Green Building Economy in China

Xuefeng Li and Lijuan Song

Abstract During the 13th Five Year Plan period, China's economic development has entered a new stage. "Green," "low carbon," and "ecology" in the construction field have become the global development trend. Based on the development status of green building economy in China, this paper expounds the meaning of green building economy and analyzes the necessity of its development. Then, combined with the current situation of China and developed countries, this paper summarizes and analyzes the shortcomings of green building economic development in China. Finally, through the method of investigation and literature review, it gives the opinions of sustainable development of green building economy. This paper has practical significance.

Keywords Green building · Economy · Sustainable development

1 Introduction

In recent years, China's construction industry has achieved rapid development. While people pursue the maximization of economic interests, environmental protection is gradually included in the category of considering the merits and demerits of buildings [1]. The practice of green building in China has also been steadily promoted (as shown in Fig. 1). As can be seen from the figure, the green building area in China was only 40 million square meters in 2011, and it has reached 1 billion square meters in 2017. According to the data comparison, the growth rate of green building area in China is obvious, and it has increased steadily year by year and has made gratifying achievements. However, because the development of green building in China is

X. Li
International College, Shenyang Jianzhu University, 25 Hunnan Middle Road, Hunnan District, Shenyang, China

L. Song (✉)
School of Management, Shenyang Jianzhu University, 25 Hunnan Middle Road, Hunnan District, Shenyang, China
e-mail: z1134407309@qq.com

Y. Li et al. (eds.), *Advances in Simulation and Process Modelling*,
Advances in Intelligent Systems and Computing 1305,
https://doi.org/10.1007/978-981-33-4575-1_13

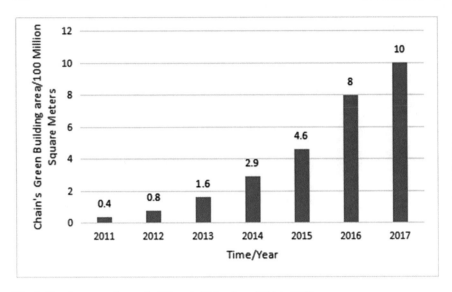

Fig. 1 Development of green buildings in China from 2011 to 2017

15 years later than that of developed countries, there are still many deficiencies. How to promote sustainable development of green building economy is still a difficult problem we need to explore [2].

2   The Meaning of Green Building Economy

The concept of green building was first proposed by American architect Paul in the 1960s. It rose in China in the 1990s and has made rapid development during the Eleventh Five Year Plan period [3]. Generally speaking, green building is to use environmental protection materials and equipment in various activities of construction, pay attention to the innovation of construction production technology, minimize the damage to the ecological environment, and save natural resources and energy to the greatest extent. In the process of planning, design, construction, operation, demolition or reuse, it puts forward high standards for architectural designers, managers, and construction operators requirements. It improves the quality of architecture, optimizes people's living environment and working environment as a whole, and realizes the harmonious coexistence of human and nature [4], which is an important embodiment of sustainable development in China.

Green building economy is an innovative concept of sustainable development. It is a building economy with comprehensive development and promotion of green building construction as the core [5].

3 The Necessity of Green Building Economic Development

With the continuous progress of society, the process of green and sustainable urbanization has become an inevitable trend. The development of green building economy promotes the development and utilization of new energy and the recycling of resources, which plays an extremely good role in promoting the process of urbanization [6].

The development of green building economy combined with ecological knowledge aims to create a comfortable living environment with energy saving and emission reduction, green and low carbon, create a good shopping experience for consumers, and improve people's quality of life. With the development of green building economy, resource saving and comfortable living are complementary to each other, showing the concept of people-oriented and sustainable development. It is an important measure to promote the harmonious coexistence between human and nature in China.

4 The Deficiency of Green Building Economic Development

4.1 Insufficient Government Support

The economic concept of green building economy was put forward relatively late in China. At present, there are some deficiencies in the government's cognition and people's recognition is not high. Based on this, China's green building economy in the rules and regulations, policies, legal system, and other aspects is not perfect, lack of good development space, in China is still in the concept and idea of the primary stage, compared with developed countries there is a big gap [7]. At the management level, due to the lack of government attention, there is no specific and perfect law to implement the protection of green building economy, resulting in a "cliff type" situation. The enthusiasm of enterprises is difficult to mobilize because of the lack of incentive measures, and there is no resultant force within the industry.

4.2 Less Practical Operation Experience in the Industry

There are many people in our country who think that green building must have high cost, first-class ventilation, lighting and heating technology at the same time [8]. Therefore, in the pursuit of the interests of the supremacy of the present, the application of green building in China is still relatively narrow, promotion is also concentrated in the city. Due to the lack of practical application, there is a lack of advanced technology for guidance, which leads to a vicious circle in which theory cannot be combined with practice. Enterprises cry for lack of capital and technical support, but high-end comprehensive talents will not rush to this industry. However, green building is a broad concept, such as some buildings with folk characteristics and perfect combination with nature, because it has the characteristics of low cost, energy saving and environmental protection, and embracing nature, it is also regarded as green building.

4.3   The Concept of Sustainable Development of Enterprises is Insufficient

Due to the lack of basic constraints of relevant laws and regulations, many enterprises in China's construction industry simply pursue economic interests. Green buildings are more like empty slogans of some enterprises, and the concept of sustainable development has not been implemented in daily work [9]. The relevant national level of publicity is not very good, incentive measures are not perfect, the concept of sustainable development of enterprises has not been implemented, resulting in the construction industry "bean curd residue" project, construction waste pollution of soil and water.

4.4   The Evaluation Mechanism of Green Economy is not Perfect

China's green building is still in the primary stage of development, and the excessive momentum of enterprises has led to some drawbacks in the green building economy. For example, some real estate developers have single green space area, but consumers have not satisfied with the green building in the later use process. The evaluation mechanism of green building economy in China cannot adapt to the current social development. So far, the mechanism is not perfect and cannot play an effective role in evaluation. Therefore, it is difficult to ensure the accuracy, consistency, and scientificity of the evaluation process [10].

5   Countermeasures for Economic Development of Green Building

5.1   Increase Policy and Legal Support

Green building economic market is an indispensable part of China's socialist market economy, and China's market economy has not yet formed a perfect system and mechanism. From the perspective of economic interests, some construction enterprises may ignore environmental protection. At this time, it is necessary for the government to improve the understanding of green building economy, pay attention to the development of long-term economic interests, formulate corresponding incentive measures, improve the enthusiasm of enterprises for innovation, increase the media publicity, and escort the development of green building economy with capital as the source power [11]. On the other hand, the government formulates corresponding laws and regulations to prevent accidents in advance and prepare for the rainy days. If the enterprises cause pollution, they should be severely punished, so as to form an atmosphere in which laws must be followed and law enforcement must be strict in the whole society.

5.2 Improve the Green Building Economic Evaluation Mechanism

The development of green building economy in China has been for a period of time, but it has not yet formed a perfect, scientific, and advanced evaluation mechanism matching with China's national conditions, which makes the promotion of green building economy lack of extensive foundation, which is also one of the important reasons why China's green building economy lags behind the western countries [12]. On the one hand, the government should develop a set of evaluation system, and then upgrade and optimize it based on China's national conditions, so as to provide a professional basis for the development of China's construction economy. On the other hand, enterprises should scientifically carry out their work according to the evaluation system, so as to make a virtuous circle of energy conservation and environmental protection, actively contribute to sustainable development, and refuse to focus on immediate economic interests and pick up small and lose big ones [13].

5.3 Cultivate Professional Talents

The development of green building economy is inseparable from the support of high-tech. The state should strengthen the training of talents, accumulate more practical operation experience through a large number of experiments and tests, and give play to the core value of green building economy. We will increase financial support and attract a large number of comprehensive talents who are proficient in the fields of architecture, law and ecology to enter the green building economy industry. Innovative technology is the key to eternal development. Enterprises should encourage employees to innovate, actively organize staff training, give full play to the industry's technical strength, reward advanced employees, and form an enterprise culture that encourages innovation [14].

5.4 Learn Foreign Advanced Experience and Technology

At present, it is necessary to enhance the exchange of advanced experience and technology with foreign countries. The development of green building economy in China lags behind that of developed countries and lacks practical operation experience. It belongs to the development mode of learning from others' experience and exploring by ourselves. In this regard, the government should strengthen cooperation and exchange, introduce technology, learn from the advanced experience and technology of developed countries, promote the healthy, stable and long-term development of green building economy in China, steadily build the environment of green building, and finally realize win–win cooperation and mutual benefit [7]. Through learning and exchange, the concept of development will be deeply rooted in the hearts of the people, and people's attention will be raised. The social foundation will be more solid, and the development of the industry will be longer.

6 Conclusion

Green building economy is the necessary trend of construction industry development, which plays a key role in promoting sustainable development and harmonious development between man and nature [15]. Compared with other countries, the development of green building economy in China is relatively late, and it is still in the primary stage in planning, design, construction, and so on. In the actual process, the effect is not obvious, and there are still many deficiencies, which need further research and exploration. With the progress of technology policy, the development of green building economy in China will be better and better, which is a coordinated project participated by the government and the people.

References

1. Wang, Z.Y., Qiu, F., Yang, W.S.: Solar water heating: From theory, application marketing and research. Renew. Sustain. Energy Rev. 01(41), 68–84 (2015)
2. Teng, J.Y., Xu, C., Ai, X.J.: Driving structure modeling and driving strategies of green building sustainable development. J. Civil Eng. Manage. 36(06), 124–137 (2019) (in Chinese)
3. Wang, Q.Q.: Review on development and standards for green buildings in China. Archit. Technol.49(04),340–345(2018)(in Chinese)
4. Huang, L., Wang, J.T.: Review on research progress of green buildings operation management. Constr. Econ. 36(11),25–28(2015)(in Chinese)
5. Huang,Q.R.: The main factors analysis on life cycle cost of green building. Constr. Manage. Modernization(04),49–51(2008)(in Chinese)
6. Luo, J., Guo, Y.F., Huang, Y.: Analysis of green building industry chain and its social and economic effects.Eng. Econ. 27(07),63–65(2017)(in Chinese)
7. Lu, B.B.: Analysis of the restricting factors in construction industry's promoting and developing green buildings. Eng. Econ. (02),81–85(2015)(in Chinese)

8. Xue, F., Shen, L., Qian, J.: Influential factors of green residential building's incremental cost.J. Civil Eng. Manage. **36**(01),194–200(2019)(in Chinese)
9. Han, Y.Y., Shen, L.Y., He, B.: Analysis of the restraining factors in promoting the development of green building based on ISM: a case study of Chongqing. Constr. Econ. **38**(02),26–30(2017)(in Chinese)
10. Wang, F., Chen, L.Q.: The analysis of problems and countermeasure in green building economic evaluation. Shanxi Archit. **41**(09),225–227(2015)(in Chinese)
11. Wang, X.L., Du, Z.F.: Study on the development countermeasures of green building under the low-carbon concept. Constr. Econ. **35**(06),80–82(2014)(in Chinese)
12. Ke, S.L.: Research on improvement and perfection of evaluation standard for green construction of building. Eng. Econ. **27**(03),69–73(2017)(in Chinese)
13. Wang, X.W., Zheng, X.X.: Research on the evaluation of green core competitiveness of construction enterprises. Eng. Econ. **27**(07),57–62(2017)(in Chinese)
14. Guo, H.D., Zhang, Y.X., Zhang, H.Y.: Mechanism construction and optimization for coordinated development of green building supply chain under the leadership of core enterprises.Constr. Econ. **40**(11),79–83(2019)(in Chinese)
15. Ye, Z.D.: Study on the macroeconomic impacts of green buildings. Urban Dev. Stud. **10**, 97–103 (2012). (in Chinese)

Research on the Balance of Automobile Mixed-Flow Assembly Line

Qi Ge, Qianqian Shao, Yunfeng Zhang, and Siqi Zhang

Abstract Mixed-flow production technology is widely used by automobile manu-facturers to cope with the fierce global competition. In order to improve the produc-tion efficiency of the mixed-flow assembly line in the assembly workshop of an automobile production company, the relevant parameters are determined. The 0–1 integer programming model with the smallest number as the objective function, linear interactive and general optimizer (LINGO), is used to deal with the model and verify the effect of the optimized balance scheme through FLEXSIM simulation software. The simulation results show that the rearrangement of each station makes the mixed-flow production line more stably, and the production efficiency is greatly improved, which provides a corresponding improvement plan for the production decision of the enterprise.

Keywords Mixed-flow assembly line · Simulation optimization · Commissioning · FLEXSIM

1 Introduction

As a State-owned key large-scale enterprise, automobile production has a large number of international advanced technology and equipment, which realizes the assembly of CNC machine tools and machining centers, automatic welding, computer-controlled online measurement and standardized inspection. However, there is still much room for improvement in the aspects of digitalization of workshop manufacturing process and scientific production scheduling of assembly line.

There are two problems to be solved in mixed assembly line optimization: produc-tion line balance problem and scheduling problem. Generally, a balanced decision is made first, and then the ranking of production is carried out according to the results of the balance. Taking the mixed-flow assembly line as the research object, priority is given to the rational distribution of tasks among the mixed-flow assembly

Q. Ge · Q. Shao (✉) · Y. Zhang · S. Zhang
School of Transportation Engineering, Shenyang Jianzhu University, Shenyang, China
e-mail: mr_Crowley@163.com

© The Author(s), under exclusive license to Springer Nature Singapore Pte Ltd. 2021
Y. Li et al. (eds.), *Advances in Simulation and Process Modelling*,
Advances in Intelligent Systems and Computing 1305,
https://doi.org/10.1007/978-981-33-4575-1_14

line workstations, and the uniform distribution of working time of each assembly line workstation in mixed-flow production is realized, making the operation of the assembly line in the mixed-flow production environment more stable and improve the output efficiency.

Based on the actual demand of the enterprise to improve the production beat and the real operation of the assembly line, combined with the balance theory, the mathematical model of the balance problem considering the mixed-flow production of different products is constructed, and the stations of the mixed-flow assembly line are rearranged. In order to better solve the practical problems, the research results provide a set of relatively complete balance optimization scheme of the mixed-flow assembly line for the final assembly workshop, and give a reasonable solution and theoretical guidance for the balance optimization of the production line, so as to improve the production efficiency. Reduce production costs, so that the variety, output, working hours, and labor load on the final assembly line to achieve a comprehensive balance.

2 Related Works

Boysen et al. [1] summarized the decision-making problems of parts logistics and distribution in automobile manufacturing industry and introduced the basic flow steps of automobile parts logistics and the main characteristics of mixed-flow assembly line logistics and distribution. Mohammed Alnahhal et al. [2] studied how to use forward-looking information to control the fluctuation of logistics in mixed-flow assembly line. Faccio et al. [3] considered the situation of multiple supermarkets supplying an assembly line and established a mathematical programming model to determine the number and location of supermarkets with the goal of minimizing transportation cost and fixed cost. In addition, Dong et al. [4] also study the problem of tractor scheduling in the case of tractor failure or the change of production sequence in the mixed-flow assembly line of the material supermarket and put forward a dynamic distribution strategy based on heuristics. Dong et al. [5] considered that in the mixed-flow assembly line, due to the different supplementary lead time, the change of production combination and the need to assemble different product models, all these pose a challenge to the traditional Kanban optimization method, so a new way to optimize the number of Kanban is designed in the literature. At the same time, Faccio et al. [6] also study an example of the application of a supermarket material distribution system driven by Kanban in the O-type assembly line.

In the current general research, the solution results cannot fully explain the various problems encountered in real life, resulting in the application of the results which is not high. In this paper, after the solution, the FLEXSIM software is used to simulate the results, which not only verifies the effectiveness of the overall optimization scheme of the mixed-flow assembly line, but also improves the overall output efficiency of the enterprise mixed-flow assembly line and optimizes the production management process.

3 Preliminary Balance Optimization of Production Line

Given the production time to solve the first kind of mixed-model assembly line balancing problems to minimize the number of workstations (MMALBP-1), the generalized mathematical programming model for this kind of problem is as follows:

Assumptions: (1) Demanding in the coming year is predictable, non-random demand;

(2) It is assumed that the number of assembly tools and the replacement time during the replacement of assembly products are not taken into account in the mixed-flow assembly line, the staffing is sufficient, and no buffer is set up; and.

(3) There are S stations on the mixed-flow assembly line, including a total of N processes, to produce M varieties of products.

Notations:

T: the effective working time in a balance cycle;

M: the number of product types;

N: the total number of jobs;

S: the total number of stations;

Dm: the predicted market demand for each type of product, non-random demand;

$D = \sum_{m=1}^{M} D_m$: the total market demand;

$C = \frac{T}{D}$: the average production beat;

$qm = \frac{Dm}{D}$: the proportion of product type m in the total demand;

t_{im}: the working time of the product m in the task i.

Decision variables:

x_{ik}: 0–1 variable

$$xik = \begin{cases} 1, \text{Indicates that job } i \text{ is allocated to station } k \\ 0, \text{Indicates that job } i \text{ is not allocated to station } k \end{cases}$$

AK: indicator variable of the station

$$AK = \begin{cases} 1, \text{Indicates that the } k\text{-th station is occupied} \\ 0, \text{Indicates that the } k\text{-th station is not occupied} \end{cases}$$

$$\min J = \sum_{k=1}^{S} A_k \tag{1}$$

St:

$$\sum_{k=1}^{S} x_{ik} = 1, \forall k \tag{2}$$

$$\sum_{k=1}^{S} k \cdot x_{ik} - \sum_{h=1}^{S} h \cdot x_{jh} \leq 0, \forall (i, j) \in P(i, j) \tag{3}$$

$$\sum_{m=1}^{M} \sum_{i=1}^{N} xik \cdot tim \cdot qm \leq C, \forall k \tag{4}$$

$$\sum_{i=1}^{k} x_{ik} \leq NAk(k \in S) \tag{5}$$

$$xik \in \{0, 1\}, \forall i, k \tag{6}$$

$$Ak \in \{0, 1\}, \forall k \tag{7}$$

Selecting the minimum number of workstations as the objective function (1), on the one hand, the reasonable distribution of workstations can reduce the number of workstations, improve assembly efficiency, and at the same time, reduce the number of assembly personnel and save labor costs. On the other hand, for the mixed assembly line with numerous assembly processes, the balance model of minimizing the number of workstations is selected with a given production beat. It can realize the segment balance of the complex large-scale mixed assembly line on the premise of ensuring the consistent production beat of the mixed-flow assembly line, accelerate the speed of solution, and make the solution of the theoretical model more practical. Constraint (2) means that each process can only be arranged in a workstation. Constraint (3) indicates the sequence constraint of job tasks. Constraint (4) indicates that the weighted load of each product on the workstation does not exceed the average production beat; constraint (5) means that if any work has been assigned to the workstation, the workstation must be turned on; constraint (6) and constraint (7) indicate that the task assignment and whether the workstation is enabled are 0–1 variables.

4 Solution and Simulation

4.1 LINGO Solution

At this stage, there are 31 processes in the preloading section, and the specific timing data are given in Table 1. $T1$ is the measurement time value of the light truck basic model, and the correction coefficient is 1, so the value of 1 is the normal time, the light truck standard time $T2$ is obtained from $T1 \times (1 + 10\%)$, the release rate is 10%, $T3$ is the measurement time value of the medium and heavy truck basic vehicle model, the correction coefficient and the relief rate are the same as the light truck, the $T4$ standard time is obtained by $T3 \times (1 + 10\%)$, and $T5$ is the weighted time of

Table 1 LINGO solution results

Number	Job number (option 1)	Station time/s	Job number (option 2)	Station time/s
Pre-installed-1	1	406	1	406
Pre-installed-2	2	386	3	653
Pre-installed-3	3	653	2	386
Pre-installed-4	4, 5	700	4, 5	700
Pre-installed-5	6, 9	700	6, 9	700
Pre-installed-6	7, 8, 11	600	7, 10, 11	718
Pre-installed-7	10, 12, 15	615	8, 12, 15, 19	663
Pre-installed-8	13, 16, 17, 18, 19, 20, 21	597	13, 14, 20, 21	363
Pre-installed-9	14, 22	635	16, 17, 18, 22	703
Pre-installed-10	23, 24, 25	600	23, 24, 25	600
Pre-installed-11	26, 27	656	26, 27	656
Pre-installed-12	28, 29	718	28, 29	718
Pre-installed-13	30, 31	612	30, 31	612

each process. The calculation equation is shown in Eq. (8). The priority relationship of each work time in the pre-installation section is shown in Fig. 1.

$$T5 = T2 \times q1 + T4 \times q2 \tag{8}$$

The LINGO 17.0 version is selected as the solving tool, and the error limit (Initial Nonlinear Feasibility Tol) of the initial nonlinear feasible solution is set to 0.001.

The LINGO operation result of the preloading section is shown in Fig. 2.

Analysis results of preloading section optimization.

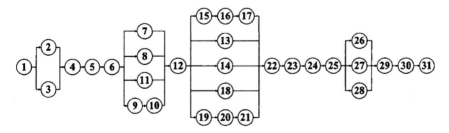

Fig. 1 Time measurement data of working procedure in preloading section

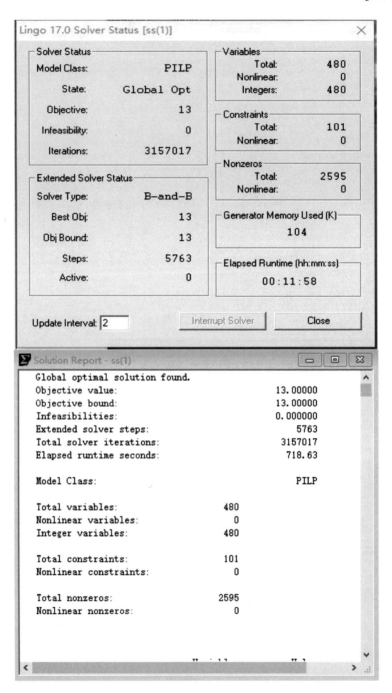

Fig. 2 LINGO running result

$$S_{min} = \left[\frac{\sum_{m=1}^{M} \left(Dm \sum_{i=1}^{N} tim \right)}{C \times D} \right] \qquad (9)$$

In the mixed-flow assembly line, the number of minimized workstations in the ideal state can be estimated by Eq. (9). The minimum number of workstations in the preloading section is 11, and the workstation set is set to 1–11. The LINGO solver shows that there is no feasible solution, and there are no feasible solutions in 12 stations. Then, we continue to increase the number of workstations, and the feasible solution of 13 workstations can be obtained after running the program for 11 min and 58 s. At this time, it can be considered that the solution of 13 workstations is the optimal solution to the balance problem of the preloading section.

Based on the LINGO software, two groups of schemes are set up, one group brings in the weighted average time, the other brings in different kinds of homework time, and the balance effect of the two groups is compared. Through the analysis of the solution results of the two sections, this paper thinks that the difference between the two data processing methods is limited to the solution difference of the iterative process within the program (LINGO integer programming model default branch and bound method to solve), and there is no difference between the advantages and disadvantages. The allocation scheme of LINGO workstations in the preloading section is given in Table 1, the station time is equal to the weighted time sum of each job, and the result satisfies the constraint condition of job sequence.

Therefore, based on the solution results of the previous section, in the solution process of the power section, it is divided into a power group (process 1–35) and a power group (process 36–76) for solution. After the program runs, both stages are obtained. The optimal solution is to minimize the number of workstations to 12 and 11, respectively. Based on LINGO's optimization results and later analysis, the balance optimization plan of mixed-flow assembly line is obtained as given in Table 2.

The power section contains 76 processes. For information on the detailed working time and priority of work, please see Appendix 3. The scheme of this paper is a 0–1 integer programming model. After analyzing the model, it is found that the number of variables in the power section is up to 1617, and the scale of the solution increases rapidly. The scale of solving this kind of equilibrium problem LINGO should not be too large, and the artificial intelligence algorithm is more suitable for solving large-scale equilibrium problems and achieving better performance. Therefore, based on the solution results of the previous section, in the process of solving the power section, it is divided into power group 1 (process 1–35) and power group 2 (process 36–76). On the one hand, it accords with the division of the actual production management organization of the enterprise. It is beneficial to the balance adjustment in the later stage, and on the other hand, it greatly speeds up the efficiency of solving the optimal solution. After the program runs, the optimal solution is obtained, and the number of minimized workstations is 12 and 11, respectively.

Table 2 LINGO solution result

Number	Job number (option 1)	Station time/s	Job number (option 2)	Station time/s
Power-1	1, 2, 4, 5, 6	653	1, 3, 6	583
Power-2	3, 7, 8	667	2, 4, 5, 7, 8	737
Power-3	9, 11	685	9, 11	685
Power-4	10, 14	674	10, 12	744
Power-5	13, 15	657	14, 15	517
Power-6	12, 17, 18	579	13, 17, 18	649
Power-7	16, 21, 22, 25	719	16, 19, 23	693
Power-8	20, 24	746	20, 21, 22	701
Power-9	19, 23, 27	730	24, 25	738
Power-10	26, 28, 29, 30	707	26, 27, 28, 29, 31, 32, 33	732
Power-11	33, 35	652	30, 35	690
Power-12	34	472	34	472
Power-13	37, 38, 39	354	36, 37, 38, 39	635
Power-14	36, 40	723	40, 41	535
Power-15	41, 42, 43, 44	741	42, 43, 44	648
Power-16	45	494	45, 46, 47, 48, 49	603
Power-17	46, 47, 48, 49, 50, 52	685	50, 52, 53	707
Power-18	51, 53, 57	736	51, 54, 57	645
Power-19	54, 55, 56, 58, 60	603	55, 56, 58, 59, 60	695
Power-20	59, 61, 64	727	61, 62, 63, 64	706
Power-21	62, 63, 65, 66, 67, 68, 69, 72	692	65, 66, 67, 68, 70, 72, 73	586
Power-22	70, 71, 73, 75	729	69, 71, 75	724
Power-23	74, 76	540	74, 76	540

4.2   Simulation Study and Analysis

Use FLEXSIM to realize the dynamic simulation of the scene and analyze all kinds of statistical data from the simulation output.

Since the assembly of a commercial vehicle requires many assemblies, parts, and components, for the sake of simplicity, the assembly process of each assembly workstation is represented by the assembly time of the vehicle on each workstation, without modeling each specific assembly. There are 34 stations in the assembly line optimization plan, including 12 stations in the preloading section (named as preloading 1–12 in turn) and 22 stations in the power section (named power 1–22 in turn). In the simulation scheme, the assembly line is arranged in the shape of "one,"

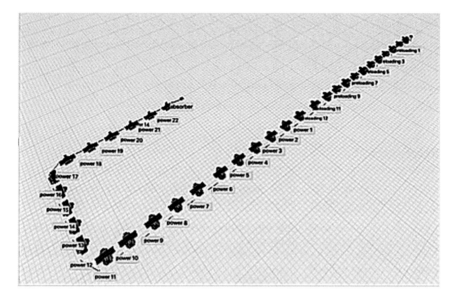

Fig. 3 FLEXSIM simulation

and each station module is connected in series to form a complete assembly line. The simulation model is shown in Fig. 3.

Under normal circumstances, the output of the simulation results is a total of 7046 vehicles, including 6342 light trucks and 704 medium and heavy trucks, with an average time of 8.51 h per vehicle from going online to assembling and going offline, although the output has increased compared with 5800 last year, but it has not yet reached the expected maximum design capacity of 10,000 vehicles.

5 Conclusions

This paper constructed a 0–1 integer programming model of mixed-flow assembly line balance with the optimal goal of minimizing the number of workstations, works out established the optimization scheme of station assignment by using LINGO software, and further adjusted and optimized the scheme by simulation software, which showed the rationality and feasibility of the optimization scheme.

For this kind of manufacturing enterprises implementing mixed-flow production mode, the research of this paper can make them realize the importance of balanced production and provide a relevant theoretical basis for them to make reasonable decisions according to the actual situation of their own enterprises. It also has certain application and promotion value to other manufacturing enterprises. It is of great significance to improve the overall efficiency of the assembly line, reduce WIP between processes, and pursue synchronous production.

Acknowledgements This research work is partially supported by the Scientific Research Project of the Educational Department of Liaoning Province 2019 of China (Project Codes: lnqn201916), Shenyang Social Science Project 2020 of China (Project Codes: SYSK2020-09-14, SYSK2020-09-32), Project Plan of China Logistics Learned Society and China Logistics and Purchasing Federation 2020 (Project Codes: 2020CSLKT3-187, 2020CSLKT3-188), and Logistics Reform and Teaching Research Project 2020 of China (Project Codes: JZW2020205).

References

1. Boysen, N., Emde, S., Hoeck, M., Kauderer, M.: Part logistics in the automotive industry: decision problems, literature review and research agenda. Eur. J. Oper. Res. **242**(1), 107–120 (2015)
2. Alnahhal, M., Noche, B.: Dynamic material flow control in mixed model assembly lines. Comput. Industrial Eng. **85**, 110–119 (2015)
3. Faccio, M., Cohen, Y., Alnahhal, M., Noche, B.: A genetic algorithm for supermarket location problem. Assembly Autom. (2015)
4. Dong, J., Zhang, L., Xiao, T.: Part supply method for mixed-model assembly lines with decentralized supermarkets. Tsinghua Sci. Technol. **21**(4), 426–434 (2016)
5. Dong, J., Zhang, L., Xiao, T., Li, H.: A dynamic delivery strategy for material handling in mixed-model assembly lines using decentralized supermarkets. Int. J. Modeling Simula. Sci. Comput. **6**(04), 1550038 (2015)
6. Faccio, M., Cohen, Y., Bortolini, M., Ferrari, E., Gamberi, M., Manzini, R., Regattieri, A.: New Kanban model for tow-train feeding system design. Assembly Autom. (2015)

Research on Scheduling Method for Uncertainty of Hit Rate of Molten Steel Based on Q Learning

Liangliang Sun, Tianyi Lu, Shuya Sha, Wanying Zhu, Qiuxia Qu, and Baolong Yuan

Abstract Aiming at the problem of uncertain scheduling of molten steel hit rate in the steel refining process, taking into account the multi-stage, multi-equipment, and multi-constrained production process conditions of refining production and the process of refining process due to the uncertainty of molten steel hit rate during the refining process, in order to obtain a scientific and feasible approximate optimal scheduling plan in a short period of time, the system state and system state transfer rules of the steel production process are defined, and the random evolution scheduling optimization system model of steel production refining based on the discrete-time Markov chain is established. At the same time, in the refining process scheduling optimization problem, the complexity of the solution will increase exponentially with the increase of the number of reprocessing processes, and a stochastic dynamic programming algorithm based on heuristic simulation strategy and improved Q learning is designed to solve the problem. Aiming at the uncertain scheduling problem of molten steel hit rate under different process production paths, simulation experiments using actual production data of a large domestic steel mill verify the effectiveness of the proposed model and algorithm.

Keywords Production scheduling · Molten steel hit rate · Markov chain · Q learning

1 Introduction

The typical steel production process includes iron making, steelmaking-continuous casting, and rolling [1]. Steelmaking-continuous casting production is the bottleneck of the entire steel production which includes three stages of steelmaking, refining, and continuous casting. The refining stage consists of RH, CAS, KIP, and other refining stations [2].

L. Sun (✉) · T. Lu · S. Sha · W. Zhu · Q. Qu · B. Yuan
Faculty of Information and Control Engineering, Shenyang Jianzhu University, Shenyang, China
e-mail: swinburnsun@163.com

© The Author(s), under exclusive license to Springer Nature Singapore Pte Ltd. 2021   149
Y. Li et al. (eds.), *Advances in Simulation and Process Modelling*,
Advances in Intelligent Systems and Computing 1305,
https://doi.org/10.1007/978-981-33-4575-1_15

As the refining process of the steelmaking-continuous casting intermediate link, from the perspective of the process, the process undertakes the ironmaking production process and closely follows the rolling production process, and its productivity is weaker than other production processes [3], so it is considered to be steel production bottleneck link; from the perspective of management level, not only Enterprise Resource Planning (ERP) and Process Control System (PCS), but also the core link of Manufacture Executive System (MES) [4]. In the refining stage, steel production enterprises usually arrange molten steel into refining equipment for different impurity removal methods for production according to customers' requirements for molten steel composition. The steel scheduling in the refined production process is based on the number of furnaces. According to the corresponding production process paths obtained by different furnaces in different attributes of the batch planning process in daily units, considering the constraints of the steelmaking-continuous casting production process, select each types of refining equipment corresponding to the production time of each furnace and the different equipment serial numbers under the same type of refining equipment [5]. As the core link of steel production, the steelmaking-continuous casting production process is disturbed by many uncertain factors [6]. After the occurrence of the molten steel hit rate that does not meet the standard with a high frequency occurs, it is necessary to determine the content of sulfur, phosphorus, and carbon in the molten iron in the unit of the furnace after the production of a certain equipment is completed at a certain stage. If the content of the component meets the standard, the next process is processed according to the established process path and static scheduling compilation rules; if the content of the component does not meet the standard, the accurate content of the component needs to be judged. However, due to the difficulty of guaranteeing the stability of the molten steel component treatment with the furnace as the production unit in the refining production equipment (RH type, CAS type, KIP type, etc.), the steel hit rate may not reach the standard (the composition of molten steel from a certain refining process in units of furnace times does not meet the established requirements). As a result, the molten steel needs to be returned to one or more of the previous processes for refining. If a batch of refining occurs due to unqualified molten steel quality, it will seriously affect the normal production of subsequent batches. As a result, the refining production process cannot be produced in strict accordance with the production scheduling plan formulated before production, which is difficult to guarantee. The entire steelmaking-continuous casting production rhythm affects the efficiency of steel production, increases the energy consumption of steel production, and increases the use of personnel costs. To this end, this paper proposes a study on the uncertain scheduling method of molten steel hit rate in the refining production process to ensure the smooth progress of steel production.

2 Scheduling Modeling

In general, when defining the system state of the refining production process scheduling problem, it is necessary to consider the number of furnaces that the steel production enterprise simultaneously performs refining production on the same set of refining equipment, the production status of each furnace at a certain time, and the number of ways for each refining process to complete the production task and the situation of each type of refining equipment being occupied at a certain time. According to the above requirements for the definition of the status of the scheduling system of the steel-refining production process, we assume that a steel production enterprise simultaneously carries out D refining production tasks on the same set of refining production equipment, and the completion of D refining production tasks requires E types refining equipment, the number of each refining equipment is $H_k (k \in \{1 \ldots E\}$ (which is a positive integer). The state of the refining production process scheduling system is defined as follows:

$$X = [s_1, s_2, \ldots, s_D, q_1, q_2, \ldots, q_D, R_1, R_2, \ldots, R_E, t]^T \tag{1}$$

where s_i is a positive integer, indicating the current refining production status of the i-th heat, $i \in \{1, \ldots, D\}$; q_i is an integer which represents the number of ways to complete production tasks in a refining production process of the i-th furnace; R_j represents the number of refining and processing equipment of the j-th type that is not occupied at the current moment, and $R_j \in \{0, \ldots, H_j\}$ is a positive integer ($j \in \{1, \ldots, E\}$, j is a positive integer).

In general, before defining the execution status of the refining production process in the refining production process, it must be ensured that at least one refining production and processing equipment required to execute this refining production process is available to execute this refining production process. According to the above requirements for the execution status of the production process in the refining production process, we define the execution status of the production process in the refining production process as:

$$U = [\beta_1, \beta_2, \ldots, \beta_D]^T \tag{2}$$

where U represents the production execution state of each furnace process in the refining production process; β_i represents the execution state of a production process of the i-th furnace, $\beta_i = 1$ or 0, ($i \in \{1, \ldots, D\}$, i is positive Integer), $\beta_i = 1$ means to execute a certain process of the i-th heat and $\beta_i = 0$ means not to execute a certain refining production process of the i-th heat.

During the refining production process, when a certain refining production process within a certain furnace is started or the task of a refining production process is completed, the state of the refining production process scheduling system will shift. If in the same state of refining production scheduling system, the production task of a production process of a certain batch of refining production is completed first, the state of refining production scheduling system will be transferred to the state of temporary refining production scheduling system. We assume that the initial state of the refining production scheduling system is:

$$x(0) = [1, \ldots, 1, 0, \ldots, 0, R_1, \ldots, R_E, 0]^T \tag{3}$$

where $x(0)$ represents the initial state of the refining production scheduling system; 1 represents the state of the production scheduling system where each furnace is ready to perform the first refining process; 0 represents the completion of each furnace performing the first refining process; $R_j \in \{0, \ldots, H_j\}$, $(j \in \{1, \ldots, E\}$, j is a positive integer, R_j' is a positive integer), which represents the number of the j-th refining and processing equipment that is not occupied at the current moment.

$$x(0) = [1, \ldots 1, 0, \ldots, 0, R_1, \ldots, R_E, 0]^T \tag{4}$$

Make reasonable arrangements for the refining equipment required for each refining production and the start processing time of each refining production process in each furnace. Therefore, in this paper, 'the minimum sum of the waiting time of the processing in each process of the refining production process' and 'the minimum difference between the ideal opening time and the actual opening time of each furnace in the steel-refining production process' are taken as the optimization goals, with 'the same refinement equipment handles adjacent furnaces at the same time without conflicting furnaces' as a constraint, and establishes the following mathematical objective function, $Q(x(k), u(k))$ means that refining is performed in the refining production state $x(k)$. The starting processing time for processing by the production action is $u(k)$, where $Q(x(k + 1), u(k + 1))$ represents the set of all refining production actions that can be selected for the refining production state $x(k + 1)$. The starting processing time for production, $g(x(k), u(k), x(k + 1))$ is the steel-refining production action $u(k)$ from the refining production state $x(k)$ to the steel-refining production. Time of state $x(k + 1)$:

$$\begin{aligned}
\text{Min}(Qx(k), u(k)) &= \alpha \text{Min}_{u(k+1)\in U^{x(k+1)}} E[Q(x(k + 1), u(k + 1))] \\
&+ (1 - \gamma)Q(x(k), u(k)) + \gamma\{[g(x(k), u(k), x(k + 1))\} \\
&+ |Q(x(k), u(k)) - T_i|
\end{aligned} \tag{5}$$

where $\text{Min} Q(x(k), u(k))$ represents the minimum processing time for the refining production system to perform the refining production action $u(k)$ in the refining production state $x(k)$; $Q(x(k), u(k))$ represents the processing time for the current refining production system to perform the refining production action $u(k)$ in the refining production state $x(k)$; $g(x(k), u(k), x(k + 1))$ represents the time from the refining production state $x(k)$ to perform the refining production action $u(k)$ for refining production to the refining production state $x(k + 1)$; $\text{Min}_{u(k+1)\in U^{x(k+1)}} E[Q(x(k + 1), u(k + 1))]$ means to perform the refining production action $u(k + 1)$ processing in the refining production state $x(k + 1)$, the minimum value of the expected value of time mathematics; γ represents the discount factor ($\gamma \in \{0, \ldots, 1\}$); α represents the learning coefficient ($\alpha \in \{0, \ldots, 1\}$); $\|Q(x(k), u(k)) - T_i\|$ means that the refining production system performs refining in the refining production state $x(k)$, the difference between the ideal opening time of the production action $u(k)$ and the actual opening time; T_i represents the furnace i ideal opening pouring time ($k \in \{1, \ldots, E\}, i \in \{1, \ldots, D\}$, k and i are positive integers), and the ideal opening time of furnace i needs to meet:

$$T_i > T_\varphi + T_\omega. \tag{6}$$

T_φ represents the processing time of the heat on the refining equipment; T_ω represents the sum of the waiting time of the adjacent processes of the heat on the refining equipment. Through Eq. (5), the performance indexes 'sum of waiting times of refining production process heat in each process' and 'difference between the ideal opening time and actual opening time of each heat in refining production process' are converted into optimization goals, Next, another performance index, 'processing adjacent furnaces on the same refining equipment within the same time without 'operation conflict,' is converted into constraint conditions by Eq. (7).

$$Q(x(k+1), u(k+1)) > Q(x(k), u(k)) + g(x(k), u(k), x(k+1)), \\ k \in \{1, \ldots, E\}; i \in \{1, \ldots, D\} \tag{7}$$

In Eq. (7), $g(x(k), u(k), x(k+1))$ is the refining production action $u(k)$ from the refined state $x(k)$ to the refined state $x(k+1)$'s production time.

3 Solution Methodology

The problem of refining production scheduling under the uncertain environment of molten steel hit rate is a large-scale flow shop scheduling problem, which is to solve the NP problem. Considering that the Q learning algorithm in reinforcement learning has adaptive, greedy search, and can quickly search for the optimal solution, but the traditional Q learning algorithm has the disadvantage that it cannot accurately select the next optimal state. Taking this problem into consideration, improvements are made on the basis of the traditional Q learning algorithm. Iterative calculation using the improved Q learning algorithm will save processing time in the refining production stage, improve the efficiency of refining production scheduling, and be able to better cope with actual refining production scheduling during the process, and a sudden situation occurs that the steel composition cannot meet the production requirements and need to be reprocessed. According to the scheduling mathematical model built in this paper, a method for solving the uncertain scheduling problem of molten steel hit rate in the refining production process using the improved Q

learning method is proposed. The following is the procedure of improving the Q learning algorithm.

Step 1: Define the state action pair of refining production process $(x(k), u(k))$, $x(k)$ represents the production state of each furnace in the refining production stage, $u(k)$ represents each furnace in the refining production stage for the same kind of refining production and processing equipment selection status $(k \in \{1, \ldots, E\}$ is a positive integer); and need to build a matrix of uncertain molten steel hit rate in the refining production process, the refining production process experience matrix R and refining production process learning matrix Q.

Step 2: Initialize the learning matrix Q of the refining production process and set the learning coefficient α;

Step 3: Select the state action pair $(x(1), u(1))$ of the first refining process in the refining production process as the initial state;

Step 4: Use the objective function formula to calculate;

Step 5: When refining production occurs on the same refining production equipment at the same time for adjacent processes of different times, select the probability by calculating the action, such as Eq. (8), arrange the order in which adjacent refining processes of different furnace times enter the refining equipment for production

$$P(a_i / S_t) = Q(x_k, u_k) / \sum_j Q(x_k, u_k) \tag{8}$$

where $Q(x_k, u_k)$ is the processing time matrix for production on the refining equipment u_k in the production state x_k; $\sum_j Q(x_k, u_k)$ means that in the j-th furnace production state x_k is selected in the refining equipment. The sum of the processing time matrices of all the processes performed on u_k, $u_k \in \{U\}$;

Step 6: Determine the update status of the learning matrix Q matrix in the refining production process. If the objective function is greater than 0, the learning matrix Q in the refining production process is updated. Otherwise, it will go back to step 4 and use the selected next production state as the initial state to perform iterative calculation using the objective function;

Step 7: Determine the convergence of the learning matrix Q in the refining production process. The conditions for the learning matrix Q to converge are: $\max_j \left| (Q_j^{i+1} - Q_j^i) / Q_j^i \leq 1 \right|$.

Step 8: Obtain the learning matrix Q for the condensed refined production process, then the algorithm ends.

At end of improved Q learning algorithm, it enters the manual online decision-making process, the dispatcher uses the Q learning algorithm to obtain the condensed refined production learning matrix Q. The molten steel produced by each process is individually selected according to the different production requirements of each furnace:

$$u^*(k) = \arg \max_{u(k) \in U^{x(k)}} Q(x(k), u(k)) \tag{9}$$

where $u^*(k)$ represents the optimal refining production equipment selection state; $\arg\max\limits_{u(k)\in U^{x(k)}} Q(x(k), u(k))$ indicates the state action pair that maximizes the value of $Q(x(k), u(k))$.

4 Simulation Study

A steel production enterprise simultaneously undertakes three refining production tasks with different process paths and furnaces. The equipment and corresponding quantities used by the enterprise to complete this refining production task are: two RH refining equipments, two KIP refining equipments, two LF refining equipments, and one CAS refining equipment, and steel production enterprises simultaneously undertake three different process paths for refining production flowcharts.

The improved Q learning algorithm solution strategy proposed in this paper is used to verify and solve the problem of uncertain scheduling optimization of molten steel hit rate in the refining production process of different process paths, and the learning matrix Q of the convergent refining production process is obtained. When it is necessary to carry out refining, the condensed refining production process learning matrix Q is used for manual online decision making to obtain a scientific and effective scheduling plan.

Using heuristic strategy simulation, the refining production process, the furnaces in each process have a small waiting time, and the small furnace production process first enters the refining production equipment for production and the furnace process with early opening time first enters the refining production equipment for production. Each time 1000 simulations are performed, and 2000 refining production process paths can be obtained. The simulation results are shown in Figs. 1 and 2.

In Figs. 1 and 2, the simulation strategy 2 is that the furnace process with the early opening time first enters the refining production equipment for production. The

Fig. 1 Simulation results 1

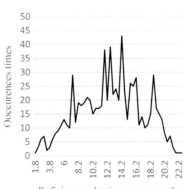

Refining production process path production time/hour

Fig. 2 Simulation results 2

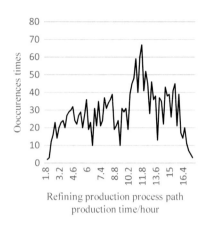

Refining production process path
production time/hour

simulation results are concentrated around the average production time of the refining production process path and the refining production process path in the simulation process. The average production time is 4.6 h ahead of simulation strategy 1.

5   Conclusion

This paper first describes the two performance indicators 'the sum of the waiting time of the furnace process in each process is the smallest' and 'the deviation between the ideal opening time and the actual opening time is small'; the Markov chain is used to describe the hit rate of molten steel uncertainty, define the refining production state, refining production process execution status, and refining production state transfer rules; establish a scheduling optimization mathematical model for the uncertain scheduling problem of molten steel hit rate in the refining production process. Then, a solution strategy for the refining production scheduling problem under uncertain environments of molten steel hit rate is proposed. First, the solution strategy is to use heuristic strategy simulation to reduce the optimal process path selection range and obtain the initial Q value of the iterative calculation of the subsequent improved Q learning method; second, using improved Q learning method to iteratively solve the refining production scheduling problem under uncertain environments of molten steel hit rate. In order to overcome the problem that the traditional Q learning algorithm cannot accurately select the optimal state action pair for each iteration calculation when solving the scheduling problem, use the Pareto solution set optimization solution idea introduces the action selection probability; finally, manual online decision making is used to obtain a scientific and feasible production scheduling scheme for the refining stage under the uncertain environment of molten steel hit rate. At the end of the paper, the solution strategy proposed in this paper is used to verify the case of molten steel hit rate uncertain scheduling under different process production paths. The verification results show that the proposed solution is feasible to solve the problem of scheduling optimization in the uncertain environment of molten steel hit rate in the refining production process, which promotes the research of industrial scheduling theory with the characteristics of large-scale mixed flow shop scheduling problems.

Acknowledgements The research is sponsored by the National Natural Science Foundation of China (61873174, 61503259), China Postdoctoral Science Foundation Funded Project (2017M611261), Science and Technology Projects of Ministry of Housing and Urban Rural Development (2018-K1-019), Liaoning Provincial Natural Science Foundation of China (20180550613, 2020-KF-11-07), Young science and technology innovation talent support plan (RC200003), and Liaoning Revitalization Talents Program (XLYC1807115).

References

1. Pang, X.F., Jiang, Y.C., Yu, S.P., Li, H.B., Gao, L., Che, Z.H.: Flexible job shop rescheduling method of steelmaking-continuous casting base on human-computer cooperation. Comput. Integr. Manuf. Syst. **24**(10), 2415–2427 (2018) (in Chinese)
2. Mao, K.: Lagrange Relaxation Level Optimization Method and its Application to Production Scheduling of the Steelmaking-Continuous. Northeastern University (2014) (in Chinese)
3. Liu, G., Luh, P.B., Resch, R.: Scheduling permutation flow shops using the Lagrangian relaxation technique. Ann. Oper. Res. **3**(4), 171–189 (1997)
4. Zheng, Z., Long, J.Y., Gao, X.Q., Gong, Y.M., Hu, W.Z.: Present situation and prospect of production control technology focusing on planning and scheduling in iron and steel enterprise. Comput. Integr. Manuf. Syst. **20**(11), 2660–2674 (2014) (in Chinese)
5. Sun, L.L.: Research on the optimal scheduling method for the productive process of steelmaking-refining-continuous casting. Northeastern University (2014) (in Chinese)
6. Tan, Y.Y., Huang, Y.L., Liu, S.X.: Two-stage mathematical programming approach for steelmaking process scheduling under variable electricity price. J. Iron Steel Res. (International) **12**(20), 1–8 (2013)

Research on Obstacle Factors of Project Operation and Maintenance Based on BIM Technology

Shengkai Zhao, Haiyi Sun, Zhe Huang, Ning Li, Mingze Guo, and Xiaohu Li

Abstract Building Information Modeling (BIM), as one of the emerging technologies in the development of the construction industry, is tried to be applied in the operation and maintenance of projects. The project operation and maintenance based on BIM technology is limited by various obstacles and has not been widely promoted. By reading relevant literature and combining with expert opinions, this paper has sorted out 13 independent influencing factors from five dimensions of industry, economy, policy, technology, and law. In this study, questionnaire survey and key interview were used to obtain preliminary data. Principal component analysis has been used to reduce the dimension of obstacle factors, and social network has been used to further analyze the centrality of obstacle factors. The data analysis results show that the deep influencing factors that hinder the operation and maintenance of BIM are insufficient technology and lack of policies, while "supervision intensity," "scientific research support," "compatibility," and "integrity" are important factors that hinder the development of BIM operation and maintenance. The results of this study can provide effective reference for the promotion and development of project operation and maintenance based on BIM technology, help improve the maintenance level of building equipment, and promote the efficient and sustainable operation of intelligent buildings.

Keywords BIM · Project operation and maintenance · Obstacle factors · Principal component analysis · Social network analysis

S. Zhao · Z. Huang · M. Guo · X. Li
Faculty of Information and Control Engineering, Shenyang Jianzhu University, Shenyang 110168, China

H. Sun (✉)
College of Science, Shenyang Jianzhu University, Shenyang 110168, China
e-mail: shy_xx@163.com

N. Li
College of Sciences, Northeastern University, Shenyang 110819, China

© The Author(s), under exclusive license to Springer Nature Singapore Pte Ltd. 2021
Y. Li et al. (eds.), *Advances in Simulation and Process Modelling*,
Advances in Intelligent Systems and Computing 1305,
https://doi.org/10.1007/978-981-33-4575-1_16

1 Introduction

The full life cycle of a project usually consists of four stages, including planning, design, construction, operation, maintenance, and abolition. If the project cost is planned according to four stages, the operation and maintenance stage usually accounts for more than 80% of the total cost, far exceeding the use cost of other stages. Thus, it can be seen that the efficient operation of project operation and maintenance can greatly save the use cost. With the continuous maturity of BIM technology in recent years, more and more researchers carry out in-depth research on BIM operation and maintenance. So far, the idea of BIM operation and maintenance has been applied in all aspects [1]. For example, BIM operation and maintenance is applied to the space management of educational office buildings to enable users to observe indoor facility layout and environment through three-dimensional layout drawings [2]. A 3D data cube model is established based on BIM, and a new operation and maintenance data mining method is adopted to improve the data-driven method of operation and maintenance management of large public buildings [3]. The bridge management system based on BIM technology realizes the collaborative management of different users and provides a beneficial platform for the maintenance and management of large bridges in China [4]. The management system of highway tunnel facilities based on BIM specification greatly improves the productivity of highway construction [5]. In spite of this, the BIM operation and maintenance system is still being continuously improved. For example, the establishment of EBS codes suitable for different stages of the project can greatly improve the compatibility of BIM operation and maintenance system [6]. Establish BIM - LCA/lCC framework for information integration and exchange in BIM environment, and overcome functional limitations caused by lack of BIM semantic information [7]. Clarifying the transformation mechanism between BIM and asset information model has also become an indispensable requirement [8]. BIM operations exposed a lot of deficiency in the process of practical application, the use of BIM operations reduced for complex interior design accurate representation, but further increased the difficulty of data processing [9], at the same time, the operational platform plus the lack of information sharing, the lack of specific information collecting channel for project and work [10], the compatibility and integration of BIM itself a limited development of BIM operations [11], and transformation mechanism and classification of BIM ops lack of unified standard, make different project information cannot be Shared to use [12]. BIM-based information modeling has changed the working mode of conventional information and created a new working mode that can change the design and construction industry [13]. However, the research on BIM operation and maintenance still needs to be further advanced, and energy management has become the main direction of BIM operation and maintenance [14]. In view of the current development status of BIM operation and maintenance, this study investigates and studies its main obstacles from five dimensions, namely industry, economy, policy, technology, and law and summarizes its biggest development weaknesses at present.

The remaining part of this paper is organized as follows. In Sect. 2, it makes an in-depth study of the principle of sample collection and sets up the obstacle factors by referring to literature. In Sect. 3, it adopts principal component analysis to reduce the dimension of obstacle factors and reduce the difficulty of data analysis. In Sect. 4, the node centrality of obstacle factors is analyzed by adjacency matrix, and the social network with obstacle factors as nodes is constructed. In Sect. 5, the conclusion is drawn.

2 Research Strategy

2.1 Questionnaire Design

Scientific principle: When promoting the index system of influencing factors of BIM technology-based project operation and maintenance application, rational analysis should be made, combining theory with practice, fully considering the characteristics of passive buildings, so as to objectively reflect the actual situation of BIM operation and maintenance application in all aspects.

Principle of feasibility: The selected evaluation indicators should be quantifiable and easy to collect data, and evaluation procedures and work should be made as simple as possible.

Principle of comprehensiveness: Through repeated comparison, the main factors affecting BIM operation and maintenance are screened out.

2.2 Pre-survey and Improvement of the Questionnaire

In order to conduct a field test on the contents of the survey plan and understand whether the survey work arrangement is reasonable, the team conducted a pre-survey test on the questionnaire. The objects of this survey are designers, and real estate developers from architectural design institutes involved in architecture-related projects. This survey is conducted in the form of offline distribution, mainly for architectural design institutes.

2.3 Obstacle Factors

Industry dimension: degree of emphasis, energy, data sharing.
Technical dimension: compatibility, coding system, integrity.
Economic dimension: research cost, training expenses, revenue cycle.
Legal dimension: laws and regulations, standardization.

Policy dimension: supervision strength, scientific research support.

Technology, economics, policy, legal, and policy obstacles are set up separately, among which obstacles indicators are defined in combination with labels. For example, the compatibility is set as F_4, and the survey results are analyzed by principal component analysis in combination with the questionnaire data.

3   Data Analysis

In practical analysis, when the Kaiser–Meyer–Olkin (KMO) statistic is above 0.8, the effect of factor analysis is generally better. Through the analysis of the questionnaire data in this paper, the KMO value was greater than 0.9 and passed the Bartley ball test with a significance level of 0.00, indicating that the questionnaire had a good structure and can be used for factor analysis (Table 1). The data of this table comes from the analysis of online questionnaire.

3.1   Principal Component Analysis

Based on the questionnaire, the two main components of the obstacle reduction factor are obtained. The detailed meaning of indicators is shown in Table 2.

$$A_1 = 0.655F_1 + 0.626F_2 + 0.738F_3 + 0.493F_4 + 0.699F_5$$
$$+ 0.445F_6 + 0.690F_{10} + 0.703F_{11} + 0.634F_{12}$$
$$+ 0.721F_{13} \tag{1}$$

$$A_2 = 0.611F_4 + 0.681F_6 + 0.715F_7 + 0.841F_8$$
$$+ 0.766F_9 + +0.356F_{12} + 0.327F_{13} \tag{2}$$

The cumulative contribution rate of the extracted principal component was more than 70%, component 1 was 60. 547%, and the total cumulative contribution rate was 85. 756%. Principal component 1 for F_1, F_2, F_3, F_5, F_{10}, F_{11}, F_{12}, F_{13}. Principal component 2 was more dependent on F_4, F_6, F_7, F_8, F_9.

Table 1 Bartley ball test

KMO sampling suitability		0.912
	Approximate Chi square	2828.884
Bartlett sphericity test	Degree of freedom	186
	Significant	0.000

Table 2 Node degree centrality

Index	Code name	Barriers	Degree	NrmDegree
1	F_1	Degree of emphasis	48.000	40.000
2	F_2	Synergy	46.000	38.333
3	F_3	Data sharing	58.000	48.333
4	F_4	Compatibility	56.000	46.667
5	F_5	Coding system	44.000	36.667
6	F_6	Integrity	54.000	45.000
7	F_7	Research cost	50.000	41.333
8	F_8	Training expenses	34.000	28.333
9	F_9	Revenue cycle	46.000	38.333
10	F_{10}	Laws and regulations	54.000	45.667
11	F_{11}	Standardization	46.000	38.333
12	F_{12}	Supervision strength	62.000	51.667
13	F_{13}	Scientific research support	62.000	51.667

4 Social Network Analysis

Different from the traditional data analysis, social network focuses on the analysis of the relationship between variables and the impact on the whole network structure. When the input variables are relatively small, the training of neural network is sometimes affected by the input order and has some defects. In order to improve the reliability of data analysis, this paper USES social network to analyze the questionnaire data again.

In this paper, the obstacle indexes that affect BIM operation and maintenance are set as nodes of the social network, and the relevant parameters between each index are set as connections of the neural network. By combining the questionnaire data and using SPSS to analyze the correlation of the questionnaire indicators, BIM operation and maintenance obstacle factors ($[F_1 \sim F_{13}]$) adjacency matrix are obtained. The adjacency matrix of obstacle factors is input into ucinet6 for data analysis. At the same time, the network structure of BIM operation and maintenance is drawn by using Netdraw software, as shown in Fig. 1.

4.1 Point Degree Centrality

The degree centrality of nodes can reflect the communication ability of nodes, the concentration degree of network relations and the communication ability of factors. The greater the point degree center of the factor node, the stronger the communication ability of the node and the closer the connection with other nodes, which is conducive to the construction of a stable network system. The point degree and center degree

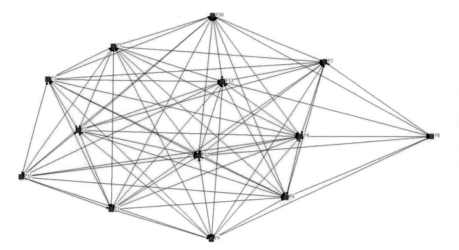

Fig. 1 BIM operation and maintenance obstacles network architecture

of obstacles to BIM + operation and maintenance platform promotion are shown in Table 2.

According to the data analysis results in the table above, the center degree of supervision and scientific research support is the highest, which is most closely related to other indicators, followed by laws and regulations, scientific research cost, data sharing, compatibility and integrity.

The center degree of the seven indexes is above 50, and the average relative point degree is also higher than other indicators, indicating that these seven indicators have a great influence on other indicators. The point degree of centrality only reflects the influence of a single node, and the obstacle factors should be analyzed in combination with other indicators.

4.2 Degree of Mediation Center

The degree of mediation center reflects that a node ACTS as an important medium for other nodes to communicate with each other, helping other nodes to transmit information. Based on the adjacency matrix, this paper analyzes the intermediary degree of obstacles to BIM + operation and maintenance platform promotion (Table 3).

According to the data analysis in the table above, the seven obstacle indicators of scientific research support, integrity, compatibility, supervision, laws and regulations, revenue cycle and research cost have the highest degree of intermediary centers with strong ability to transmit information. The overall data shows that the overall degree of intermediary center is not high, and it still needs to be analyzed in combination with other indicators.

Table 3 Node intermediary centrality

Index	Code name	Barriers	Betweenness	nBetweenness
1	F_1	Degree of emphasis	2.929	1.408
2	F_2	Synergy	2.929	1.408
3	F_3	Data sharing	2.929	1.408
4	F_4	Compatibility	2.929	1.408
5	F_5	Coding system	2.652	1.270
6	F_6	Integrity	2.652	1.270
7	F_7	Research cost	1.634	1.004
8	F_8	Training expenses	0.988	0.289
9	F_9	Revenue cycle	0.934	0.273
10	F_{10}	Laws and regulations	0.912	0.255
11	F_{11}	Standardization	0.561	0.121
12	F_{12}	Supervision strength	0.191	0.021
13	F_{13}	Scientific research support	0.191	0.031

4.3 Proximity Centrality

Proximity centrality refers to the reciprocal of the sum of the shortest distance between a node and other nodes in the network. The greater proximity centrality is, the closer its relationship with a node is, the easier it is to establish a connection. The proximity to the center of obstacles in BIM + operation and maintenance platform promotion is shown in Table 4.

Table 4 Node approaching centrality

Index	Code name	Barriers	Farness	nCloseness
1	F_1	Degree of emphasis	22.000	96.623
2	F_2	Synergy	22.000	96.623
3	F_3	Data sharing	22.000	96.623
4	F_4	compatibility	22.000	96.623
5	F_5	Coding system	28.000	92.308
6	F_6	Integrity	28.000	92.308
7	F_7	Research cost	28.000	92.308
8	F_8	Training expenses	28.000	92.308
9	F_9	Revenue cycle	28.000	92.308
10	F_{10}	Laws and regulations	28.000	92.308
11	F_{11}	Standardization	28.000	92.308
12	F_{12}	Supervision strength	36.000	85.714
13	F_{13}	scientific research support	39.000	70.588

Table 5 Identification of key obstacle indicators

Index	Code name	Barriers	Point degree centrality	Intermediary centrality	Approaching centrality
13	F_{13}	Scientific research support	48.000	2.929	96.623
6	F_6	Integrity	46.000	2.929	96.623
4	F_4	Compatibility	58.000	2.929	96.623
12	F_{12}	Supervision strength	56.000	2.929	96.623

According to the data analysis in the table above, the four indicators of scientific research support, integrity, compatibility, and supervision are close to the maximum centrality.

4.4 Identification of Key Obstacle Indicators Based on Centrality Analysis

Combining the data analysis results of point degree centrality, intermediate degree centrality and proximity degree, the key obstacle indexes were identified. The selected key indicators are in the center of the network structure, which can effectively maintain the stability of the network structure. The key indicators are shown in Table 5.

According to the data in the table above, scientific research support, integrity, compatibility, and supervision are the key obstacle indicators for the development of BIM operation and maintenance. This indicates that the establishment of BIM operation and maintenance platform is especially inseparable from the macro-control of the state, and active scientific and technological research can effectively promote the development of BIM operation and maintenance platform. Netdraw software was used to conduct statistical analysis of the results as shown in Fig. 2.

As shown in Fig. 2 [F_4, F_6, F_7, F_9, F_{10}, F_{12}, F_{13}], the larger the index nodes, it means that these obstacle have a larger weight in the overall core and have a greater impact on the entire network architecture. In network architecture, the connection between each node represents the relevant parameters between each node, and the deeper the connection, the greater the degree of correlation between the two.

5 Conclusion

According to the collected samples, the types of projects managed through BIM + operation and maintenance platform are limited, most of them are public places close

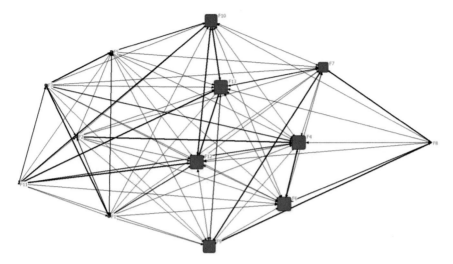

Fig. 2 Analysis of key obstacle indicators

to people's living needs, but seldom involved in the types of using advanced technology such as comprehensive pipe gallery, bridge, and tunnel. More than 80% of the respondents think that the basic operation of BIM + operation and maintenance platform is more complex. Combined with the questionnaire data, this paper analyzes the adjacency matrix of questionnaire data based on social network and analyzes the centrality from three aspects: point centrality, intermediary centrality, and proximity centrality. The three aspects of centrality analysis can take into account the individual and the overall multi angle analysis. From the analysis data of point centrality, intermediary centrality, and proximity centrality, it can be concluded that supervision, scientific research support, integrity and compatibility are the main obstacle indicators for the development of BIM operation and maintenance. The results of data analysis show that the lack of policy and technology has great restrictions on the promotion of BIM operation and maintenance platform.

Acknowledgements This work was partially supported by Liaoning Provincial Department of Education Scientific Research Fund Project-Basic Research Project (2020018), Liaoning BaiQianWan Talents Program (grant no. 2017076), and the Natural Science Foundation of Liaoning Province (grant no. 20170540769).

References

1. Liu, R., Issa, R.R.A.: Survey: common knowledge in BIM for facility maintenance. J. Performance Constr. Facilities **30**(3), 04015033 (2016)
2. Ma, G.F., Song, X., Shang, S.S.: BIM-based space management system for operation and maintenance phase. J. Civil Eng. Manage. **26**(1), 29–42 (2020)

3. Wen, Q., Zhang, J.P., Hu, Z.Z., Xiang, X.S., Shi, T.: A data-driven approach to improve the operation and maintenance management of large public buildings. IEEE Access **7**, 176127–176140 (2019)
4. Wan, C.F., Zhou, Z.W., Li, S.Y., Ding, Y.L., Xu, Z., Yang, Z.G., Xia, Y.F., Yin, F.Z.: Development of a bridge management system based on the building information modeling technology. Sustainability **11**(7), (2019)
5. Chen, L.J., Shi, P.X., Tang, Q., Liu, W., Wu, Q.L.: Development and application of a specification-compliant highway tunnel facility management system based on BIM. Tunn. Undergr. Space Technol. **97**, 103262 (2020)
6. Zhang, L.Y., Dong, L.J.: Application study on Building Information Model (BIM) standardization of Chinese Engineering Breakdown Structure (EBS) coding in life cycle management processes. In: Advances in Civil Engineering, 1581036 (2019)
7. Santos, R., Costa, A.A., Silvestre, J.D., Pyl, L.: Integration of LCA and LCC analysis within a BIM-based environment. Autom. Constr. **103**, 127–148 (2019)
8. Heaton, J., Parlikad, A.K., Schooling, J.: Design and development of BIM models to support operations and maintenance. Comput. Ind. **111**, 172–186 (2019)
9. Jung, J., Hong, S., Jeong, S., Kim, S., Cho, H., Hong, S., Heo, J.: Productive modeling for development of as-built BIM of existing indoor structures. Autom. Constr. **42**, 68–77 (2014)
10. Park, C.S., Lee, D.Y., Kwon, O.S., Wang, X.: A framework for proactive construction defect management using BIM, augmented reality and ontology-based data collection template. Autom. Constr. **33**, 61–71 (2013)
11. Ding, Z.K., Zuo, J., Wu, J.C., Wang, J.Y.: Key factors for the BIM adoption by architects: a China study. Eng. Constr. Archit. Manage. **22**(6), 732–748 (2015)
12. Ahmed, A.L., Kassem, M.A.: Unified BIM adoption taxonomy: conceptual development, empirical validation and application. Autom. Constr. **96**, 103–127 (2018)
13. Hautala, K., Jarvenpaa, M.E., Pulkkinen, P.: Digitalization transforms the construction sector throughout asset's life-cycle from design to operation and maintenance. STAHLBAU **86**(4), 340–345 (2017)
14. Gao, X.H., Pishdad-Bozorgi, P.: BIM-enabled facilities operation and maintenance: a review. Adv. Eng. Inform. **39**, 227–247 (2019)

Simulation of Shenyang Pork Supply Chain System Based on System Dynamics Model

Qianqian Shao, Xiaojing Zhang, Yunfeng Zhang, and Siqi Zhang

Abstract In most of the leading research on pork supply chain (PSC) analysis, only a certain part of PSC is studied qualitatively or quantitatively. While they analyze the transmission of price in PSC, connection between price and operation of PSC has not been established. In this paper, system dynamics (SD) and supply chain theory are used to analyze the system behavior of PSC. The model takes into account the natural elimination of slaughtered pigs during feeding and the perishable nature of pork as fresh food. Based on the analysis of the logical relationship and data transfer between the links of PSC, the SD model of Shenyang pork supply chain system is formulated, and the corresponding flow diagram and the equations expressing the relationship between variables are developed. The simulation experiments are designed with step demand as the boundary condition of the system. The fluctuation trend of pork price in Shenyang is analyzed, and the robustness of Shenyang PSC is further discussed.

Keywords Pork price · Pork supply chain · System dynamic

1 Introduction

As an important part of daily consumption, pork industry is closely related to people's life and economic development. With the development of economy, the pork industry is gradually changing from retail farming and scattered related enterprises to large-scale breeding management, processing, transportation, and sales. The concept of pork supply chain (PSC) is gradually introduced for overall management with the development of economy. Aiming at the lack of combining qualitative and quantitative methods, the dynamic change and trend prediction of PSC are modeled and analyzed. System dynamics (SD), as a qualitative analysis of correlation feedback and a quantitative analysis of system flow diagrams, is suitable for analyzing complex systems of PSC. At present, with the rapid development of logistics in China, the research on SD applied to supply chain-related aspects in the existing literature

Q. Shao (✉) · X. Zhang · Y. Zhang · S. Zhang
School of Transportation Engineering, Shenyang Jianzhu University, Shenyang 100168, China
e-mail: mr_Crowley@163.com

© The Author(s), under exclusive license to Springer Nature Singapore Pte Ltd. 2021　　169
Y. Li et al. (eds.), *Advances in Simulation and Process Modelling*,
Advances in Intelligent Systems and Computing 1305,
https://doi.org/10.1007/978-981-33-4575-1_17

focuses on cold chain logistics and inventory recently. A few literatures related to PSC analyze the composition structure and nodes of supply chain. Therefore, based on the existing supply chain literature, the PSC is modeled.

As a simulation method to analyze complex systems, SD has been widely used in logistics and supply chain modeling and simulation with the increase of supply chain level. Qiu and Chen [1] put forward the dynamic models of dual-channel supply chain inventory independent system, single-stage inventory cooperative system, and multi-stage inventory cooperative system by using the theory and method of system dynamics. These are used to study the characteristics of fresh agricultural products dual-channel supply chain system. He and Liu [2] proposed an SD simulation model for information transfer of aquatic product supply chain. Chen et al. [3] proposed taking banana supply chain as an example, analyzed, and sorted out the composition and characteristics of fresh agricultural products supply chain and drew the inventory flow diagram of four-level supply chain. Hu and Liu [4] proposed to determine the boundary of the supply chain system of fruits, vegetables, and agricultural products in western Hunan based on risk factors, analyze the system structure, determine the causal relationship between each subsystem, and finally build an SD model for simulation. In order to analyze the influence of various factors on fresh agricultural products logistics inventory, Wei and Liu [5] proposed to use SD method to establish the flow chart of fresh agricultural products logistics inventory system and used VENSIM software for simulation.

As a simulation method to analyze complex systems, SD has been widely used in logistics and supply chain modeling and simulation with the increase of supply chain level. The rest of this paper is organized as follows: Sect. 2 described A causal diagram of the PSC. The dynamic model of PCS is established, and the main equations are described in Sect. 3. Section 3 describes SD model of Shenyang PSC system. Section 4 analyzes the simulation and robustness of Shenyang PSC model. The price delay sensitivity analysis is also introduced. The last section provides conclusions and future work.

2 PSC Causal Diagram

The causal cycle diagram is used to represent the feedback relationship between variables in the supply chain system and qualitatively represents the behavior of each node in the supply chain. This paper mainly studies the PSC system and only focuses on the inventory transfer and information transfer feedback between key nodes of the supply chain, without considering other meat substitutes, breeding industry consumption, feed cost, pig breeding capital quota, and other factors. In the simulation process, there is a time lag between the purchase decision of pig farm and the supply of pig from the upstream supplier. The cause-and-effect diagram of the PSC is shown in Fig. 1.

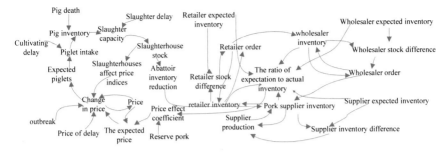

Fig. 1 PSC causal diagram

3 SD Model of Shenyang PSC System

3.1 Upstream Production Subsystem

The upstream production subsystem, as the upstream node, consists of two parts: piglet inventory and slaughterhouse. Slaughterhouses are connected to downstream pork suppliers, so the pigs slaughtered are affected by the productivity of the pork suppliers. First, the supplier determines the supplier's production demand based on the supplier's sales forecast. Then, production demand affects supplier productivity. Finally, according to the supplier's productivity feedback, the slaughterhouse will supply production. In the real world, slaughterhouse orders need to be verified and stocked, so they cannot be delivered immediately. Therefore, the supplier order cycle is set. The main equations of pork supplier inventory subsystem are shown in Table 1.

Table 1 Main equations of pork supplier inventory subsystem

Sequence number	Equation
1.	Pig inventory = INTEG(Piglet intake –Slaughter capacity-Pig death, Initial pig inventory)
2.	Slaughterhouse stock = INTEG(Slaughter capacity-Abattoir inventory reduction, Initial pig inventory)
3.	Piglet intake = IF THEN ELSE((Expected piglets- Pig inventory)/Cultivating delay ≤ 0,0,(Expected piglets- Pig inventory)/ Cultivating delay + Slaughter capacity)

Table 2 Main equations of the wholesaler inventory subsystem

No	Equation
1.	Pork supplier inventory = INTEG(+Supplier productivity-Supplier delivery rate-Pork supplier inventory quality rate change, Initial quantity of pork supplier inventory)
2.	Supplier productivity = DELAY3(Supplier production demand, Meet slaughterhouse order cycle time)
3.	Supplier delivery rate = MIN(Pork supplier inventory, DELAY3(Wholesaler order, transport delay))

3.2 Downstream Sales and Supply Subsystem

3.2.1 SD Model Analysis of Pork Supplier Inventory Subsystem

The slaughterhouse is the intermediary between the upstream supplier and the downstream wholesaler, whose demand information is also passed to the slaughterhouse by order. After slaughtering pigs in the slaughterhouse, the slaughterhouse supplies according to the wholesaler's order. At the same time, the slaughterhouse should also predict the demand of wholesalers and make replenishment in time. In reality, due to the periodicity of pig fattening, slaughterhouses do not slaughter pigs at any time, and they have the characteristics of periodic ordering. Therefore, the abattoir order cycle is added into the factor. After ordering, the supplier needs to transport the pigs, and there is a transportation time delay, so factor 1 of transportation delay is set. In addition, considering that the slaughterhouses have the preparation time after receiving the order from the wholesaler, the moving smoothing time of the slaughterhouse is set. The main equations of the wholesaler inventory subsystem are shown in Table 2.

3.2.2 SD Model Analysis of Wholesaler Inventory Subsystem

In reality, retailers have a large sales volume, but their inventory is small, so the demand should be from wholesalers with certain storage conditions to order. As the intermediate node between downstream retailers and upstream pork suppliers, wholesalers need to consider the transportation delay and inventory adjustment time in the process, so as to set the transportation delay factor. In addition, after receiving the order from the retailer, the wholesaler prepares and transports the goods to the retailer. Therefore, the ordering period of the wholesaler is set to express the material and information delay in the transportation process. The main equations of the wholesaler inventory subsystem are shown in Table 3.

Table 3 Main equations of the wholesaler inventory subsystem

No	Equation
1.	Wholesaler inventory = INTEG(Supplier delivery rate- Wholesaler delivery rate-Wholesalers change the quality of inventory, Wholesaler stock initial quantity)
2.	Supplier delivery rate = MIN(Pork supplier inventory,DELAY3(Wholesaler order, transport delay))
3.	Wholesaler delivery rate = MIN(Wholesaler inventory,DELAY3(Retailer order, Meet wholesaler order cycle time) + Reserve pork)

Table 4 Main equations of the retailer inventory subsystem

No	Equation
1.	retailer inventory = INTEG(+Wholesaler supply rate- Market demand rate- Retailer inventory quality change, Retailer initial inventory)
2.	Wholesaler delivery rate = MIN(wholesaler inventory,DELAY3(Retailer order, Meet wholesaler order cycle time) + Reserve pork)

3.2.3 SD Model Analysis of Retailer Inventory Subsystem

The retailer is a downstream node connecting the terminal consumer demand of the wholesaler, and the demand information of the retailer is transmitted to the wholesaler by order. The wholesaler supplies goods according to the retailer's order forecasts the retailer's sales and makes replenishment in time. In reality, after ordering, the wholesaler needs to prepare and transport the products after receiving the retailer's order. There is a delay in transportation time, so the factor of satisfying the ordering cycle of the wholesaler is set. The main equations of the retailer inventory subsystem are shown in Table 4.

3.3 Price Response Subsystem

The price will affect the purchase intention of piglets, thus affecting the expected piglet quantity. The price is mainly affected by the change in price and the expected price. In the absence of external influence conditions, the expected price is affected by the price effect coefficient, and the price change is affected by the price delay. The main equations of the price response subsystem are shown in Table 5.

3.4 Global Flow Diagram of Shenyang PSC System

After analyzing the above upstream production subsystem, inventory subsystem and price response subsystem of three nodal enterprises, a complete SD model of

Table 5 Main equations of the price response subsystem

No	Equation
1.	Price = INTEG(Change in price, initial price)
2.	Price effect coefficient = ABS(IF THEN ELSE(ABS(The ratio of expected to actual inventory) < 3, The ratio of expected to actual inventory/3,RANDOM NORMAL(0.7, 0.9, 0.8, 4.4)))

Shenyang PSC system were obtained. The global system flow diagram is shown in Fig. 2.

4 Simulation Study and Result Analysis

4.1 Simulation Study

After testing the model, the simulation experiment was carried out on the PSC system model. Set the simulation time step to 1 week and the simulation time to 100 weeks. In this paper, we observe the change of inventory level under the condition of step demand fluctuation. Through the analysis of the demand of Shenyang pork market, this situation is a special holiday before and after the Spring Festival, the pork market demand will suddenly increase. The simulation model assumes the consumer function in the pork market for the above situation and the related system dynamics research on the supply chain inventory. The function is used for two simulations.

STEP function: STEP function describes the sudden increase of the initial value to the final value, which can be used to represent the sudden increase of pork demand due to special holidays in reality, and is an important situation for research.

4.2 Results and Analysis

(1) Simulation results output under STEP demand: in the condition of STEP demand, $1000 + STEP(1800, 30)$ was input into the market demand formula, that is, consumer demand increased by 1800 in the 30th week. The simulation output results are shown in Fig. 3.

(2) Simulation results output under STEP demand: in the condition of STEP demand, $1000 + STEP(2000, 30)$ was input into the market demand formula, that is, consumer demand increased by 2000 in the 30th week. The simulation output results are shown in Fig. 4.

In Figs. 3 and 4, it shows that early step demand for flat demand fluctuations, so prices fluctuate at first is in a stable state, when a step change, after a sudden increase demand remains at a relatively high demand, the original inventory to meet

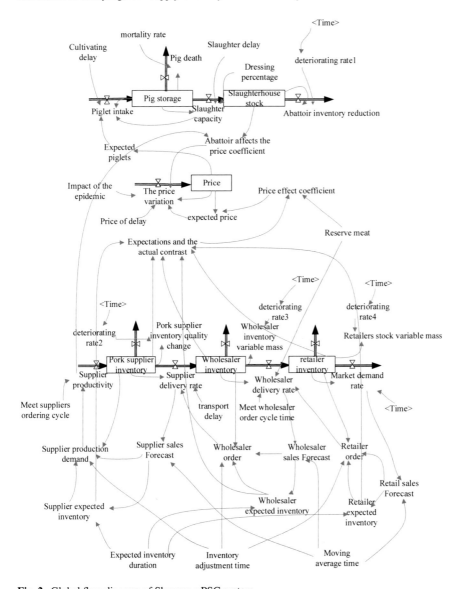

Fig. 2 Global flow diagram of Shenyang PSC system

the sudden increase in demand, added a large number of products, one could be down at the next higher level at the same time increase the order. However, due to the delay of transportation and stock preparation, the original stock was in short supply after the maximum supply was allowed, which led to a sudden increase in the price later. By increasing the arrival of supplementary products to adjust the market, slow down the price rising trend. Compare Fig. 3 with Fig. 4, it can be concluded that there is a

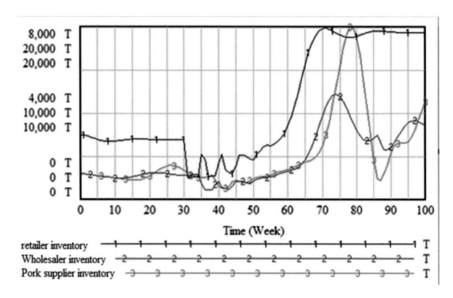

Fig. 3 Inventory changes under step demand 1

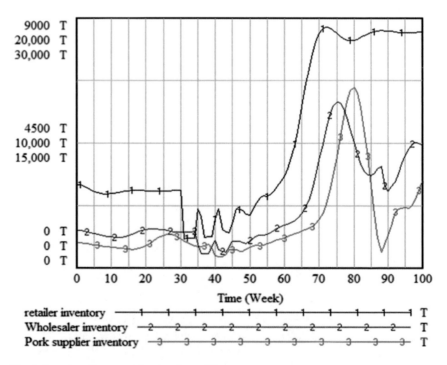

Fig. 4 Inventory changes under step demand 2

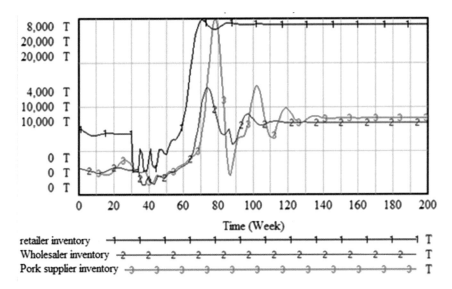

Fig. 5 Inventory changes under 200 weeks of simulation

significant improvement in both weeks around 80. However, the fluctuation of pork supplier inventory in Fig. 3 is relatively large, while that in Fig. 4 is relatively small. The results showed that when the demand for pork increased suddenly, the increase of 2000 could reduce the inventory of pork suppliers and alleviate bullwhip effect.

By observing Figs. 3 and 4, it can be found that in the case of step demand, the images of inventory levels at all levels and replenishment at all levels have obvious fluctuations in the case of a sudden increase in demand, but with the increase in time, the fluctuation amplitude begins to be significantly reduced compared with the initial sharp oscillation. To further analyze this phenomenon, the simulation time was extended to 200 weeks and 300 weeks, respectively, for simulation, and the inventory fluctuation in Figs. 5 and 6 are obtained.

Comparison of inventory changes after changing the simulation time, volatility can be observed after a period of time, amplitude recovery from higher volatility weakened until smooth, thus manifests the rapid growth in demand suddenly the dangerous situation, system has a certain robustness, volatility, when the system can adjust themselves in order to maintain the stability of the system.

5   Conclusion

According to design relevant suggested four-point optimization simulation results for the optimization of supply chain, respectively, from the integration of simplified nodes to reduce the bullwhip effect of cascade multilevel demand amplification and

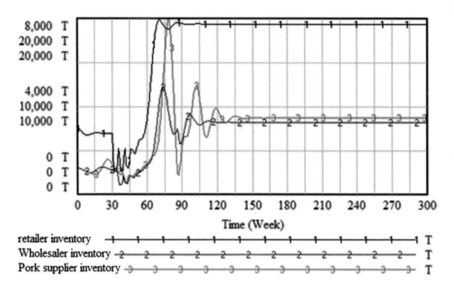

Fig. 6 Inventory changes under 300 weeks of simulation

distortion, to support the breeding reduce delay cultivation, and strengthen the cooperation mechanism to eliminate bullwhip effect brought by the information transfer is
not transparent and national reserve effectively on the application of the four aspects
of the meat.

The model in this paper is to simplify the nodal enterprises involved in the actual
PSC and extract the relevant nodal links that are obvious to the supply and demand
changes, so as to analyze the relationship between the supply and demand changes
and the price. However, part of the data used in the simulation is difficult to obtain
accurate values, which is mainly obtained through system debugging by referring
to relevant data and simulation, or through reasonable analysis based on the actual
situation. There are certain errors, which affect the accuracy of the model.

Acknowledgements This research work is partially supported by the Scientific Research Project
of the Educational Department of Liaoning Province 2019 of China (Project Codes: lnqn201916),
Shenyang Social Science Project 2020 of China (Project Codes: SYSK2020-09-14, SYSK2020-09-
32), Project Plan of China Logistics Learned Society and China Logistics and Purchasing Federation 2020 (Project Codes: 2020CSLKT3-187, 2020CSLKT3-188), Logistics Reform and Teaching
Research Project 2020 of China (Project Codes: JZW2020205)

References

1. Qiu, H.Q., Chen, W.Y.: Study on the inventory cooperative strategy of dual-channel supply chain in fresh agricultural products. Math. Pract. Theor. **49**(09), 1–10 (2019). (in Chinese)
2. He, J., Liu, W.X.: Simulation of aquatic products supply chain information transfer based on system dynamics. Food Industry **40**(09), 302–306 (2019). (in Chinese)
3. Chen, W., Shen, Y., Tian, Y.: Modeling and simulation of fresh agriculture products supply chain based on system dynamics. Logistics Eng. Manage. **40**(12), 101–104 (2018). (in Chinese)
4. Hu, L.H., Liu, T.: Research on supply chain risk for fruit and vegetable products in big Xiangxi. J. Huaihua Univ. **38**(07), 74–79 (2018). (in Chinese)
5. Wei, J., Liu, W.H.: Fresh agricultural products logistics inventory simulation analysis based on system dynamics. Storage Process **18**(02), 130–134 (2018). (in Chinese)

Experimental Study on Grinding Surface Roughness of Full-Ceramic Bearing Ring

Songhua Li, Kechong Wang, and Jian Sun

Abstract To solve the problems of low efficiency, high cost, and high difficulty of full-ceramic bearing ring grinding machining, the effect of process parameters such as grinding wheel grain size, grinding wheel speed, workpiece speed, grinding wheel feed rate, and Z-axis oscillation rate on the value of ceramic ring surface roughness Ra was studied by MK2710 NC grinding machine. The results show that grinding parameters have different influences on the value of Ra. The value of ceramic ring surface roughness Ra decreases with the grain size becoming smaller. As the grinding wheel speed and workpiece speed increases, the value of Ra decreases. The value of ceramic ring surface roughness Ra increases with the increase of grinding wheel feed rate and Z-axis oscillation rate. The determination of the best grinding process will provide the basis for full-ceramic bearing ring precision machining.

Keywords Bearing ring · Grinding parameters · Surface roughness · Single-factor test · Engineering ceramics

1 Introduction

Engineering ceramics with lightweight, high strength, wear resistance, corrosion resistance, high-temperature resistance, small coefficient of thermal expansion, and other excellent properties [1–3] have been widely used in bearing manufacturing industry, aerospace, cutting tool, instrumentation, ceramic armor, biological medicine, and other fields, and the material properties of bearing steels and engineering ceramics are shown in Table 1. There is no doubt that engineering ceramics have many advantages compared with bearing steel. However, it is hard and brittle, which also cause a lot of trouble in processing and manufacturing [4]. Therefore, it is of great significance to study the machinability of engineering ceramics to obtain high-quality ceramic parts.

S. Li · K. Wang (✉) · J. Sun
School of Mechanical Engineering, Shenyang Jianzhu University, Shenyang, China
e-mail: Charles_WKC@163.com

© The Author(s), under exclusive license to Springer Nature Singapore Pte Ltd. 2021
Y. Li et al. (eds.), *Advances in Simulation and Process Modelling*,
Advances in Intelligent Systems and Computing 1305,
https://doi.org/10.1007/978-981-33-4575-1_18

Table 1 Performance comparison between bearing steel and engineering ceramics

Indicators	Unit	Steel	SiC	ZrO$_2$	Si$_3$N$_4$
Density	g/cm^3	7.85	3.10–3.20	5.7–6.05	3.20–3.30
Coefficient of thermal expansion	10^{-6}/K	10.0	4.0–4.8	7.0–10.5	3.1–3.3
Modulus of elasticity	GPa	208	410–450	180–210	300–320
Poisson's ratio	–	0.30	0.14	0.30	0.26
Hardness	HV	700	2100–2400	800–1500	1300–1800
Bending strength	MPa	2400	350–450	900–1200	800–1000
Compressive strength	MPa	–	2250–3500	1000–3000	2000–3500
Bending toughness	MPa·m$^{1/2}$	25	3.5–4.5	8.0–10.0	5.0–7.0
Thermal conductivity	W/m·K	30–40	150	2–3	29–35

Up to now, the main reason for restricting the wide application of ceramic materials is the high processing cost, in which the grinding processing cost accounts for more than 80% [5]. Scholars worldwide have done a lot of research on engineering ceramics, but there is little research on the grinding surface roughness of ceramic bearing rings. Wu et al. [6] proposed a new surface roughness model for brittle materials based on a series of experiments on grinding of SiC ceramics, and their results indicate that the predicted values of the model have the advantage of high goodness of fit with the experimental results and satisfactory effect. Ma et al. [7] studied the removal mechanism of hard–brittle material and developed a cutting force model of hard–brittle material based on fracture mechanics. The results will reflect the fracture removal process of brittle material. Wan et al. [8] created a finite element model of silicon nitride grinding based on virtual abrasives, simulated the subsurface damage depth, and analyzed the influence of the grinding parameters, such as the wheel speed, workpiece speed, and grinding depth on the subsurface damage depth. It is obvious that the research on high-efficiency and low-cost processing methods needs further research in actual production.

For this purpose, this paper takes the outer ring of zirconia ceramic bearing as the research object, and MK2710 NC grinding machine is selected for the experiment to discuss the effect of processing parameters on the value of zirconia ceramic ring surface roughness Ra.

In this paper, Sect. 2 describes the environment and equipment used for the experiments, Sect. 3 provides an analysis and discussion of the results of the experiment, and Sect. 4 shows the main conclusions obtained from the experiment. Hence, this paper has a positive effect on the rational selection of process parameters, improving surface quality and reducing production costs during the production and processing of zirconia ceramic ring.

2 Experimental Work

2.1 Experimental Materials and Conditions

In the grinding experiments, our research objects are zirconia ceramic outer rings, which are provided by the Shanghai Institute of Ceramics of the Chinese Academy of Sciences. The size of the zirconia ceramic ring sample is $\varphi 75$ mm $\times$ $\varphi 63$ mm $\times$ 16 mm. Table 2 lists the main mechanical properties of the zirconia ceramics used in the experiment.

In this experiment, resin-bonded diamond grinding wheels are used as shown in Fig. 1, and its specifications are showed in Table 3. All experiments are carried out under wet conditions, which is a 5% solution of water-based cooling of fluid. In addition, the flow rate of grinding fluid is controlled at the level of 100 L/min.

Table 2 Performance of zirconia ceramic rings

Density (g cm^{-3})	Elasticity modulus (GPa)	Thermal expansivity (10^{-6} K^{-1})	Poisson's ratio	Fracture toughness (MPa m$^{1/2}$)	Hardness (HRC)	Bending strength (MPa)
5.88	195	8.75	0.30	10.50	78	750

Fig. 1 Diamond grinding wheel

Table 3 Performance indicators of the diamond grinding wheel

Indicators	Value
Outer diameter (mm)	50
Thickness (mm)	5
Concentration (%)	100
Bond	Resin

Table 4 Parameters of the single-factor experiment

No.	Grain size (#)	Grinding wheel speed (m/s)	Workpiece speed (r/min)	Grinding wheel feed rate (μm/min)	Z-axis oscillation rate (mm/min)
1	40, 60/70, 80/100, 120/140, 140/170, 230/270, W20	35	100	20	600
2	80/100	20, 25, 30, 35, 40, 45, 50, 55, 60, 65	100	20	600
3	80/100	35	100, 200, 300, 400, 500, 600, 700, 800	20	600
4	80/100	35	100	4, 6, 8, 10, 12, 14, 16, 18, 20	600
5	80/100	35	100	20	100, 200, 300, 400, 500, 600, 700, 800, 900

2.2 Experimental Design

In order to investigate the effect of processing parameters of zirconia ceramic ring on the value of surface roughness Ra, single-factor experiments are used to analyze the influence of grinding wheel grain size, grinding wheel speed, workpiece speed, grinding wheel feed rate, and Z-axis oscillation rate on the value of zirconia ceramic ring surface roughness Ra. The parameters of the single-factor experiment are shown in Table 4.

2.3 Experimental Apparatus and Measuring Instruments

Due to the good performances of the high-precision MK2710 NC grinder, it is selected as our experimental platform. Its minimum resolution does not exceed 0.001 mm and the maximum grinding spindle speed can be up to 36,000 r/min. Our experimental platform is shown in Fig. 2. As shown in Fig. 3, we can measure the value of zirconia ceramic ring surface roughness Ra by the Surtronic 25 Taylor Hobson roughness measuring instrument, and the measurement error does not exceed 0.001 μm. At the same time, we can use the Hitachi S-4800 Cold Field Emission Scanning Electron Microscope to observe the surface topography of the zirconia ceramic rings, which allows us to see more clearly.

Fig. 2 Grinding experiment system

Fig. 3 Taylor Hobson roughness measuring instrument

In addition, a number of measures have been taken with data processing in order to make our test results more accurate. Six sets of data are measured along the circumference direction for each ceramic ring, and after deleting the maximum and minimum values, the average of remaining four sets of data is the final experimental result.

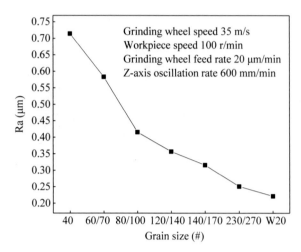

Fig. 4 Effect of the grinding wheel grain size on the value of Ra

3   Analysis of Experimental Results and Discussions

3.1   Effect of the Grinding Wheel Grain Size on the Value of Ra

As can be seen in Fig. 4, the value of Ra decreases with the decrease of grinding wheel grain size. At the initial stage, the roughness changes dramatically, but as the decrease of grain size, the roughness changes tend to be flat, the surface roughness gets decreased, and the surface quality becomes better. The reason for this phenomenon is that the larger the grain size, the wider the scratches produced by the grain on the surface, the greater the height of the grooves and bulges, and the worse the surface quality. Furthermore, it can be clearly seen from Fig. 5 that there are many visible grinding grooves on the grinding surface of the ceramic bearing ring. And grinding wheel grain size is smaller, and grinding grooves are showing a narrow, smooth state.

3.2   Effect of the Grinding Wheel Speed on the Value of Ra

The effect of grinding wheel speed on the value of Ra of ceramic bearing ring is shown in Fig. 6. As can be seen in Fig. 6, the value of Ra decreases with the increase of the grinding wheel speed. When the grinding wheel speed is up to 55 m/s, the grinding wheel speed has little effect on the value of Ra of ceramic bearing ring and the value basically tends to be stable. As is known to all, the maximum undeformed cutting thickness of single grain decreases with the increase of the grinding wheel speed, the grinding force of single grain gets decreased and specific grinding energy gets

Fig. 5 SEM image

Fig. 6 Effect of the grinding wheel speed on the value of Ra

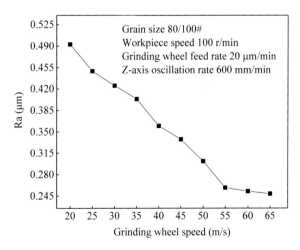

Grain size 80/100#
Workpiece speed 100 r/min
Grinding wheel feed rate 20 μm/min
Z-axis oscillation rate 600 mm/min

Ra (μm)

Grinding wheel speed (m/s)

increased. The proportion of plastic removal of the ceramic material gets increased and the surface quality of the ceramic ring will become better [9, 10]. Therefore, the grinding wheel speed can be increased in the grinding process, which will be more conducive to improving the surface quality of the ceramic ring.

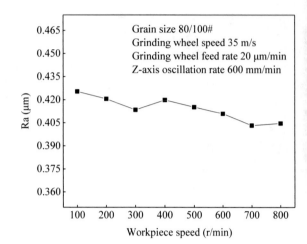

Fig. 7 Effect of the workpiece speed on the value of Ra

Grain size 80/100#
Grinding wheel speed 35 m/s
Grinding wheel feed rate 20 μm/min
Z-axis oscillation rate 600 mm/min

3.3 Effect of the Workpiece Speed on the Value of Ra

We can clearly see from Fig. 7 that the effect of the workpiece speed on the value of Ra of ceramic bearing ring is not very obvious. But as the workpiece speed increases, the surface roughness generally shows a slight downward trend. Compared with the high-speed rotating grinding wheel, the workpiece speed has little effect on the test results. However, with the continuous increase of workpiece speed, the scratches produced by the abrasive on the surface of the ceramic ring become denser and denser. Consequently, the roughness of ceramic bearing ring tends to decrease.

3.4 Effect of the Grinding Wheel Feed Rate on the Value of Ra

The result shows that the value of Ra of ceramic ring generally increases with the increase of the grinding wheel feed rate, and the variation of the value of Ra with the grinding wheel feed rate is as shown in Fig. 8. When the grinding wheel feed rate does not exceed 10 μm/min, the surface roughness is small and stable. With the further increase of grinding wheel feed rate, the cutting thickness of single grain is up to the critical cutting depth of plastic and brittle removal, thus the proportion of brittle removal of the ceramic material gets increased [11]. As a result, surface roughness increases sharply and it results in rapid deterioration of surface quality.

Fig. 8 Effect of the grinding wheel feed rate on the value of Ra

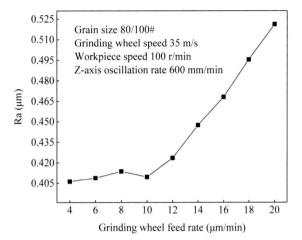

Grain size 80/100#
Grinding wheel speed 35 m/s
Workpiece speed 100 r/min
Z-axis oscillation rate 600 mm/min

3.5   Effect of Z-axis Oscillation Rate on the Value of Ra

The variation of the value of Ra of zirconia ceramic ring is as shown in Fig. 9. We can clearly see from Fig. 9 that the trend of Ra varies with the variation of Z-axis oscillation rate. With the increase of Z-axis oscillation rate, the surface roughness increases continuously. The reasons for this trend are as follows: when the Z-axis oscillation rate is low, the scratches produced by abrasive on the surface of the ceramic ring are dense and the value of Ra is small. With the continuous increase of the Z-axis oscillation rate, the scratches on the surface of the workpiece become more and more sparse, and the surface roughness increases significantly.

Fig. 9 Effect of Z-axis oscillation rate on the value of Ra

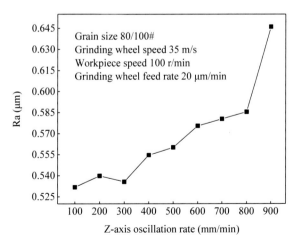

Grain size 80/100#
Grinding wheel speed 35 m/s
Workpiece speed 100 r/min
Grinding wheel feed rate 20 μm/min

4 Conclusions

During the grinding process of the zirconia ceramic bearing ring, the value of Ra varied with processing parameters. The surface quality of zirconia ceramic ring was closely related to the removal mode of surface material. Hence, a reasonable selection of grinding parameters could predict and control the surface roughness of zirconia ceramic ring, which would be of great significance for production and processing.

Acknowledgements This research work is partially supported by the National Natural Science Foundation of China (Project Codes: 51975388), Liaoning Provincial Natural Science Foundation of China (Project Codes: 2019-ZD-0666, 2020-BS-159).

References

1. Kalin, M., Vižintin, J., Novak, S.: Wear mechanisms in oil-lubricated and dry fretting of silicon nitride against bearing steel contacts. Wear **210**(1–2), 27–38 (1997)
2. Wang, L., Wood, R.J.K., Harvey, T.J.: Wear performance of oil lubricated silicon nitride sliding against various bearing steels. Wear **255**(1–6), 657–668 (2003)
3. Kang, J., Hadfield, M., Ahmed, R.: The effects of material combination and surface roughness in lubricated silicon nitride/steel rolling contact fatigue. Mater. Des. **24**(1), 1–13 (2003)
4. Li, Z.P., Zhang, F.H., Luo, X.C.: Material removal mechanism of laser-assisted grinding of RB-SiC ceramics and process optimization. J. Eur. Ceram. Soc. **39**(4), 705–717 (2019)
5. Baraheni, M., Amin, S.: Predicting subsurface damage in silicon nitride ceramics subjected to rotary ultrasonic assisted face grinding. Ceram. Int. **45**(8), 10086–10096 (2019)
6. Wu, C.J., Li, B.Z., Liu, Y.: Surface roughness modeling for grinding of silicon carbide ceramics considering co-existence of brittleness and ductility. Int. J. Mech. Sci. **133**, 167–177 (2017)
7. Ma, L.J., Li, C., Chen, J.: Prediction model and simulation of cutting force in turning hard-brittle materials. Int. J. Adv. Manuf. Technol. **91**(1–4), 165–174 (2017)
8. Wan, L.L., Liu, Z.J., Deng, Z.H.: Simulation and experimental research on subsurface damage of silicon nitride grinding. Ceram. Int. **44**(7), 8290–8296 (2018)
9. Liu, W., Deng, Z.H., Shang, Y.Y.: Effects of grinding parameters on surface quality in silicon nitride grinding. Ceram. Int. **43**(1), 1571–1577 (2017)
10. Wang, K.C., Li, S.H., Sun, J.: Study on surface roughness of zirconia ceramics in high efficient and precision grinding process. In: IOP Conference Series: Materials Science and Engineering (2020)
11. Ding, Z.S., Jiang, X.H., Guo, M.X.: Investigation of the grinding temperature and energy partition during cylindrical grinding. Int. J. Adv. Manuf. Technol. **97**, 1767–1778 (2018)

Transportation and Traffic Systems

Research on Traffic Impact Assessment of Project Under Construction Based on TransCAD

Lei Zhang and Zhu Bai

Abstract Traffic impact assessment is to evaluate the potential effects of a particular development on the traffic network in the affected area. It aims to test the interaction degree of various traffic and discuss whether the new contradictions and the negative effect of traffic are brought. This article first determined the basic parameters of the traffic impact assessment. After determining the study area, the traffic flow of each intersection within the scope of the study was counted, and the specific conditions of the roads involved were investigated. On the basis of the data obtained from the survey, TransCAD software was used in the estimation of The OD matrix to predict the background traffic, and then superimposed the traffic volume to obtain the traffic assignment result. Finally, a traffic impact assessment was carried out according to the relevant regulations, and proposed relevant improvement measures for each specific problem.

Keywords Traffic impact assessment · Estimation of OD matrix · Improvement measures

1 Introduction

A traffic impact assessment is an evaluation of the potential impact that a particular development's traffic will have on the road network in its scope of impact [1]. The study of the effect of new development on the current urban traffic is now a significant topic in terms of sustainable development of urban traffic [2]. It was studied in [3, 4] the management and methods of urban road construction on road traffic impacted area.

In terms of traffic impact assessment, after experiencing the climax of urban construction and development from the 1920s to 1970s, the USA has carried out relevant thinking on the construction of transportation facilities. By analyzing the impact of construction projects on the road network and other traffic systems in

L. Zhang · Z. Bai (✉)
School of Transportation Engineering, Shenyang Jianzhu University, Shenyang, China
e-mail: baizhu@sjzu.edu.cn

the influenced regions of projects, the USA decided whether to modify its scale, and from the middle of 1980s, it made a more in-depth and extensive study of the traffic impact assessment system [5, 6]. To study the relationship between traffic impact assessment and city construction, Chen et al. [7] summarized the traffic impact assessment project development trend and characteristics by project count and scale, geography position, and project type. In addition, TIA was applied in many fields, such as incident management [8], moving work zone operations [9], etc. Kaparias et al. identified and proposed related models and metrics linking traffic characteristics with traffic safety effect [10].

With this background of the research, this paper carries out research on TIA of project under construction with TransCAD software, then the capacity and level of service of each road are analyzed in detail. Finally, some useful measures are proposed.

2 The Determination of Parameter

In this paper, the Yujingwan District in Shenyang, China, is regarded as the object, which is located in the west of Hunnan District, the Olympic Sports Center to the east, museums to the south, Changbai Island to the west, and Hun River to the north. There are more than ten bus lines. Youth Avenue, Shenying Street, Hunnan West Road, Danfu Expressway, Metro Line 2, and bus lines run through the city, and Metro Line 9 is under construction. There are 21 intersections around the project, 22 open parking lots, 2 main roads, 2 secondary roads, and 10 branches, totals 14 roads.[1] The traffic is complex. Therefore, it is very reasonable to select this communities as the research object.

According to *Technical Standards for Traffic Impact Assessment of Construction Projects*, the influenced scope of assessment was set to 100,000 m², which is the area enclosed by Nandi West Road, Sanyi Street, Shenying Street, Hunnan West Road and Jinyang Street, while the third year after the completion of the project, and morning rush hour of the working day were set as the evaluation period of the project.

3 Traffic Survey and Data Processing

The survey involved 21 intersections and 14 roads. Intersections include Sanyi Street-Nandi West Road, Caixia Street-Linbo Road, Qixia Street-Linbo Road, Jiahe Street-Linbo Road, Qixia Street-Hunnan West Road, Linbo Road Lane 2-Linbo Road, and Linbo Road Lane 2-Mingbo Road, etc., while roads include Nandi West Road,

---

[1]The transport data for this research work was obtained from the actual intersection of Qixia Road, Yujingwan District, Shenyang, China.

Table 1 Capacity of a single lane

Type of the road	The capacity of a single lane
Main road	1350
Secondary road	1300
Branch	1250

Xiandao South Road, Changbai South Road, Hunnan West Road, Jinyang Street, and Qixia Street, etc.

The bidirectional capacities of the roads are shown in Table 1.

The bidirectional capacity is equal to the capacity of a single lane multiplied by the number of the lanes. The capacity of the roads is shown in Table 2.

Table 2 Capacity of road

No.	Road name	Number of lanes	Road nature	Bidirectional capacity (pcu/h)
1	Nandi West Road	Bidirectional four lanes	Secondary road	5200
2	Xiandao South Road	Bidirectional four lanes	Branch	5000
3	Changbai South Road	Bidirectional four lanes	Branch	5000
4	Hunnan West Road	Bidirectional ten lanes	Main road	13,500
5	Jinyang Street	Bidirectional eight lanes	Secondary road	10,400
6	Linbo Road	Bidirectional four lanes	Branch	5000
7	Qixia Street	Bidirectional six lanes	Branch	7500
8	Jiahe Street	Bidirectional four lanes	Branch	5000
9	Caixia Street	Bidirectional six lanes	Branch	7500
10	Mingbo Road	Bidirectional two lanes	Branch	2500
11	Linbo Road 2	One-way street	Branch	1250
12	Sanyi Street	Bidirectional four lanes	Branch	5000
13	Shenying Street	Bidirectional six lanes	Main road	8100
14	Youth Avenue Branch	Bidirectional six lanes	Branch	7500

4 Traffic Demand Forecast Based on TransCAD

4.1 Background Traffic Forecast

The Estimation of OD Matrix

With the help of TransCAD software, the data obtained is divided into each specific road section. And then, a new basic matrix is built with a value of 1, and the interior of the communities is 0. Therefore, the estimation result of the OD matrix is obtained through the traffic assignment method and user equilibrium (see Fig. 1).

According to the estimation result of the OD matrix obtained in Fig. 1, the traffic flow, traffic saturation, and the level of service of each road are obtained by traffic assignment for the traffic area and the estimation result of the OD matrix (see Table 3).

Taking Hunnan West Road as an example, one-way traffic capacity is equal to half of the bidirectional capacity, while maximum in one-way section is equal to the maximum flow of one-way section divided by one-way traffic capacity, and the level of service is calculated by *Technical Standards for Traffic Impact Assessment of Construction Projects.*

From the above table, we can see that the saturation of each road section in the scope of the assessment is uneven, and the level of service is different, which are four degrees of A, B, C, and F. The two roads with a lower level of service are Shenying Street and Sanyi Street, which are connected to Youth Avenue and close to the Olympic Center with large traffic volume. Therefore, there are many traffic problems with a low level of service. While the main reason for the other roads with a high level is that wide road, and less traffic demand of the surrounding traffic districts. Thus, the function of districts has not been fully developed and utilized, this causes the traffic produced and attracted small, and the level of service is high.

Background Traffic and Traffic Attraction Forecast

In the scope of the study, the utilization of each construction projects should be investigated, which are occupancy of residential projects, utilization of commercial and official projects, and then the situation of the target year can be predicted based on

	1	2	3	4	5	6	7	8	9	11	12	13	14	15	16	17	18	19	20	21	22	23
1	0.00	199.16	65.62	136.60	17.46	3.12	0.00	40.04	49.26	146.34	89.80	134.24	21.73	29.07	19.66	23.33	19.87	21.47	38.62	21.42	27.20	45.46
2	199.16	0.00	25.58	95.82	6.70	0.94	0.00	21.98	32.71	132.82	70.41	122.58	16.21	23.18	12.69	16.72	13.90	16.03	29.20	12.45	21.56	37.82
3	65.62	25.58	0.00	46.18	10.15	0.16	0.00	21.98	37.65	363.77	21.26	29.81	14.96	22.06	10.50	15.33	12.22	14.79	30.23	9.64	21.02	40.52
4	136.60	95.82	46.18	0.00	1.87	2.54	0.00	89.90	95.04	121.00	61.28	133.43	45.57	61.16	55.53	56.98	45.85	43.97	93.78	91.99	55.80	87.90
5	17.46	6.70	10.15	1.87	0.00	1.90	0.00	14.59	22.41	34.49	11.77	26.49	22.91	31.19	20.37	24.36	20.72	22.41	28.50	22.02	29.09	37.47
6	3.12	0.94	0.16	2.54	1.90	0.00	0.00	1.14	2.30	6.18	3.62	6.90	4.31	6.57	2.34	3.76	3.21	4.34	5.08	1.62	6.21	9.09
7	1.00	1.00	1.00	1.00	1.00	1.00	0.00	1.00	1.00	1.00	1.00	1.00	1.00	1.00	1.00	1.00	1.00	1.00	1.00	1.00	1.00	1.00
8	40.04	21.98	21.88	89.90	14.59	1.14	0.00	0.00	107.76	35.96	21.69	53.54	14.20	24.42	7.56	14.09	10.44	13.90	99.35	4.53	21.99	87.06
9	49.26	32.71	37.65	95.04	22.41	2.30	0.00	107.76	0.00	22.14	59.58	104.14	42.68	73.18	83.30	72.38	42.00	39.90	91.87	33.57	62.17	81.40
11	146.34	132.82	363.77	121.00	34.49	6.18	0.00	35.96	22.14	0.00	39.61	60.05	18.05	30.62	33.68	39.31	13.93	17.51	4.96	27.18	27.41	16.90
12	89.80	70.41	21.26	61.28	11.77	3.62	0.00	21.69	59.58	39.61	0.00	1217.34	39.65	61.19	50.86	54.77	38.38	37.80	57.27	221.78	53.94	58.80
13	134.24	122.58	39.81	133.43	26.49	6.90	0.00	53.54	104.14	60.05	1217.34	0.00	65.22	80.77	113.19	96.59	72.12	62.03	89.74	447.56	79.10	96.02
14	21.73	16.21	14.96	45.57	22.91	4.31	0.00	14.20	42.68	18.05	39.65	65.22	0.00	747.63	31.18	39.29	44.45	39.69	97.28	24.20	258.56	943.31
15	29.07	23.18	22.86	61.16	31.19	6.57	0.00	24.42	73.18	30.62	61.19	80.77	747.63	0.00	70.31	67.90	143.44	83.66	168.80	45.67	422.87	800.53
16	19.66	12.69	10.50	55.53	20.37	2.34	0.00	7.56	83.30	33.68	50.86	113.19	31.18	70.31	0.00	63.68	22.59	28.56	64.81	15.07	56.55	66.35
17	23.33	16.72	15.33	56.98	24.38	3.76	0.00	14.09	72.38	39.31	54.77	96.59	39.29	67.90	63.68	0.00	37.32	36.97	64.26	30.05	57.93	64.77
18	19.87	13.90	12.22	45.85	20.72	3.21	0.00	10.44	42.00	13.93	38.38	72.12	44.45	143.44	22.59	37.32	0.00	36.33	236.68	18.39	98.06	135.56
19	21.47	16.03	14.79	43.97	22.41	4.34	0.00	13.90	39.90	17.51	37.80	62.03	39.69	83.66	28.56	36.97	36.33	0.00	84.13	22.96	67.01	78.80
20	38.62	29.20	30.23	93.78	28.50	5.08	0.00	99.35	91.87	4.96	57.27	89.74	97.28	168.80	64.81	64.26	236.68	84.13	0.00	38.39	130.13	78.23
21	21.42	12.45	9.64	91.99	22.02	1.62	0.00	4.53	33.57	27.18	221.78	447.56	24.20	45.67	15.07	30.05	18.39	22.96	38.39	0.00	39.20	43.48
22	27.20	21.56	21.02	55.80	29.09	6.21	0.00	21.99	62.17	27.41	53.94	79.10	258.56	422.87	56.55	57.93	98.06	67.01	130.13	39.20	0.00	427.80
23	45.46	37.82	40.52	87.90	37.47	9.05	0.00	87.06	81.40	16.90	58.80	96.02	943.31	800.53	66.35	64.77	135.56	78.80	78.23	43.48	427.80	0.00

Fig. 1 Estimation result of OD matrix

Table 3 Traffic flow and the level of service

No	Road name	One-way traffic capacity (pcu/h)	Maximum flow of one-way section (pcu/h)[a]	Maximum in one-way section V/C	The level of service
1	Hunnan West Road	6750	3105	0.46	B
2	Shenying Street	4050	2714	0.67	C
3	Nandi West Road	2600	1300	0.5	B
4	Jinyang Street	5200	1768	0.34	A
5	Qixia Street	3750	712	0.19	A
6	Caixia Street	3750	1125	0.3	A
7	Jiahe Street	2500	875	0.35	A
8	Changbai South Road	2500	958	0.38	A
9	Linbo Road	2500	700	0.28	A
10	Mingbo Road	1250	423	0.34	A
11	Sanyi Street	2500	2932	1.17	F
12	Xiandao South Road	2500	400	0.16	A
13	Linbo Road 2	625	149	0.24	A
14	Youth Avenue Branch	7500	2181	0.29	A

Letter "a" represents each or one

above results to obtain the traffic production and attraction; while in the surrounding scope of the study area, virtual traffic districts should be investigated according to its economic growth, and then traffic production and attraction of each district in the target year can be determined according to the growth.[2] The results are shown in Table 4.

4.2 Traffic Volume of Superimposed Projects

Based on the original the OD matrix, the OD matrix obtained by the gravity model method is used to superimpose it, and then the superimposed OD matrix obtained by traffic distribution is used to traffic assignment for the road network. Thus, a diagram of the desire line is drawn (see Figs. 2 and 3; Table 5).

It can be seen in Table 4 that after superimposing the traffic volume within the scope of the assessment, the saturation of each road section is uneven, and the level

---

[2]The data in Table 4 was calculated by the simulation software-TransCAD.

Table 4 Forecast results of background traffic demand

Communities number	Status quo		Target year		Prediction criteria	Assignment ID
	Production pcu/h	Attraction pcu/h	Production pcu/h	Attraction pcu/h		
1	1150	1151	1533	1535	Occupancy rate 75%	3
2	908	910	1816	1820	Occupancy rate 50%	20
3	824	825	1177	1179	occupancy rate 70%	8
4	1421	1423	1776	1779	Occupancy rate 80%	2
5	406	407	1015	1018	Utilization rate 40%	11
6	75	76	750	760	Occupancy rate 10%	1
7	21	35	210	350	Utilization rate 10%	21
8	706	707	1176	1178	Occupancy rate 60%	4
9	1155	1156	1283	1284	Occupancy rate 90%	23
10	1192	1193	1703	1704	Occupancy rate 70%	5
11	2271	2272	2672	2673	Occupancy rate 85%	6
12	3099	3100	4714	4715	Average annual growth 15%	13
13	2531	2532	3849	3851	Average annual growth 15%	18
14	3003	3004	4567	4569	Average annual growth 15%	14
15	829	830	1842	1844	Occupancy rate 45%	19
16	880	881	2200	2203	Occupancy rate 40%	17
17	1065	1067	1521	1524	Occupancy rate 70%	15
18	772	773	1930	1932	Occupancy rate 40%	16

(continued)

Table 4 (continued)

Communities number	Status quo		Target year		Prediction criteria	Assignment ID
	Production pcu/h	Attraction pcu/h	Production pcu/h	Attraction pcu/h		
19	1531	1532	2041	2043	Occupancy rate 75%	24
20	1171	1172	1464	1465	Occupancy rate 80%	22
21	1964	1965	2311	2312	Occupancy rate 85%	25
22	3227	3228	4908	4909	Average annual growth 15%	12

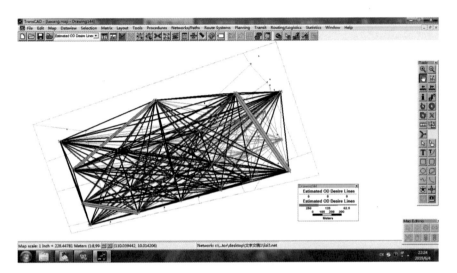

Fig. 2 Diagram of travel desire line after superposition

of service is different, which are five degrees of A, B, C, D, and F, but the overall level of service is high, and there are more A-levels. Except for Nandi West Road and Sanyi Street, the levels of service of other roads are relatively high. Hunnan West Road, Sanyi Street, and Nandi West Road are adjacent to Youth Avenue with large traffic volume. Hence, the road saturation is high, and the burden is heavy, especially Sanyi Street. Qixia Street is located in the center of the scope of the study, and the project newly built is besides, most of the newly produced traffic will be loaded on this road, thus the level of service is low. Of course, its assignment results may not be consistent with the truth. Firstly, the traffic flow of the roads is obtained by artificial observation and statistical analysis; hence, there are some errors. The second is that in the process of analysis with TransCAD, the centroid of traffic districts may not

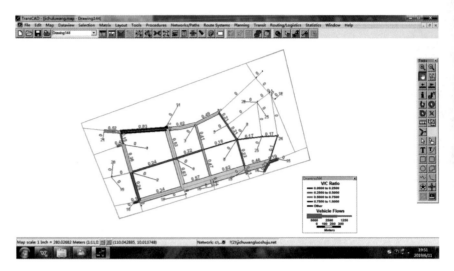

Fig. 3 Diagram of traffic assignment results

Table 5 Results of superimposed traffic assignment in target year

Serial number	Road name	One-way traffic capacity (pcu/h)	Maximum flow of one-way road section(pcu/h)	Maximum in one-way section V/C	The level of service
1	Hunnan West Road	6750	4185	0.62	C
2	Shenying Street	4050	1013	0.25	A
3	Nandi West Road	2600	2158	0.83	D
4	Jinyang Street	5200	1872	0.36	A
5	Qixia Street	3750	2363	0.63	C
6	Caixia Street	3750	863	0.23	A
7	Jiahe Street	2500	1175	0.47	B
8	Changbai South Road	2500	950	0.38	A
9	Linbo Road	2500	600	0.24	A
10	Mingbo Road	1250	425	0.34	A
11	Sanyi Street	2500	2925	1.17	F
12	Xiandao South Road	2500	1000	0.4	A
13	Linbo Road 2	625	150	0.24	A
14	Youth Avenue Branch	7500	2175	0.29	A

be the origin and destination of transportation, and the traffic demand produced by traffic districts directly reaches the intersection, which is inconsistent with the truth, therefore, there are errors.

5 Traffic Impact Assessment and Improvement Measures

5.1 Assessment Results

Based on the above traffic impact assessment, the following results are obtained:

(1) The construction of the project has increased the load on the surrounding traffic network, and the level of service of all roads has changed to a certain extent. Except for Qixia Street, the impact of each road is less than the maximum allowable impact, but all are at the lowest acceptable degree, and some roads have even improved. The project under construction has no significant impact on the traffic of the surrounding roads; therefore, it is acceptable.

(2) The construction of the project has declined the level of service of the surrounding intersections to a certain extent, but it is still higher than the minimum acceptable level of service. The project under construction has no significant impact on the surrounding intersections; therefore, it is acceptable.

(3) The scheme of project construction has planned an entrance and exit in the east, west, south, and north four directions. Therefore, traffic and pedestrians can pass in a safe and orderly manner, and the entrances and exits are not set on the main roads of the city; therefore, it has a little direct impact on it.

(4) According to the mass traffic impact assessment, after the project is finished and put into use, the new public traffic demand is less than its remaining capacity, and public traffic system can satisfy its demand. Thus, there is no significant impact and it can be accepted.

(5) According to the static traffic impact assessment, parking facilities for the project cannot satisfy the parking demand of the project itself. Therefore, the project under construction has a significant impact on the surrounding static traffic system, and it is unacceptable.

From the above assessment results, it can be known that there is no significant impact on the traffic system within the scope of assessment due to the finished project, and relevant improvement measures should be proposed.

5.2 Improvements

According to the relevant requirements, if the new project has a significant impact on the surrounding traffic system, it is necessary to propose corresponding improvement measures. Improvement measures should be improved from two aspects of the project itself and the surrounding traffic system.

(1) For roads, channelization should be implemented to regulate the traffic to improve the traffic capacity. Overpasses can also be built on roads where the traffic volume is close to saturation to reduce traffic delays caused by pedestrians crossing the road and improve traffic capacity of the road.
(2) For intersections, it is possible to optimize the traffic light timing, to increase the green time of the road with heavy traffic, and improve the level of service of intersections.
(3) For the entrance and exit of the project, it is recommended that the properties should implement traffic control of right-in and right-out at the entrance and exit of the project in peak hours to reduce the impact of its traffic volume on adjacent roads, thus the traffic of the communities can be integrated into the traffic of the road more quickly and reduce delays.
(4) For public transport, the departure frequency should be adjusted in peak hours to satisfy travel demand.
(5) For parking facilities, parking areas should be divided, and ground parking facilities should be added.

6 Conclusion

This paper carries out research on traffic impact assessment of the project under construction with TransCAD and analyzes the traffic impact of the project under construction. Firstly, parameters of traffic impact assessment, which are start threshold, the scope of influence, and the evaluation period were determined. Then, the estimation of the OD matrix was carried out with TransCAD software, and the forecast results of traffic demand in target year without project were obtained. Based on this, the new traffic demand of the new project was predicted and loaded it into the district, and the gravity model was used for traffic distribution of production and attraction, and the user equilibrium model was used to traffic distribution. Finally, traffic assessment results were obtained by a traffic impact assessment. The results show that the completion of the project has no significant impact on the transportation system within the evaluation fields.

Acknowledgements This research work is partially supported by Researcher of Economic and Social Development Project of Liaoning Province in 2021 by Provincial Federation of Social Sciences of China (Project Codes: 2021lslybkt-023).

References

1. Sharmeen, N., Sadat, K., Zaman, N., et al.: Developing a generic methodology for traffic iImpact assessment of a mixed land use in Dhaka City. J. Bangladesh Inst. Plan. **5**, 119–128 (2012)
2. Xiao, X.Q.: A study on the major problems of urban traffic impact analysis. Appl. Mech. Mater. **178–181**, 2619–2622 (2012)
3. Richards, S.H., Dudek, C.L.: Field evaluation of traffic management strategies for maintenance operation in freeway middle lanes. Traffic Res. Rec. **703**, 31–36 (2010)
4. Dudek, C.L., Richards, S.H.: Traffic capacity through urban freeway work zones in Texas. Traffic Res. Rec. **869**, 14–18 (2012)
5. Franco, V., Garrain, D., Vidal, R.: Methodological proposals for improved assessments of the impact of traffic noise upon human health. Int. J. Life Cycle Assess **15**(8), 869–882 (2010)
6. Chartasa, C., Sexton, K.G., Gibson, M.D.: Traffic impacts on fine particulate matter air pollution at the urban project scale: a quantitative assessment. J. Environ. Prot. **4**(12), 49–62 (2013)
7. Chen, Y.L., Du, H.B.: Relationship between traffic impact analysis and city construction—a case study in Beijing. J. Transp. Syst. Eng. Inf. Technol. **9**(6), 21–25 (2009)
8. Dia, H., Gondwe W., Panwai S.: Traffic impact assessment of incident management strategies. In: 2008 11th International IEEE Conference on Intelligent Transportation Systems, pp. 441–446. IEEE, Beijing (2008)
9. Edara, P., Brown, H., Sun, C.: Traffic Impact Assessment of Moving Work Zone Operations. Tech Transfer Summary, pp. 1–3 (2017)
10. Kaparias, I., Liu, P., Tsakarestos, A., et al.: Predictive road safety impact assessment of traffic management policies and measures. Case Stud. Transp. Policy **8**(2), 508–516 (2020)

Research on Influencing Factors of Combined Transportation System Based on Analytic Hierarchy Process

Yan Xing, Yingwen Xing, Weidong Liu, Zongze Li, and Shanshan Fan

Abstract In order to improve the efficiency of an urban combined transportation travel system and solve the problem of urban residents' "last mile" travel, this paper used the analytic hierarchy process to study the factors that affect travelers' choice of bicycle-sharing and urban public transportation combined travel systems. We chose twelve factors that may have an impact on travelers' choice behaviors based on the past experience. Taking into account of three characteristics of personal attributes, travel attributes and transfer traffic attributes. We sorted the twelve possible influencing factors according to the three characteristics to conduct an in-depth analysis based on the ranking results obtained. It is found that transfer characteristics have the greatest impact on travelers' travel choice. Based on the above research, we provided suggestions on the configuration of combined transportation and travel systems to improve the attractiveness of combined travel system and increase the probability of travelers choosing combined travel system. The research results have guiding significance for guiding urban residents to choose travel modes and can improve the current status of urban traffic congestion.

Keywords Analytic hierarchy process · Combined travel system · Influencing factors

1 Introduction

China's economic and social development is in a period of rapid development. However, due to the multiple constraints of road conditions, funds and other aspects, the construction of traffic infrastructure cannot meet the needs of motor vehicle growth. Nowadays, giving priority to the development of green transportation and

Y. Xing (✉) · Y. Xing · W. Liu
School of Transportation Engineering, Shenyang Jianzhu University, Shenyang, China
e-mail: xingyan@sjzu.edu.cn

Z. Li · S. Fan
Security Department, Shenyang Jianzhu University, Shenyang, China

slow traffic system has become the common goal of alleviating traffic congestion and improving the travel efficiency of residents.

The urban public transport system is mainly composed of buses, subways, taxis and other modes of travel. It has the advantages of large passenger capacity and small environmental pollution and can reduce lane occupancy and per capita possession of road resource, so it is favored by urban residents. However, the development of urban public transport system in China started late, which may not meet the expectations of urban residents in terms of punctuality and comfort. At the same time, due to its fixed route, it cannot solve the "last kilometer" travel problem and needs to be combined with other travel modes. Bicycle-sharing system has the characteristics of energy saving, low cost and environment-friendly. When the riding distance is appropriate and the riding time is within the acceptable range of the traveler, it can also improve the rider's physical fitness and achieve the effect of exercise.

In the previous studies, the factors affecting residents' travel mode selection are generally summarized into three aspects: Traveler characteristics, travel characteristics and vehicle characteristics [1–3], Kuby [4], Narisra [5] and Loo [6] found that land is an important factor affecting residents' travel choice; with the accelerating process of urbanization in China, the emergence of "separation of work and residence" and "spatial dislocation" makes [7] the influence of urban land use characteristics on residents' travel in China is more and more significant; Yang et al. take the trip chain as the analysis unit to construct the structural equation model of residents' travel mode selection in the process of rapid urbanization, and finally get the influence direction and degree of the personal and family attributes, land use, travel chain, etc. on travel mode [8]. Liu et al. discussed the measures to reduce the environmental impact of tram traffic noise in 2015 [9], and then he put forward the countermeasures and management of parking planning in Shenyang [10].

On the basis of previous studies, this paper analyzes the influencing factors of the configuration of the combined travel system of bicycle-sharing and public transport by using analytic hierarchy process (AHP), so as to solve the "last mile" travel problem and provide suggestions for easing traffic congestion.

2 Basic Principle of Analytic Hierarchy Process

The analytic hierarchy process (AHP) divides the problem into different levels. Then, get the judgment matrix according to the importance degree of each two elements at the same level. Finally, calculate the weight of each element, and the best scheme is determined according to the principle of maximum combination weight [11].

2.1 Establish Hierarchical Structure Model

When we use AHP to solve problems, it generally includes the following levels: the highest level: the purpose of solving the problem, that is, the goal to be achieved; the middle level: the link needed to take a certain method to achieve the highest level; the lowest level: the solution that can be directly used to solve problems.

2.2 Construct Judgment Matrix

Before constructing the judgment matrix, it is necessary to make clear how to describe the comparison results of the importance between the elements. At present, the commonly used method is to use the 1–9 scale to describe the important relationship between the two elements. The larger the numerical value is, the higher the importance degree is. In the process of AHP, the number of elements in each level and the comparison result of the importance degree will change with the natural environment and economic development. When the judgment matrix to solve the problem is constructed, the maximum eigenvalue and maximum eigenvector of the judgment matrix need to be calculated to pave the way for consistency test.

2.3 Conformance Test

In order to make the calculated weight vector reflect the importance of each factor, it is required that the weight vector should be consistent or the deviation degree should not be too large. Therefore, the consistency test of the judgment matrix should be carried out before calculating the weight. The common inspection objectives are as follows:

Consistency index:

$$CI = \frac{\lambda_{\max}}{n - 1} \tag{1}$$

Random consistency index:

$$CR = \frac{CI}{RI} = \frac{\lambda_{\max} - n}{(n - 1)RI} \tag{2}$$

where CI is the consistency index, n is the order of matrix A, and RI is the average random consistency index.

The common values are shown in Table 1.

Table 1 Average random consistency index

n	RI	n	RI	n	RI
1	0	6	1.26	11	1.54
2	0	7	1.41	12	1.56
3	0.52	8	1.46	13	1.58
4	0.89	9	1.49	14	1.59
5	1.12	10	1.52		

When the random consistency index CR is less than 0.1, it is generally considered that the obtained judgment matrix can be accepted, and the above method can be used to calculate the weight. Otherwise, the judgment matrix A will be readjusted until it meets the requirements of consistency test.

2.4 Hierarchical Total Ranking

The importance of all the factors at the lowest level is ranked by the weight of the highest level. If the middle level A contains n elements and the lowest level B contains m elements, it is necessary to sort the weights of the intermediate elements first, and then multiply the weight value of the factors at the lowest level by the weight value of the middle level, in order to obtain the combination weight of the lowest level B relative to the top layer.

3 Study on Influencing Factors of Combined Transportation Travel System Based on AHP

3.1 Analysis of Influencing Factors

First, we analyzed the influencing factors of combined travel system, and the main influencing factors were taken as the bottom consideration factors, as shown in Table 2.

In terms of personal characteristics, the individual factors that affect travelers' choice of combined travel system include gender, age, education level, bicycle-sharing preference, etc. In terms of gender, the social and family roles of different gender travelers are different, which leads to different transportation modes, daily travel times and travel service requirements. Male travelers pay more attention to the efficiency of travel in the purpose of study and work; women travelers prefer to choose one mode of transportation to avoid transfer in shopping and other activities. Age and education level are also the key factors influencing the choice of combined

Table 2 Influencing factors of behavior choice of combined travel system

Individual factors A_1	Travel factors A_2	Transfer factors A_3
Gender B_{11}	Bus waiting time consumption B_{21}	Convenience of storing bicycles B_{31}
Age B_{12}	Bus operation time consumption B_{22}	Transfer type B_{32}
Education level B_{13}	Purpose of travel B_{23}	Transfer time consumption B_{33}
Bicycle-sharing preference B_{14}	Total travel expenses B_{24}	Transfer safety B_{34}

travel system. Health status has been greatly affected by the difference of age, and it results in different travel mode preferences. Students and office workers tend to choose the combined travel system. Middle aged and elderly people may not have enough understanding of the transfer system, and they will directly choose public transportation or walking mode in most cases. Travelers do not consider other modes of travel before they travel, unless their habits are broken by the environment.

In terms of travel characteristics, the travel factors affecting travelers' choice of combined travel system include bus operation status, waiting time consumption, travel purpose, total travel cost, etc. Travelers will focus on the waiting time and operation time consumption of public transport when they choose the travel mode. They will choose more efficient transportation; as for travel purpose, in the usual rigid travel, travelers need to arrive at the destination quickly and on time. In flexible travel, travelers pay more attention to the comfort of travel and pay less attention to travel time consumption and cost. Generally speaking, taxi and walking travel account for a large proportion; travel cost is the purpose of travel for travelers The cost is in the process. When the cost of this mode of travel is high, travelers will choose other ways to travel instead.

In terms of transfer characteristics, the factors that affect travelers' choice of combined travel system include the convenience of bicycle storage, transfer type, transfer time consumption, transfer safety, etc. The convenience of bicycle storage refers to whether the traveler can quickly and safely store the bicycle in the parking area; the transfer type refers to whether the traveler prefers to use the combined travel mode at the starting end or at the terminal end, or uses the group in the whole travel process. When the transfer time is within the acceptable range of the traveler's walking, the traveler is more inclined to walk to the bus station; when the transfer time is beyond the tolerance range, the longer transfer time will consume more physical strength, which will greatly reduce the travel comfort and increase the travel cost, and the travelers will choose other more comfortable transportation modes. Safety factors will also be taken into account when choosing the travel mode. Whether the road is suitable for bicycle, whether there are protective measures, and the safety of bicycle itself will affect the choice of travelers.

Table 3 Judgment matrix of the middle layer relative to the highest level

G	A_1	A_2	A_3
A_1	1	1/5	1/7
A_2	5	1	1/3
A_3	7	3	1

3.2 Generating Judgment Matrix

According to the development of traffic transfer system, the weights of various factors in the middle layer are determined by relevant experts. The judgment matrix is shown in Table 3.

According to the judgment matrix, the maximum eigenvalue and the eigenvector corresponding to the maximum eigenvalue are calculated, then the consistency test is carried out as follows:

Maximum eigenvalue: $\lambda_{g\max} = 3.0647$;

Consistency index:

$$CI = \frac{\lambda_{\max}}{n - 1} = \frac{3.0647}{3 - 1} = 1.5324 \tag{3}$$

Random consistency index:

$$CR = \frac{CI}{RI} = \frac{\lambda_{\max} - n}{(n - 1)RI} = \frac{3.0647 - 3}{2 \times 0.52} = 0.0622 < 0.1 \tag{4}$$

The assumed judgment matrix holds $W = (0.1108, 0.4297, 1.000)^T$; it is obtained by normalization $W = (0.0719, 0.2789, 0.6492)^T$.

3.3 Get the Hierarchical Total Ranking Results

The weight results of each influencing factor calculated by AHP are shown in Tables 4, 5, 6 and 7. The bigger the total ranking weight value, the more important the influencing factor.

Table 4 Hierarchical ranking results of the middle level A

	A_1	A_2	A_3
Value	0.0719	0.2789	0.6492

Table 5 Hierarchical ranking results of the lowest level B_1

B_1	A_1	Total ranking R of layer B_1
B_{11}	0.0745	0.0054
B_{12}	0.1022	0.0073
B_{13}	0.1679	0.0121
B_{14}	0.6554	0.0471

Table 6 Hierarchical ranking results of the lowest level B_2

B_2	A_2	Total ranking R of layer B_2
B_{21}	0.1751	0.0488
B_{22}	0.0991	0.0276
B_{23}	0.4206	0.1173
B_{24}	0.3052	0.0851

Table 7 Hierarchical ranking results of the lowest level B_3

B_3	A_3	Total ranking R of layer B_3
B_{31}	0.1142	0.0741
B_{32}	0.0878	0.0570
B_{33}	0.3113	0.2021
B_{34}	0.4867	0.3161

3.4 Result and Discussion

According to the results of hierarchical total ranking, it can be found that transfer characteristics have the greatest impact on travelers' travel choice. Small transfer time consumption, safety and convenience of transfer can improve the attractiveness of combined travel system and increase the probability of travelers choosing combined travel system.

4 Conclusion

The conclusion for the analysis results are as follows:

(1) Transfer time consumption

It is necessary to optimize the bicycle riding road to reduce the transfer time and draw up the appropriate bicycle-sharing transfer lines or use the existing lines to determine the transfer planning layout of bicycle lanes.

(2) Transfer safety

The vicinity of the transfer area, the non-motorized vehicle lane and the motor vehicle lane should be separated as far as possible. If there is no green belt or railing, the method of increasing the longitudinal height of the bicycle lane can be adopted to realize the separation between the motor vehicle and the non-motor vehicle.

(3) Transfer convenience

The setting of the bicycle-sharing parking space should consider the factors such as the passenger flow of the bus station, the traffic conditions around the bus station and the land use nature around the station. The scale of the transfer site should be planned according to the demand and the above factors, and the bicycle-sharing reservation system can be established to improve the convenience of bicycle storage. The setting of bicycle-sharing parking space should avoid the impact on the surrounding roads and intersections as far as possible, so as to facilitate the collection and distribution of the bicycles and take into account the relationship with other modes of transportation in road facilities and operation organization.

Acknowledgements This research work is partially supported by the Natural Science Foundation of Liaoning Province, China (Project Codes: 2019-ZD-0658), Shenyang philosophy and social science planning, China (Project Codes: SC19007).

References

1. Ben-Akiva, M., Lerman, S.R.: Discrete Choice Analysis: Theory and Application to Travel Demand. MIT Press, Cambridge, Massachusetts (1985)
2. Bowman, J.L., Ben-Akiva, M.E.: Activity-based disaggregate travel demand model system with activity schedules. Transp. Res. Part A **35**, 1–28 (2000)
3. Zong, F.: Activity-based travel choice model and traffic demand management strategy. J. Jilin Univ. (Engineering and Technology Edition) **1**(37), 48–53 (2007) (in Chinese)
4. Michael, K.: Factors influencing light—rail station boardings in the United States. Transp. Res. Part A **28**, 223–247 (2004)
5. Narisra, L.: The influence of socioeconomic characteristics, land use and travel time considerations on mode choice for medium - and longer distance trips. J. Transp. Geogr. **14**, 327–341 (2006)
6. Loo, B.P.Y., Chen, C.: Rail-based transit—oriented development: lessons from New York City and Hong Kong. Lands. Urban Plan. **97**, 202–212 (2010)
7. Zhang, Y.: Research on Beijing City commuting based on comparison of residential district. Geogr. Res. **5**(28), 1327–1337 (2009) (in Chinese)
8. Yang, L.Y.: Choice of residents' travel mode in the process of rapid urbanization. China Soft Sci. **02**, 71–79 (2012). (in Chinese)
9. Liu, W.D.: Environmental impact analysis of traffic noise on Shenyang modern tram. Arch. Budget **02**, 39–42 (2015). (in Chinese)
10. Liu, W.D.: Analysis of parking planning and management countermeasures in Shenyang. J. Shenyang Jianzhu Univ. (Social Science Edition) **20**(04), 395–400 (2018) (in Chinese)
11. Xu, S.B.: Principles of analytic hierarchy process. Tianjin University Press, Tianjin (1988). (in Chinese)

Simulation on Two Types of Improved Displaced Left-Turn Intersections Based on VISSIM

Yingcheng Zheng, Qianqian Shao, Yunfeng Zhang, and Siqi Zhang

Abstract Displaced left-turn (DLT) intersections, which solve the conflict between left-turn traffic and opposing-through traffic are currently the most efficient innovative intersection design. There are no T-shaped DLT intersections in use due to few researches have proposed the implementation of DLT design at T intersections. This paper proposes an improved design method for DLT intersections suitable for two types of intersections, which considering the demand of pedestrian and non-motorized vehicles crossing the street, the phase scheme of the improved DLT, the length of the storage section of the left-turn vehicle and lane-changing length is designed. The operational performance of two types of the improved DLT intersection is validated using the VISSIM simulation. Besides, a sensitivity analysis is demonstrated to compare two types of DLT with conventional intersections under different left-turn ratios. The results demonstrate the effectiveness of the proposed method and provide a basis for the design of DLT intersections geometric layout and signal timings.

Keywords Traffic engineering · Displaced left-turn · Signal design · Geometric layout · Simulation evaluation

1 Introduction

Congestion at intersections continues to worsen in numerous cities throughout the world. As the demand for transportation has gradually increased, urban traffic has become increasingly congested. However, while demands on the transportation system continue to grow, the conventional treatment of intersections through the provision of left-turn bays with protected left-turn phases may not be sufficient to avoid long delays. Consequently, various unconventional intersection designs and control strategies have been proposed by researchers to help enhance the efficiencies of intersections.

Y. Zheng · Q. Shao (✉) · Y. Zhang · S. Zhang
School of Transportation Engineering, Shenyang Jianzhu University, Shenyang, China
e-mail: mr_Crowley@163.com

© The Author(s), under exclusive license to Springer Nature Singapore Pte Ltd. 2021
Y. Li et al. (eds.), *Advances in Simulation and Process Modelling*,
Advances in Intelligent Systems and Computing 1305,
https://doi.org/10.1007/978-981-33-4575-1_21

213

Among these innovative intersection designs, the displaced left-turn (DLT) intersection is quickly becoming prevalent worldwide [1–3]. Wu et al. [4] analyzed the influence of the DLT design on the trajectory of left-turning vehicles, and established a left-turn vehicle capacity calculation model and a left-turn vehicle delay calculation model at the DLT intersections, which demonstrated the effectiveness of the DLT. Yang et al. [5] conducted a comprehensive analysis of the delay of the entire DLT intersection, identified potential queue overflow locations, and developed a set of DLT design geometry planning stage models. Chang et al. [6] in order to further improve the traffic capacity of DLT intersections, the signal coordination relationship between the main intersection and sub-intersection was analyzed, and a signal timing optimization model with general adaptability was proposed. You et al. [7] proposed a model for a full continuous flow intersection (CFI), which is a type of the DLT intersection, with the objective of cycle length minimization to obtain signal timings based on the analysis and formulation of the queue length. Sun et al. [8] further proposed a simplified continuous flow intersection (called CFI-Lite) design, which uses the existing upstream intersection, instead of newly constructed sub-intersection, to allocate left-turn traffic to DLT lane.

The current studies suggest that determining optimal control strategies and suitable geometric layout for unconventional intersection designs remain a significant challenge in practice. In this paper, the design of intersection that combines the type of DLT intersection selection, the length of the DLT lane, and signal timings is presented. This paper innovatively applied the DLT intersection design to different types of intersections, including T-intersections and cross intersections, and further hopes to provide a basis for the design of DLT intersections geometric layout and signal timings.

2 Improved Layout Design of the DLT

Displaced left-turn intersections as a new type of intersection, also referred as continuous flow intersections (CFI) and parallel flow intersection (PFI). As shown in Fig. 1, the improved DLT requires the relocation of the left-turn movement to the other side of the opposing roadway. Drivers who need to turn left at the main intersection should pass the opposite lane first turn left at the intersection upstream of the main intersection. Subsequently, left-turning traffic and through traffic on a roadway proceed simultaneously, without conflict, at the primary intersection [9]. Especially, the right-turning traffic can also leave the intersection without conflict under the unconventional layout of the improved DLT intersection.

The DLT intersection effectively circumvents the conflict between left-turning traffic and through traffic of opposite. Through the coordination of several road sections and intersection signal lights, through volume and left-turning volume are continuously passed through the intersection, reducing the number of phases and vehicle delays.

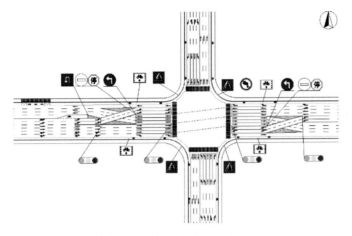

(a) Improved layout of across intersection

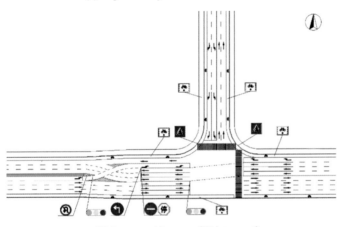

(b) Improved layout of T-intersection

Fig. 1 Improved layout of DLT at different types of intersections. **a** Improved layout of across intersection, **b** improved layout of T-intersection

Fig. 2 Lane-changing length of left-turn vehicle

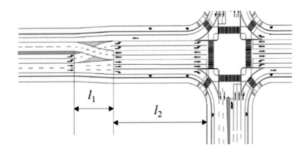

As shown in Fig. 2, for the length of lanes at the left-turn crossover needs to be limited to meet the normal traffic flow of the intersection. For the DLT intersection, the left-turning vehicle enters the shifted left-turn lane from the stop line of the road section to change lanes, and the road section l_1 must meet the vehicle turning needs and safety distance, as constraint (1) illustrated. Constraint (2) requires that the length of the DLT lane should be long enough to accommodate the queuing length of left-turn vehicles.

$$l_1 \geq 2\sqrt{r^2 - [r - (w_1 + 0.5w_c)]^2} \tag{1}$$

$$l_2 \geq \frac{3600L_{el}}{mn S_{el}} \tag{2}$$

where l_1 is the length of lane-changing, as in Fig. 2; r is the minimum turning radius of vehicle; w_1 is the width of a lane; w_2 is the width of double amber lines; l_2 is the length of the storage section of DLT, as in Fig. 2; L_{el} is the length of all left-turn vehicles in line; m is the number of left-turn lanes; n is the number of signal cycles at intersections within one hour; S_{el} is the saturated flow rate of left-turn lanes.

3 Improved Design for DLT Intersection

The optimization objective is determined to minimize the total vehicle delay at the intersection. The objective function is as follows:

$$\min P = \left(d^{sub} + \sum_{i=1}^{n} d_i^{main}\right) \Bigg/ \left(q^{sub} + \sum_{i=1}^{n} q_i^{main}\right) \tag{3}$$

where d^{sub} is the vehicle delay of sub-intersection; d_i^{main} is the vehicle delay of phase i of main intersection; q^{sub} is volume of sub-intersection; q_i^{main} is volume of phase i of main intersection.

Constraints (4)–(6) are general signal constraints for all types of intersections. For the DLT intersection design, in order to reduce the number of stops for vehicles that have left the intersection, ensure that the vehicles that left the intersection in the previous phase pass through leg l_1 smoothly before allowing a left-turning vehicle to change lanes. At the same time, the duration of green of the sub-intersection needs to consider the length of section l_2 and the volume of left-turn traffic. Constraints (7)–(9) show the special requirements of main signal and pre-signal at DLT intersection.

$$C = L + \sum_{i=1}^{3} g_i^{main} \tag{4}$$

$$g_i^{main} \leq g_i^{max} \tag{5}$$

$$g_i^{main} \geq g_i^{min} \tag{6}$$

$$g^{sub} - g_1^{main} \geq \frac{l_2}{v_{sl}} + t_1^{main} \tag{7}$$

$$g_2^{main} \geq \frac{3600 Q_{el}}{nm S_{el}} \tag{8}$$

$$g^{sub} - g_2^{main} \geq \frac{l_1 + l_2}{v_{es}} + t_2^{main} \tag{9}$$

where C is the cycle length; g_1^{main} is the duration of green for the left-turn phase of the main intersection without the sub-intersection; g_2^{main} is the duration of green for the through phase of the main intersection without the sub-intersection. There is no g_2^{main} and the phase corresponding to g_2^{main} if the type of intersection is T-shaped; g_3^{main} is the duration of green for left-turn and through phase of the main intersection with the sub-intersection; L is the loss time of signal cycle; g_i^{min} is the minimum duration of green for each phase; g_i^{max} is the maximum duration of green for each phase; g^{sub} is the duration of green of the sub-intersection; t_1^{main} is the time that the traffic flow corresponding to g_1^{main} leaves the intersection; v_1^{main} is the average speed of the traffic flow corresponding to g_1^{main}; t_2^{main} is the time that the traffic flow corresponding to g_2^{main} leaves the intersection; v_2^{main} is the average speed of the traffic flow corresponding to g_2^{main}.

4 Simulation and Analysis

4.1 Simulation Data

In order to verify the effectiveness of the proposed method, the existing two types of intersections in Lanzhou as an example.[1] The data acquisition time is from 17:30 to 18:30 in the evening peak of ordinary working days. The traffic demand for the test is summarized in Tables 1 and 2.

---

[1] The traffic data for this research work was obtained from the actual intersection of Xijin West Road, Anning District, Lanzhou City, China.

Table 1 Traffic demand of cross intersection

Across intersections	East import (pcu/h)	West import (pcu/h)	South import (pcu/h)	North import (pcu/h)
Left	370	336	760	178
Through	1379	1816	625	327
Right	100	554	616	256
Total imports	1849	2706	2001	761
Total intersections	7314			

Table 2 Traffic demand of T intersection

T intersections	East import (pcu/h)	West import (pcu/h)	South import (pcu/h)
Right	688	None	252
Through	1148	1475	None
Right	None	488	251
Total imports	1836	1963	503
Total intersections	4302		

4.2 Traffic Simulation—VISSIM

In this section, the performance of two types of the improved DLT intersections are evaluated through simulation tests. The intersection where DLT is set and the current situation is executed in VISSIM 6.0. The phase diagram scheme of intersection entered in VISSIM is shown in Fig. 3. The green light duration of each phase of the cross-shaped main intersection and sub-intersection of the improved cross-shaped DLT is 0–38 s, 41–68 s, 71–104 s, and 48–69 s, respectively. The green light duration of each phase of the T-shaped main intersection and sub-intersection of the improved

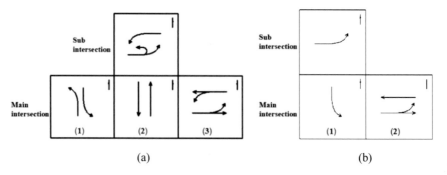

Fig. 3 Phase diagram scheme of intersection. **a** Scheme of cross DLT, **b** scheme of T-shaped DLT

Fig. 4 VISSIM simulation process

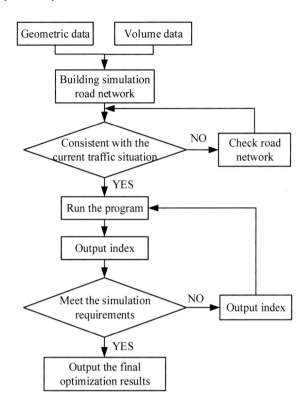

cross-shaped DLT is 0–18 s, 21–57 s, and 7–18 s, respectively. All signals use 3 s amber.

The default driver behavior parameters in VISSIM are used in the simulation of the two types of the improved DLT intersections, and considering that software simulation requires a certain amount of traffic loading time to ensure the accuracy of the collected data, the starting time of the simulation data collection is set to 100 s, and the simulation is run 10 times. The simulation steps of VISSIM are shown in Fig. 4. The geometric layout of the intersections is shown in Fig. 1, where l_2 is set to 50 m.

4.3 Simulation Results

The performance evaluation results are illustrated in Table 3. Results from analyses show that the DLT intersections outperform the conventional design in terms of improving the intersection capacity.

Table 3 Comparison of simulation results

Simulation index	Cross intersection	Cross DLT	Optimization rate (%)	T intersection	T-shaped DLT	Optimization rate (%)
Average stop delay (s)	41.3	26.9	34.87	17.9	9.9	44.69
Average vehicle delay (s)	33.4	20.6	38.32	25.6	17.5	31.64
Maximum queue length (m)	22.3	13.9	37.67	107.7	66.4	38.35
Stops	14.4	7.8	45.83	5.2	4.8	7.69

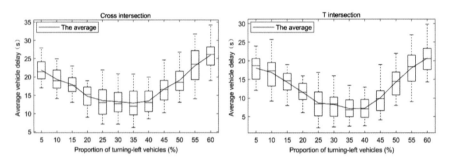

Fig. 5 Influence of left-turn ratio on delay of vehicles at intersections

Figure 5 displays the result of the sensitivity test, when the proportion of left-turn volume reaches 30–40%, the benefits of optimized design in reducing vehicle delays begin to manifest.

The indicators of the two types of intersections after setting up as DLT intersections have been greatly improved compared with the current situation, which validates that the improved DLT design method proposed in this paper is also suitable for T intersections.

5 Conclusion

It is developed, in this paper, an improved setting method for displaced left-turn (DLT) intersections suitable for two types of intersections. The contribution of this paper is comprehensively considered that there are many T intersections in the actual situation. Meanwhile, aiming at improving the general applicability of unconventional intersection design to different types of intersections. Established a VISSIM simulation model based on the actual intersection as a prototype, and the situation before and after the intersection was set as DLT intersection was simulated. The results show that the design of DLT intersection can directly and effectively reduce vehicle delays in the straight direction and enhance the traffic efficiency of the entire intersection.

However, it should be noted that DLT intersections may result in reduce driving comfort and safety and increase air pollution. Moreover, it leads to more complex intersections and requires the driver to have a strong ability to cope with unfamiliar driving environment. Therefore, future work will be introducing the evaluation of the safety performance of the intersection to ensure the DLT operation effect, while ensuring the operation efficiency of DLT intersection, the safety of the intersection will be improved to a greater extent.

Acknowledgements This research work is partially supported by the Scientific Research Project of the Educational Department of Liaoning Province 2019 of China (Project Codes: lnqn201916), Shenyang Social Science Project 2020 of China (Project Codes: SYSK2020-09-14, SYSK2020-09-32), Project Plan of China Logistics Learned Society and China Logistics and Purchasing Federation 2020 (Project Codes: 2020CSLKT3-187, 2020CSLKT3-188), Logistics Reform and Teaching Research Project 2020 of China (Project Codes: JZW2020205).

References

1. Zhao, J., Ma, W.J., Head, K.L., Yang, X.G.: Optimal operation of displaced left-turn intersections: a lane-based approach. Transp. Res. Part C. **61**, 29–48 (2015)
2. Abdelrahman, A., Abdel-Aty, M., Lee, J., Yue, L., Al-Omari, M.M.A.: Evaluation of displaced left-turn intersections. Transp. Eng. **1**, 1–8 (2020)
3. Jagannathan, R., Bared, J.G.: Design and performance analysis of pedestrian crossing facilities for continuous flow intersections. Transp. Res. Rec.: J. Transp. Res. Board **1939**(1), 133–144 (2005)
4. Wu, J.M., Liu, P., Tian, Z.Z., Xu, C.C.: Operational analysis of the contraflow left-turn lane design at signalized intersections in China. Transp. Res. Part C: Emerg. Technol. **69**, 228–241 (2016)
5. Yang, X.F., Chang, G.L., Rahwanji, S., Lu, Y.: Development of planning-stage models for analyzing continuous flow intersections. J. Transp. Eng. **139**(11), 1124–1132 (2013)
6. Chang, Y.T., Wang, Y.T.: An optimal timing model for continuous flow intersection. J. Highw. Transp. Res. Dev. **35**(4), 93–101 (2018) (in Chinese)
7. You, X.M., Li, L., Ma, W.J.: Coordinated optimization model for signal timings of full continuous flow intersections. Transp. Res. Rec.: J. Transp. Res. Board **2366**(1), 23–33 (2013)
8. Sun, W.L., Wu, X.K., Wang, Y.P., Yu, G.Z.: A continuous-flow-intersection-lite design and traffic control for oversaturated bottleneck intersections. Transp. Res. Part C: Emerg. Technol. **56**, 18–33 (2015)
9. Suh, W., Hunter, M.P.: Signal design for displaced left-turn intersection using Monte Carlo method. KSCE J. Civ. Eng. **18**(4), 1140–1149 (2014)

Simulation and Optimization of Multi-UAV Route Planning Based on Hybrid Particle Swarm Optimization

Xuezhao Peng, Feng Guan, Zhongpu Wang, and Shushida Gao

Abstract The multiple traveling salesman problem (mTSP) is an obviously NP-Hard problems in the optimization of complex planning. By adding much more conditions and meanings to the objective function of the mTSP, the vehicle routing problem (VRP) and trajectory optimization problem of the drones can be further evolved. This paper focuses on the multi-UAV route planning problem which is one of the typical mTSP. But traditional algorithms include genetic algorithm and particle swarm optimization are premature in varying degrees and easy to fall into local optimization. Therefore, this paper adds the crossover operator and mutate operator in the genetic algorithm to the particle swarm optimization, and proposes a hybrid particle swarm optimization to solve the multi-UAV route planning problem. The algorithm can make the particles cross the optimal value, and the particles can mutate themselves to strengthen the interchange between the particles. So it can enrich the diversity of the population. Meanwhile, it can improve the global optimality of the algorithm. Finally, it proves that the hybrid particle swarm optimization has better performance when compared with the simulation results of genetic algorithm.

Keywords Multi-UAV route planning problem · mTSP · Genetic algorithm · Hybrid particle swarm optimization

1 Introduction

The multi-unmanned aerial vehicle (multi-UAV) route planning problem is basically consistent with the description of the mTSP [1, 2]. Its application scenario can be described as: There are n locations and m drones. m drones start from the same location and return to the same departure location. In the process, each location is required to be visited and can only be visited one time. Each drone visits at least one location. Most importantly, it should ensure that the total distance of the entire route selection plan is the shortest.

X. Peng · F. Guan (✉) · Z. Wang · S. Gao
School of Transportation Engineering, Shenyang Jianzhu University, Shenyang, China
e-mail: benguan@126.com

© The Author(s), under exclusive license to Springer Nature Singapore Pte Ltd. 2021   223
Y. Li et al. (eds.), *Advances in Simulation and Process Modelling*,
Advances in Intelligent Systems and Computing 1305,
https://doi.org/10.1007/978-981-33-4575-1_22

In recent years, a variety of heuristic algorithms and evolutionary algorithms proposed by scholars can effectively solve mTSP. Kencana [3] uses ant colony optimization to solve complex mTSP problems. Jiang et al. [4] put forward an ant colony partheno genetic algorithm (AC-PGA), which includes partheno genetic algorithm and ant colony algorithm, to solve the large-scale mTSP. Lu and Yue [5] gave an effective way for solving mTSP. They use a task-oriented form to assign ants to the population to optimize the ant colony algorithm. Zhou et al. [6] proposed two methods to solve mTSP. One is the PGA using the roulette model and elite strategy. The other is called IPGA, which combines selection and mutation operators to improve population diversity. Dhein et al. [7] gave a label to describe the difference in dispersion between different paths, and found a genetic algorithm that can realize local search as a new method to settle the matter. The simulation results obtained in the article are very satisfactory.

This article proposes a hybrid particle swarm optimization; it can solve the problems that may occur when the particle swarm optimization is applied to multi-UAV route planning. These problems include the shortcomings of particles easily falling into local optima and slow convergence speed. Finally, the result will be compared with the result of genetic algorithm.

2 Model of Multi-UAV Route Planning Problem

The development of drone technology has brought great convenience to the logistics of remote mountainous areas where vehicles cannot reach. The drone flight does not need to consider the factors of the road network. They can directly use the coordinates to calculate the distance between two points precisely [8, 9]. As a classic mTSP, the multi-unmanned aerial vehicle(multi-UAV) route planning problem is to find the shortest path traversing all locations. Its mathematical model can be described as [10–12].

Location set $A = (1, 2, \ldots, n)$ contains n locations, and location 1 is the start and end of all drones. Drone set $B = (1, 2, \ldots, m)$ contains m drones. $S_{i,k}$ represents the collection of locations visited by the kth drone.

Define variables:

$$x_{ijk} = \begin{cases} 1, \text{The } k\text{th drone goes from location } i \text{ to } j \\ 0, \text{Or} \end{cases} \tag{1}$$

$$y_{ki} = \begin{cases} 1, \text{The } k\text{th drone visits location } i \\ 0, \text{Or} \end{cases} \tag{2}$$

Objective function:

$$L = \min \left(\sum_{k=1}^{m} L_k \right) \tag{3}$$

$$L_k = \sum_{i=1}^{n} \sum_{j=1}^{n} d_{ij} \cdot x_{ijk}, \quad k = 1, 2, \ldots, m \tag{4}$$

Subject to:

$$\sum_{k=1}^{m} y_{ki} = \begin{cases} m, i = 1 \\ 1, i = 2, \ldots, n \end{cases} \tag{5}$$

$$\sum_{i=1}^{n} x(i, i + 1, k) = y(k, i), \quad k = 1, \ldots, m \tag{6}$$

$$\sum_{i=1}^{n} x(i, i + 1, k) = y(k, i + 1), \quad k = 1, \ldots, m \tag{7}$$

$$S_{i,k} = \{x(i, i + 1, k) = 1\} = \left\{ s_{1k}, s_{2k}, \ldots, s_{qk} \right\} \tag{8}$$

$$\sum_{k=1}^{m} \sum_{i=1}^{q} s_{ik} = n \tag{9}$$

$$a \leq q \leq b \tag{10}$$

where d_{ij} d is the length of the drone from one point i to another point j. Equation (3) indicates the shortest flight distance of all drones; Eq. (4) represents the flight distance of a drone; Eq. (5) represents that all drones start at location 1, and each location has one and only one drone to visit; Eq. (6) indicates that the secondary destination city of any route has only one starting city connected to it; Eq. (7) indicates that the starting point city of any path has and only one secondary destination city is connected to it; Eqs. (6) and (7) work together to avoid sub-loops; Eq. (8) represents q locations visited by the kth drone; Eq. (9) represents the total number of locations visited by all drones; Based on Eqs. (8) and (9), Eq. (10) indicates that each drone visits at least a location and at most b locations. According to the number of locations in this paper, Eq. (10) is determined as $8 \leq q \leq 14$.

3 Hybrid Particle Swarm Algorithm for Multi-UAV Route Planning

Genetic algorithm has strong robustness and global search capabilities. It is widely used to solve mTSP and VRP. The particle swarm optimization completes the optimization process of the entire algorithm by continuously seeking individual extreme values and group extreme values. Although the algorithm principle is relatively simple and it can complete convergence in a short time, with the accumulation of the number of particle swarm iterations, the population gradually converges and concentrates and the individual particles gradually become more and more similar which may cause the particle swarm to fall into local optimal solutions.

In our article, we propose a hybrid particle swarm algorithm to simulate and solve the multi-UAV route planning problem [13]. It changes the PSO that uses the method of tracking extreme values to give each particle a new position. The crossover and mutation operations are added to the particle swarm optimization, so that particles can cross individual extreme values and group extreme values. At the same time, the particles can mutate themselves. In this way, the shortest route can be calculated. Hybrid particle swarm optimization algorithm uses the crossover operator and mutation operator to strengthen the information exchange between particles. It improves the abundance of particle swarms, optimizes individual and group, and improves the old algorithm that is easier to fall into the dilemma of locally optimal solutions.

3.1 Individual Code

In this algorithm, the individual coding of the particles adopts an integer coding method, and the location is represented by numbers 1, 2, ... Mark all locations and assign a serial number to each location. There are 2 drones that traverse 9 locations, starting and ending at location 1, and the encoding method is shown in Fig. 1.

Decoding means that the first drone departs from location 1. It follows the path 1-2-3-4-5-1 to location 5 and then returns to location 1; the second drone departs from location 1. It follows path 1-6-7-8-9-1 to location 9 and returns to location 1.

Fig. 1 Individual code

1	2	3	4	5	1	6	7	8	9	1

Fig. 2 Cross operation

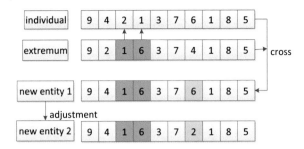

3.2 Fitness Value

In this algorithm, the fitness value is expressed as the total length of the route that all drones traverse all locations. The calculation formula is:

$$\text{fitness} = \sum_{i=1}^{m} L_i \tag{11}$$

where m represents the number of drones, i is the route of the ith drone, and L_i is the length of the ith route. The smaller the fitness, the better the individual fitness.

3.3 Cross Operation

The update of the individual is realized by the crossover between the individual and the individual extreme value or the group extreme value [14]. The crossover method in this algorithm adopts the integer crossover method. We cross the individual with the individual extreme value or the group extreme value and randomly select two intersection positions. If there is a duplicate position in the new individual, we need to replace the duplicate place with a place not included in the individual. The operation method is shown in Fig. 2.

3.4 Mutate Operation

Flip mutation operator
Any two pieces of code are interchanged [15]. The mutation is only carried out within one individual code such as the inversion of the individual code.

Swap mutation operator
Part of any two individual codes is exchanged to form a simple "swap" mutation method, as shown in Fig. 3.

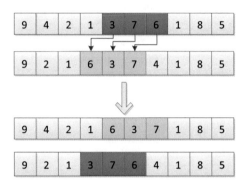

Fig. 3 Swap mutation operator

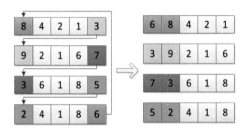

Fig. 4 Slide mutations operator

Fig. 5 "Swap + Slide" mutation operator

Slide mutations operator

The last bit of code is transferred from one individual to another to constitute a "sliding" mutation, as shown in Fig. 4.

"Swap + Slide" mutation operator

It is a hybrid mutation operator that combines exchange mutation and sliding mutation which can significantly improve the performance, as shown in Fig. 5.

3.5 Procedure of Algorithm

The algorithm flowchart based on the optimization we proposed in the article to solve the problem is indicated in Fig. 6.

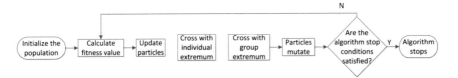

Fig. 6 Algorithm flowchart

The procedure of algorithm is as follows:

Step1: Initialize the particle swarm population, encode the individual particles, set the particle swarm parameters, and initialize the particle position;

Step2: Use the total length of the route as the fitness function to get the fitness value of the algorithm;

Step3: Update the individual optimal particle and the group optimal particle according to the particle fitness value and compare the obtained fitness value with the previous best fitness. It will be accepted instead of the best fitness value previously accumulated if the latest fitness is more suitable;

Step4: Use the crossover operator to cross the individual and individual optimal particles to obtain new particles. Cross the individual and the group optimal particles to obtain new particles;

Step5: Make the particles themselves mutate through the mutation operator to get better particles; and

Step6: Determine whether the optimal solution is obtained or the algorithm has reached the specified maximum number of iterations. If the termination conditions are met, the algorithm terminates; otherwise, go to Step 2.

4 Result and Analysis of the Simulation

4.1 Simulation Result

The coordinate data of 51 locations is used as the simulation data to solve the problem in our article. The genetic algorithm and the hybrid particle swarm optimization are, respectively, used for simulation optimization [16]. This paper uses MATLAB-R2017b to make a program to implement the algorithm. It sets the number of drones to 5 and the number of detection locations to 51. The genetic algorithm sets the total group number to 100, and the maximum iteration round is 1000. We set the crossover and mutation probabilities to 0.9 and 0.1, respectively. The particle swarm size of the optimization is 200, and the maximum iteration round is 1000. Figures 7 and 8 show the results of the two algorithms.

Fig. 7 Hybrid particle
swarm optimization

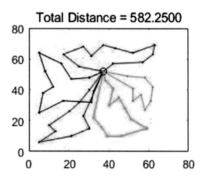

Fig. 8 Genetic algorithm

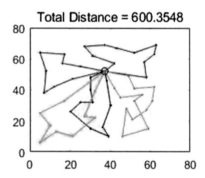

Table 1 Comparison of simulation results

Algorithm	Optimal value	Worst value	Average value	Average number of iterations
Genetic algorithm	570.5684	627.3180	596.2746	751.24
Hybrid particle swarm optimization	581.8723	649.2876	624.2186	862.17

According to the simulation results, we can know that the calculation result obtained by the hybrid particle swarm optimization is 582.25. The calculation result obtained by the genetic algorithm is 600.3548. The two algorithms were run 30 times. We can see the comparison results in Table 1.

4.2 Result Analysis

According to the simulation calculation results, it can be seen that, compared with genetic algorithm, hybrid particle swarm algorithm can greatly speed up the iteration speed of the algorithm. The crossover and mutation operations are added on the basis of the particle swarm algorithm, so that the optimal solution can be obtained

Fig. 9 Hybrid particle swarm optimization

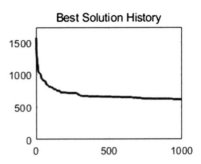

Fig. 10 Genetic algorithm

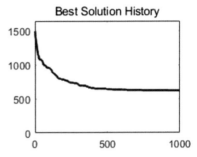

faster. From the perspective of the searched route length, the calculation result of hybrid particle swarm optimization is better than genetic algorithm. It is found that increasing the population size of the genetic algorithm will significantly increase the calculation amount and computing time of the algorithm in the process of parameter adjustment. However, increasing the number of particles has a limited impact on the computing time of the algorithm in the optimization. The iteration diagrams of the two algorithms are as follows. The iterative diagrams of the two algorithms are shown in Figs. 9 and 10.

After many tests, the results show that the optimal value of hybrid particle swarm algorithm is better than the result of GA. The GA usually converge in advance, while hybrid particle swarm algorithm has a wider search range and better diversity. It can also effectively overcome premature convergence and has faster convergence speed and better results.

5   Conclusion

In this paper, it gives a hybrid particle swarm algorithm that adds genetic algorithm ideas to the standard PSO. It is applied to solve the problem of multi-UAV route planning. Its results are compared with the simulation results of genetic algorithm. We can know from the results given by the algorithm that the particles in the algorithm cannot restricted by local optimal solutions. It can overcome the defect of early convergence to a certain extent. So it has better performance. However, the individual coding method of this algorithm has room for further optimization and adjustment. So subsequent research can further improve the performance of the algorithm in this respect.

Acknowledgements This research was supported in Liaoning Provincial Nature Fund (2019-ZD-0661), Project of Liaoning Province Social Science Circle (2020lslktqn-058).

References

1. Koerkamp, N.W., Borst, C., Mulder, M.: Supporting humans in solving multi-UAV dynamic vehicle routing problems. IFAC-PapersOnLine **52**(19), 359–364 (2019)
2. Coutinho, W.P., Battarra, M., Fliege, M.: The unmanned aerial vehicle routing and trajectory optimisation problem, a taxonomic review. Comput. Ind. Eng. **120**, 116–128 (2018)
3. Kencana, E.: The performance of ant system in solving multi traveling salesmen problem. Procedia Comput. Sci. **124**, 46–52(2017)
4. Jiang C., Wan Z., Peng Z.: A new efficient hybrid algorithm for large scale multiple traveling salesmen problems. Expert Syst. Appl. **139**, 112867(2020)
5. Lu, L.C., Yue, T.Y.: Mission-oriented ant-team ACO for min-max MTSP. Appl. Soft Comput. **76**, 436–444 (2019)
6. Zhou, H.L., Song, M.L., Pedrycz, W.: A comparative study of improved GA and PSO in solving multiple traveling salesmen problem. Appl. Soft Comput. **64**, 564–580 (2018)
7. Dhein, G., Serrano, M.: Olinto César Basside. minimizing dispersion in multiple drone routing. Comput. Oper. Res. **109**, 28–42 (2019)
8. Müller, S., Rudolph, C., Janke, C.: Drones for last mile logistics: Baloney or part of the solution? Transp. Res. Procedia **41**, 73–87 (2019)
9. Pugliese, L.D.P., Guerriero, F., Macrina, G.: Using drones for parcels delivery process. Procedia Manuf. **42**, 488–497 (2019)
10. Coelho, B.N., Coelho, V.N., Coelho, I.M.: A multi-objective green UAV routing problem. Comput. Oper. Res. **88**, 306–315 (2017)
11. Dasdemir, E., Ksalan, M.E., Öztürk, D.T.: A flexible reference point-based multi-objective evolutionary algorithm: an application to the UAV route planning problem. Comput. Oper. Res. **114**, 104811 (2020)
12. Ning, Q., Tao, G.P., Chen, B.C.: Multi-UAVs trajectory and mission cooperative planning based on the Markov model. Phys. Commun. **35**, 100717 (2019)
13. Yang, X.D., Li, G.Y.: GA-PSO based decomposition of distributed model predictive control systems. Sci. Technol. Eng. **19**(25), 262–267 (2019). (in Chinese)
14. Xu, J., Pei, L., Zhu, R.Z.: Application of a genetic algorithm with random crossover and dynamic mutation on the travelling salesman problem. Procedia Comput. Sci. 131, 937–945 (2018)
15. Hacizade, U., Kaya, I.: GA based traveling salesman problem solution and its application to transport routes optimization. IFAC-PapersOnLine **51**(30), 620–625 (2018)
16. Mu, X.C., Xie, D.L., Yan, W., Nie, J., Li, X.: Research based on genetic algorithm traveling sealer problem of trajectory optimization. J. Syst. Simul. **25**, 2013 (2013). (in Chinese)

Simulation Research of Arctic Route and Traditional Route Based on the Logit Model

Feng Guan, Yi Cao, Xuezhao Peng, and Yongbao Wang

Abstract At present, the global climate continues to warm. The Arctic shipping routes have become a reality. The opening of the Arctic route will greatly reduce the sea distance between Asia-Europe and Asia-America. It will effectively reduce the shipping cost and achieve the goal of mutually promoting regional economic development. The routes have quite high economic benefit value for all users. Take Dalian port to Rotterdam port for example. Calculate route costs and benefits. Collect relevant data through the SP survey. The discrete selection model—the Logit model is established. The transport sharing rate between the Arctic route and the traditional route is obtained. The Monte Carlo method was used to simulate the data, and the decision probability was obtained, and the error analysis was carried out with the Logit model results. It is believed that Arctic routes will increase their share of the freight market.

Keywords Arctic route · Logit model · Transport sharing rate · Monte Carlo

1 Introduction

With the acceleration of melting ice in the Arctic [1], the improvement of navigation technology, the improvement of shipping environment, the relocation of the World Trade Center, the congestion of the Panama and Suez Canals, the Arctic shipping route will surely become the new maritime Silk Road. Therefore, it is necessary to study a method that can calculate the selection probability of the Arctic route based on multi-dimensional data. The Logit model is the most widely used discrete choice model currently. Its probability expression is simple. Its solution speed is fast. Its application is convenient. And its predicted results can be compared and tested. So this paper uses the Logit model to calculate the transportation sharing rate of Arctic routes and traditional routes.

F. Guan · Y. Cao (✉) · X. Peng · Y. Wang
Faculty of Transportation Engineering, Shenyang Jianzhu University, Shenyang, China
e-mail: 1171746022@qq.com

Y. Li et al. (eds.), *Advances in Simulation and Process Modelling*,
Advances in Intelligent Systems and Computing 1305,
https://doi.org/10.1007/978-981-33-4575-1_23

233

Fig. 1 Arctic route and the
traditional route

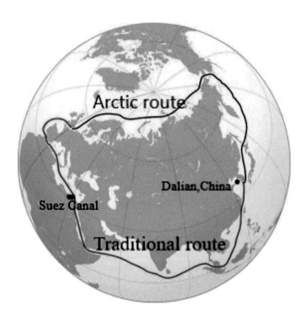

This paper discusses the sailing status of the Arctic route and the choice of ship type. Taking a trip from Dalian to Rotterdam as an example, the costs and benefits of the Arctic route and traditional route are calculated and compared, respectively. The data samples are obtained by the SP survey, and the Logit binomial discrete model is used to calculate the probability of each decision maker in the data sample choosing two routes. The decision-making behavior is simulated by the Monte Carlo method, and the error analysis is performed between the simulation results and the results obtained by the Logit model. It is predicted that the transportation sharing rate of the Arctic route under the condition of the continued decline in international fuel prices is about 31.79%. This research provides a basis for the improvement and sustainable development of commercial navigation on Arctic routes (Fig. 1).

2 Literature Review

Liu and Kronbak [2] compared full-year profits made through the Suez Canal with those made by ships through the North Sea route (NSR). It is concluded that NSR is competitive with the Suez Canal in some conditions. Verny and Grigentin [3] verified the technical and economic feasibility of conventional container transport along the NSR and found that the NSR could be an alternative to the Suez Canal transport. Dai and Yin [4] take the example of a container being transported on the Arctic route. The economic comparison between the Arctic route and the traditional Suez route is made. The questionnaire was designed, the weights were analyzed by the fuzzy analytic hierarchy process, and the data were processed by MATLAB. The results show that NSR has no direct economic advantage. From the perspective of economic feasibility, Masahiko and Furuichi [5] studied whether it was feasible to complete container transportation between East Asia and Northwest Europe through NRS/SCR(Suez Canal Route) combination transportation. Gao and Lu [6] developed a mathematical model based on a programming equation that measures the impact of increased shipping costs caused by strait or canal congestion on the Chinese fleet. Alghamedii et al. [7] determined the theoretical basis of the Logit model and derived the logarithmic probability equation. This method is used to analyze the actual market situation. And Maria et al. [8] have analyzed and modeled and gathered data and gotten preliminary results, and got airline passengers' preferences for maximum comfort. In the existing studies, the economic efficiency of the Arctic route is analyzed mainly by calculating the shipping cost quantitatively or the feasibility of the Arctic navigation qualitatively. There is little comparison with traditional shipping lanes. Based on the Logit model, Monte Carlo simulation results are used for error analysis in this paper. The transport share ratio of the Arctic route and the traditional route is obtained, and the future market share of Arctic route is predicted. The research results provide the development direction for enterprise development.

3 Route Selection Model Establishment

3.1 Problem Description

According to the deficiencies of existing studies, this paper takes the comparison of Arctic shipping routes and traditional shipping routes as the research object and takes transport sharing rate as the research purpose. Determining the transport share ratio is a common problem in traffic planning. The purpose of analyzing the sharing rate is to establish a reasonable sharing relationship by analyzing the current situation and the future. Many factors are affecting the sharing rate. Based on references and subjective judgments, this paper designs a questionnaire to obtain the original data.

3.2 Voyage Cost Analysis

Shipping cost can be divided into capital cost, operating cost, and voyage cost.

(1) **Cost of capital**
   The cost of capital is the hire of the ship

Table 1 Cost–benefit comparison (Unit/US $10,000)

	The Arctic route	The traditional route
Freight	1425	1068.75
Cost	948.65	770.97
Earning	476.35	297.78
Days	22.08	29.10

(2) **Operating costs**

1. Crew fees
2. Insurance premium
3. Maintenance fees
4. Management fees
5. Other expenses.

(3) **Voyage cost.**

The cost directly incurred in transportation is called voyage cost. These include port, fuel, canal, and other navigation costs (the Arctic route is ice-breaking pilotage).

By building a cost model, the costs and benefits of the Arctic route and the traditional route are calculated as shown in Table 1.

Although the cost advantage of the Arctic route is not obvious during the navigation period. However, due to the shortened voyage time, it can be transported more than once during the limited voyage period. So the payoff is higher. This suggests that, with the development of Arctic shipping routes, the economy of Arctic shipping routes will become higher and higher.

3.3 Discrete Selection Model

The Logit model theory is based on the stochastic utility theory and the behavioral utility maximization hypothesis [7, 9, 10]. Assume that the decision maker is under perfectly reasonable conditions. Faced with all options, decision makers often choose the strategy that works best for them. Binomial selection model is a widely used model in discrete selection model [10]. It assumes that all selection actions are at the same level, and there was no correlation between each selector limb.

Based on stochastic utility theory, different choices will produce different utility values. Under given conditions, shipping decision makers will choose the strategic plan with the greatest utility. The utility value of choosing a particular shipping route depends on a number of factors. The utility function is:

$$U_{in} = V_{in} + \varepsilon_{in} \quad i = 1, 2 \tag{1}$$

where U_{in} is the actual utility of limb i, V_{in} is the utility determination (obtained by direct observation of the data), ε_{in} is the utility error, the effects that cannot be directly observed; A_n is the collection of all shipping route options; i is the choice and n is the decision maker.

For the fixed term V_{in}, it is usually expressed as:

$$V_{in} = \beta_1 X_{in1} + \beta_2 X_{in2} + \beta_k X_{\text{ink}} \tag{2}$$

where X_{ink} is the kth utility variable in the ith route option, β_k is the coefficient corresponding to the kth variable.

The probability calculation equation is:

$$P_{in} = \frac{1}{1 + e^{-V_{in}}} \tag{3}$$

Simplified equation is:

$$\ln\left(\frac{p_i}{1 - p_i}\right) = V_{in} \tag{4}$$

Through derivation, the calculation equation is as follows:

$$P_i = \frac{1}{N} \sum_{n=1}^{N} P_{in} \tag{5}$$

where P_i is the probability of the selecting limb i, P_{in} is the probability of decision maker n choosing shipping route i.

3.4 Survey Methods

Adopt the method of stated preference (SP) survey, design the questionnaire. SP surveys collect the decision makers' choice intentions or results in hypothetical situations. It is widely used in traffic demand forecasting [8]. Using the network platform, the questionnaire was filled out. The Statistical Product and Service Solutions (SPSS) is used to solve each parameter in the model.

According to the five types of shipping service attributes concerned by decision makers and combined with existing market conditions, determine the level of shipping service attributes and select the orthogonal experimental design method of partial factorial design to determine the final factor level combination. SPSS is a commonly used data analysis software using SPSS to carry out orthogonal experiment design, choose a suitable and effective orthogonal table for design. Thirty-two combinations of service attribute levels are obtained. Distribute the questionnaire to fifty practitioners in related industries. Each person fills in ten randomly, and a total of 500 valid questionnaires are obtained. SPSS was used to process all the data in the questionnaire.

3.5 Model Analysis

SPSS was used for regression analysis of the survey data. The analysis results were as follows: Table 2 is the Hosmer–Lemeshow test in the regression analysis. This test is used to detect whether the data fit the model to meet the requirements.

This is the H-L test sheet, $P = 0.919 > 0.05$. It is considered that the model can fit the numbers well (Table 3).

In the final result, the P is all <0.05, which means that the requirements are met. P is the significance result.

By substituting parameters into the utility function, the utility function of the model is

$$V_i = 1.695X_{i9} - 2.512X_{i1} - 1.962X_{i2} + 1.666X_{i3} \tag{6}$$

According to the utility function and Eq. (6), the selection probability of two paths for each decision maker in the data sample can be calculated. The results are shown in Table 4. This paper only gives the path choice probability of 10 decision makers (Fig. 2).

Equations (4) and (5) are used to calculate the probability of each selector limb in the population sample, and then:

$$P_1 = \frac{1}{N} \sum_{n=1}^{N} P_{1n} = 0.6021$$

Table 2 Hosmer–Lemeshow (H-L) test

Procedure	Chi-square	Degree of freedom	Significance
1	3.226	8	0.919

Table 3 Regression analysis results

Attributes	B	Error	Wald	P	Exp(B)
Value	1.695	0.582	8.481	0.004	5.449
Cost	-2.512	0.391	41.305	0.000	0.081
Time	-1.962	0.532	13.581	0.000	0.141
Safety	1.666	0.444	14.065	0.000	5.293
Type	7.010	2.269	9.542	0.002	1108.078

Table 4 Select probability

Decision makers	Select probability	
	Traditional route	Arctic route
1	0.755839	0.244161
2	0.569791	0.430209
3	0.577617	0.422383
4	0.503500	0.422393
5	0.570527	0.429473
6	0.570527	0.429473
7	0.755839	0.244161
8	0.570527	0.429473
9	0.577617	0.422383
10	0.569791	0.430209

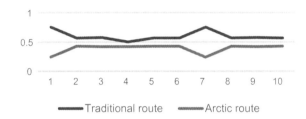

Fig. 2 Comparison chart of transport sharing rate

$$P_2 = \frac{1}{N} \sum_{n=1}^{N} P_{2n} = 0.3979$$

This paper realizes visualization based on transportation cost model and the Logit discrete selection model. Select three longitudinal parameters: time, cost, and safety parameters and two lateral parameter types of cargo and value of goods. The data were randomly divided into several parts for analysis. It can be seen that different types of goods have different requirements on time, cost, and safety. The higher the value of goods, the higher the requirements on time cost and safety. After many simulations, it can be clearly concluded that there are differences in the transport sharing rate of the Arctic route. This shows that with the increase of global temperature, the improvement of navigation technology, port facilities and other conditions change. The share of shipping on the Arctic route will gradually increase [11, 12].

Table 5 Survey data statistical indicators

Attribute	Quantity	Min	Max	Average	Standard	Variance
Cost	500	3	5	4.08	0.925	0.855
Time	500	2	5	3.76	0.699	0.488
Safety	500	2	5	3.65	0.816	0.666
Type	500	1	4	2.34	0.783	0.613

4 Monte Carlo Simulation and Analysis

4.1 Monte Carlo Simulation Process

According to the results of SPSS regression analysis, cost, time, safety, and goods value are the main attributes that affect decision makers' choice. The basic steps of Monte Carlo simulation are as follows:

(1) Determine the number of simulations. Let the total number of simulation runs be N. Based on the trial algorithm and the stability of the simulation results, $N = 5000$ is finally set in this paper.

(2) Firstly determine the probability distribution of the data. After regression analysis of data by SPSS, the Logit model meets the fitting standard. So the data is close to an exponential distribution.

(3) Generate random numbers according to probability distribution. Random numbers here use a computer to generate "pseudo-random numbers." Mathematical transformation is performed according to the probability distribution function of the data. Finally, the random number evenly distributed between [0, 1] is substituted into the transformed function. The sampling value of each random number can be obtained. This paper USES the inverse function of the data probability distribution function to generate random numbers. This method is called inverse transformation.

(4) After the random sampling is determined, the random sampling values of this set of data were substituted into the model. The final selection probability value of N groups can be obtained.

(5) N groups of probability values obtained by simulation. The expected value, variance, and standard deviation are calculated as shown in Table 5. The simulation results are compared with the values obtained from the Logit model.

4.2 Simulation Result Analysis

See Fig. 3.

According to the theory of large numbers based on the Monte Carlo method, the available simulation data results are:

Fig. 3 Simulation results

$$P_1 = 0.584964$$

$$P_2 = 0.415036$$

Compared with the calculation result of the Logit model, the error of the transportation sharing rate of traditional routes is 2.8%, and the error of the transportation sharing rate of Arctic routes is 4.1%. The simulation results show that the error of the probability results calculated by establishing the Logit model and using Monte Carlo simulation is within the allowable range. It proves that the Logit model has high applicability for solving such problems.

5 Conclusions

Navigation technology is improving as global temperatures rise and Arctic sea ice melts. The opening of Arctic shipping routes and commercial navigation is an inevitable trend. The Arctic shipping route could become a new core. As far as the current situation is concerned, the northeast Arctic route has been navigable. However, the navigation conditions of the Arctic route are still restricted by many factors. From the perspective of shipping enterprises, this paper takes Dalian to Rotterdam as an example to calculate shipping costs and benefits. This paper is based on the Logit discrete selection model. SPSS was used to process the data obtained from the SP survey, and the rate of transport sharing of Arctic routes was about 39.79%, and that of traditional routes was about 60.21%. It is expected that the Arctic route will account for about 31.79% and the traditional route for about 68.21% in the case of the continued decline in international oil prices. The Monte Carlo method was used to simulate the data, and error analysis was carried out with the data obtained by the Logit model. It is concluded that the Arctic route will take more share in the shipping market. The research results provide a new development direction for enterprises.

Acknowledgements This research was supported in Liaoning Provincial Nature Fund (2019-ZD-0661), Project of Liaoning Province Social Science Circle (2020lslktqn-058).

References

1. Fedi, L., Faury, O., Gritsenko, D.: The impact of the Polar Code on risk mitigation in Arctic waters: a "toolbox" for underwriters. Marit. Policy Manag. **45**(4) (2018)
2. Liu, M., Kronbak, J.: The potential economic viability of using the Northern Sea Route (NSR) as an alternative route between Asia and Europe. J. Transp. Geogr. **18**(3), 434–444 (2010)
3. Verny, J., Grigentin, C.: Container shipping on the northern sea route. Int. J. Prod. Econ. **122**(1), 107–117 (2009)
4. Dai, J., Yin, M.: Technical economic analysis of container ship transportation on the Arctic Northern Sea Route **5**(8), (2019)
5. Furuichi, M., Otsuka, N.: Economic feasibility of finished vehicle and container transport by NSR/SCR combined shipping between East Asia and Northwest Europe. In: Proceedings of the IAME 2014 Conference, Norfolk, VA, USA, pp. 15–18 (2014)
6. Gao, T., Lu, J.: The impacts of strait and canal blockages on the transportation costs of the Chinese fleet in the shipping network. Policy Manag. **46** (2019)
7. Ateq, A., Melfi, A.: The application of discrete choice models in marketing decisions. Math. Theory Model. **4**(9), 53–62 (2014)
8. Maria, G.B., Luigi, O., Laura, E.: Passengers' expectations on airlines' services: design of a stated preference survey and preliminary outcomes. Sustainabiility **12**(11) (2020)
9. Dohee, K., Byung-Jin (Robert), P.: The moderating role of context in the effects of choice attributes on hotel choice: A discrete choice experiment. Tour. Manag. **63** (2017)
10. Yu, P.L.H., Lee, P.H., Cheung, S.F.: The Logit tree models for discrete choice data with application to advice-seeking preferences among Chinese Christians. Comput. Stat. **31**(2), 799–827 (2016)
11. Alireza, R., Ghazaleh, A.., Xia, J.: Examining human attitudes toward shared mobility options and autonomous vehicles. Transp. Res. Part F: Psychol. Behav. **72** (2020)
12. Lee, T., Kim, H.J.: Barriers of voyaging on the Northern Sea route: a perspective from shipping companies. Mar. Policy **62**, 264–270 (2015)

Airport Taxi Driver Decision and Ride Area Scheme Design Based on Hybrid Strategy

Jingxuan Yang, Longsheng Bao, Qicheng Xu, Junjie Liu, and Yang Wang

Abstract With the development of the civil aviation market, how to improve the operating efficiency of airport taxis while ensuring a balanced income for drivers has become an important task. This paper recommends using a hybrid strategy model for analysis. It is analyzed, in this paper, the factors influencing the driver's decisions to obtain the qualitative relationship between the driver's income and costs such as time cost and no-load cost. We simplified the problems that are a game between driver groups whether or not to wait to establish a hybrid strategy model and to provide references in the selection of the two schemes for drivers. A scheme to give priority to short-distance passenger-carrying drivers is proposed. The two sides of the game are re-determined as short-distance passenger-carrying drivers and long-distance passenger-carrying drivers and income equilibrium of both parties are obtained. The results showed that under the new scheme, both parties have equal expectations of income so the new scheme is feasible.

Keywords Hybrid strategy · Income expectation · Income equilibrium · Queuing system

J. Yang (✉) · L. Bao
School of Traffic Engineering, Shenyang Jianzhu University, Shenyang 110168, China
e-mail: yjx_xuaner@163.com

L. Bao
e-mail: 13516094255@163.com

Q. Xu
School of Science, Shenyang Jianzhu University, Shenyang 110168, China

J. Liu
School of Management, Shenyang Jianzhu University, Shenyang 110168, China

Y. Wang
Faculty of Information and Control Engineering, Shenyang Jianzhu University, Shenyang 110168, China

© The Author(s), under exclusive license to Springer Nature Singapore Pte Ltd. 2021   243
Y. Li et al. (eds.), *Advances in Simulation and Process Modelling*,
Advances in Intelligent Systems and Computing 1305,
https://doi.org/10.1007/978-981-33-4575-1_24

1 Introduction

With the development of the air passenger transport market, the capacity of diversified transportation modes at airports is facing more challenges. Among them, the efficient operation of airport taxi pick-up points is a key issue, but scholars are more inclined to study service level in ride area. Few people have paid attention to the driver's decision-making before picking up passengers, and the issue of the income balance of different operating distances. Therefore, the correct establishment of a mathematical model, a selection strategy for drivers, and a reasonable arrangement of planning pick-up points are significant for improving the efficiency of taxi operations and maintaining the benefits of taxi drivers.

In terms of taxi decision and profit maximization, Zhang et al. established driver's choice decision model based on time periods and multi-objective programming model based on queuing theory [1]. Wang established a judgment formula based on the comprehensive supply and demand relationship and profit relationship and achieved the long-distance and short-distance taxi driver income balance by dividing the level [2]. Lv et al. established a multi-objective programming model, which was solved using genetic algorithms to obtain a reasonable distribution scheme in airport with the highest riding efficiency [3]. Zheng established a fitting model by collecting relevant airport data and analysis and calculation methods, and a reasonable scheme is designed for the allocation of taxi resources [4].

The revenue of airport taxis is related to both whether the taxi carrying a long-distance or a short-distance passenger and whether a return taxi carrying a long-distance or a short-distance passenger. When the airports set the pick-up points, they should consider giving a "priority" to those drivers whose latest trip was a short distance, so that the revenue of these taxis is as balanced as possible. Secondly, the existing multi-point side-by-side taxi queuing service system [5] has not maximized the riding efficiency, so taxi boarding points should be set more reasonably and efficiently.

The structure of this paper is as follows. Section 2 introduces the establishment of hybrid strategy model, which simplifies the problems to the game process of waiting or not among driver groups. Section 3 gives the scheme to make short-distance passenger-carrying drivers get priority to make their income equal, and correspondingly gives the scheme of a pick-up point, which has two parallel lanes, to improve the efficiency of the pick-up point, and shows the effectiveness of this scheme.

2 Hybrid Strategy Model

2.1 Hybrid Strategy

The research of traffic psychology shows that factors such as the driver's driving age, gender, risk perception ability, emotion, and decision-making style will have an impact on the driver's driving decision [6]. Combining with the taxi drivers in the target airport of this article, we observed that the changes in the number of passengers at the airport and the driver's expected income will also affect the driver's decision.

To facilitate the following description, the following concepts are introduced here:

- Income (W): Refers to the driver's final income under the different options.
- Estimated income (I): Refers to the income that the driver may obtain from entering the waiting area and successfully carrying passengers.
- Time cost (C_1): Refers to the revenue lost during the waiting period from when the driver enters the waiting area to successfully carry passengers.
- No-load cost (C_2): Refers to the no-load cost (gas fee) paid by the driver when he chooses an empty vehicle to return to the urban area and the possible loss of passenger income.
- Other cost (C_0): additional costs incurred in other time periods.

For the taxi drivers, they have only two decision-making schemes. One is waiting in line for passengers (hereinafter referred to as Scheme A), and the other is to empty the taxi and return to the city to carry passengers (hereinafter referred to as Scheme B).

Therefore, from the standpoint of the taxi driver (self), maximizing revenue is the key to the decision. The driver's competitors are the biggest distractions affecting the driver's earnings. Thus, the problem can be simplified as a game process of choice between Scheme A and Scheme B among taxi drivers.

Figure 1 reflects the income calculation principle of different decision-making schemes [7].

The qualitative relationship between the income and cost is as follows:

$$\text{Scheme A}: W = I - C_1 - C_2 - C_0 \tag{1}$$

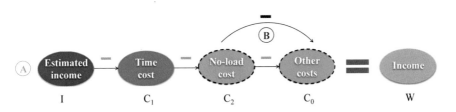

Fig. 1 Principle of income calculation

Table 1 Hybrid game matrix between "self" and "competitor"

Probability		q	$1 - q$
	Self/competitor	Scheme A	Scheme B
p	Scheme A	(W_{11}, W'_{11})	(W_{12}, W'_{12})
$1 - p$	Scheme B	(W_{21}, W'_{21})	(W_{22}, W'_{22})

$$\text{Scheme B}: W = C_2 - C_0 \tag{2}$$

Also, due to the objective existence of tourist peak-season and tourist off-season, the income and cost of drivers are not fixed, but fluctuate periodically with time.

In this step, we supposed that the income of "oneself" and the "competitor" after the game are W_{xy}, W_{xy}, where "x" represents "oneself" and "y" represents the "competitor". Then, we supposed that the probability of "oneself" choosing scheme A is p ($0 < p < 1$), so the probability of choosing scheme B is $1 - p$. Similarly, a "competitor" chooses scheme A with probability q ($0 < q < 1$), so the probability of choosing scheme B is $1 - q$. Thus, the hybrid game matrix of "self" and "competitor" is shown in Table 1 (the content of the matrix is represented in brackets).

2.2 Utility Analysis

We assumed that the probability of the "competitor" choosing Scheme A is q, then the utility function $U(x, y)$ of "self" Scheme A and Scheme B are chosen (3) and (4) as follows, respectively:

$$U(1, q) = W_{11}q + W_{12}(1 - q) \tag{3}$$

$$U(0, q) = W_{21}q + W_{22}(1 - q) \tag{4}$$

Let $U(1, q) = U(0, q)$, get the probability:

$$q = \frac{W_{22} - W_{12}}{W_{11} - W_{12} - W_{21} + W_{22}} \tag{5}$$

So as to get the income expectation of "self" as:

$$E = W_{11}q + W_{12}(1 - q) \tag{6}$$

In the same way, we assumed that the probability of "self" choosing Scheme A is P. Let $U(p, 1) = U(p, 0)$, then the probability can be obtained:

$$p = \frac{W'_{22} - W'_{12}}{W'_{11} - W'_{12} - W'_{21} + W'_{22}} \tag{7}$$

Decision-making suggestions are provided according to the above processes:

- When the probability of the "competitor" choosing Scheme A is equal to q, "self" can choose Scheme A or B. When the probability of the "competitor" choosing Scheme A is greater than q, the "own" Scheme A is more dominant. On the contrary, the "self" Scheme B is more dominant.
- When the ideal income of "self" is less than the income expectation E, Scheme A should be selected, that is, queuing up passengers to obtain greater income. When the ideal income is greater than the income expectation E, Scheme B should be selected, that is, empty the car without carrying passengers. When the two are equal, either Scheme A or B will work.

3 Income Equilibrium

3.1 Principle of Income Equilibrium

The types of passenger-carrying drivers can be divided into long-distance passenger-carrying and short-distance passenger-carrying. Among them, the revenue and cost of the two types of drivers over time are shown in Figs. 2 and 3.

It can be seen from Fig. 3 that the profit level of long-distance passenger-carrying drivers is much higher than that of the short-distance passenger-carrying drivers. Besides, since the no-load cost and time cost of the long-distance passenger-carrying driver is lower than the short-distance passenger-carrying driver, the cost level is also lower than that of the short-distance passenger transportation.

As time goes by, the polar differences between the two sides will become larger and larger, which could disrupt the stability of the taxi economy market. At this time,

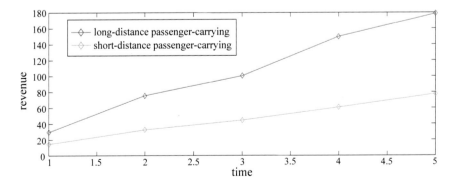

Fig. 2 Revenue of the two types of drivers change over time

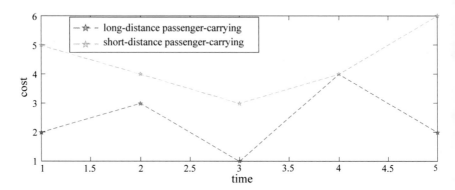

Fig. 3 Cost of the two types of drivers change over time

the airport management department must provide a certain "priority" to the drivers whose latest trip is a short distance to ensure that the benefits of both parties are balanced [8], **this is the income equilibrium**.

The two sides of the game in this problem are **the short-distance passenger driver** and **the long-distance passenger driver**. To analyze this problem, the hybrid decision-making model established in Sect. 2 can be obtained. The mathematical expression of the equilibrium of the two parties' income [7] is:

$$W_{11}q + W_{12}(1 - q) = W'_{11}p + W'_{21}(1 - p) \tag{8}$$

It means that the income expectation of "self" is equal to the income expectation of the "competitor", namely:

$$E = E' \tag{9}$$

3.2 "Priority" Scheme Design

We analyzed a model, which is a double-sided multi-point cross-tandem queue service system. In the scenario, we set there are two parallel lanes in the taxi ride area with this model. The model is a double-sided queuing system, and the pick-up points are cross-distributed, providing detour space for the vehicles from the rear. After entering the riding area, taxis could still enter any pick-up point and wait in line. When picking up passengers, they could choose to leave the riding area on the original road or take a detour. Passengers are diverted to the two sides of the two parallel lanes in the ride area through the dedicated passage, forming a line respectively, and the first passenger in the line could choose different pick-up points. The model diagram is shown in Fig. 4.

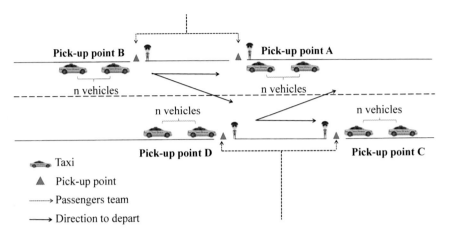

Fig. 4 Double-sided multi-point cross-tandem queue service system

To give a priority for short-distance passenger-carrying drivers, before the vehicles enter the pick-up points, they should be classified into four kinds: ordinary and priority with a long and short distance, then be arranged on both sides of the ride area respectively, as shown in Figs. 5 and 6.

When a taxi arrived at the airport, the driver should select a long or short distance to go for the next trip. Then, if the latest trip is short, the driver could enter the priority lane. If not or this is the first trip from the airport today, the driver should enter the ordinary lane. Therefore, all vehicles can be divided into four categories: ordinary vehicles A that will carry long-distance passengers, priority vehicles B that

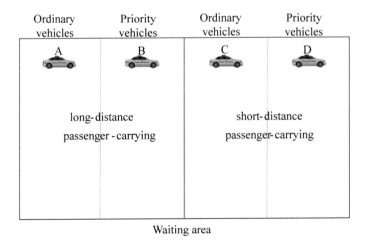

Fig. 5 Classification

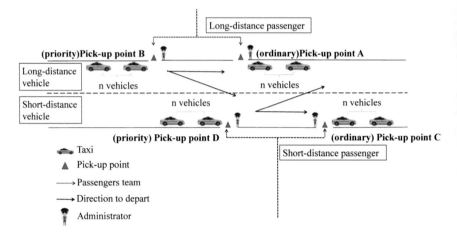

Fig. 6 Priority queuing scheme for short-distance passenger transportation-ride area

will carry long-distance passengers, ordinary vehicles C that will carry short-distance passengers, and priority vehicles D that will carry short-distance passengers.

The priority of short-distance passenger-carrying return vehicles is reflected in the separate queuing channel.

Passengers, before entering the ride area, are divided into two types: long-distance passengers and short-distance passengers. The two types of passengers enter the corresponding ride area in a line and follow the instructions of the administrator to enter the pick-up point and wait for taxis.

3.3 Effect of Hybrid Strategy Model with "Priority"

Based on the priority queuing schemes in Figs. 5 and 6, the hybrid strategy model in Sect. 2.1, the hybrid game matrix of "self" and "competitor" is re-established [9] as Table 2.

Obtaining $E = E' = \frac{7}{3}$ from Eq. (9) shows that the income expectation of "self" is equal to the income expectation of "competitor", which means the income of short-distance passenger-carrying drivers and long-distance passenger-carrying drivers are balanced. It proves the scheme of priority queuing designed in this paper is effective.

Table 2 Hybrid game matrix between yourself and the vehicle in front (with priority)

Probability		q	$1-q$
	Self/competitor	Scheme A	Scheme B
p	Scheme A	$(-3, 2)$	$(3, 0)$
$1-p$	Scheme B	$(0, -1)$	$(2, 1)$

4 Conclusions

This paper takes airport taxis as the research object and establishes a hybrid strategy model to be used into two problems. First, simplify the problem to the game process of whether the driver is waiting. Second, calculate the effectiveness of a scheme to balance the income of short-distance and long-distance passenger-carrying drivers. By analyzing above, the following conclusions can be drawn: through the hybrid strategy model, drivers can intuitively compare the final revenue expectation with their own income expectation, so as to choose the most favorable decision. The hybrid strategy model has good portability, it can adapt to different conditions by changing the main body of the game. Meanwhile, changing the scheme of the ride area, distinguishing taxi types, and giving priority to short-distance passenger-carrying drivers on an existing basis can effectively balance the benefits of the driver group.

References

1. Zhang, H.R., Zheng, Z., Cao, Y.: Airport taxi management based on dynamic decision and multi-objective programming model. Int. J. Serv. Sci. Manag. **3**(7) (2020)
2. Wang, Z.C.: Airport taxi decision and management model based on maximum benefits. SSMI **2019**(375), 412–418 (2019)
3. Lv, B.Y., Tian, Y.S., Liu, B.T.: Airport taxi dispatching based on VISSIM and multi-objective programming model. Am. J. Comput. Sci. Appl. **3**(23) (2020)
4. Zheng, Y.S.: Research on airport taxi resource allocation based on information asymmetry. Open J. Bus. Manag. **08**(02), 763–769 (2020)
5. Sun, J., Ding, R.J., Chen, Y.Y.: Modeling and simulation of single lane taxi boarding system based on Queuing theory. J. Syst. Simul. **29**(5), 996–1004(2017)
6. Chen, X.C.: The Study of Driving Decision-Making on Characteristics and Influencing Factors. Liaoning Normal University (2013) (in Chinese)
7. Antoniou, A., Lu, W.S.: Practical Optimization: Algorithms and Engineering Applications. Springer Publishing Company, Incorporated (2010)
8. Prajit, K.D.: Strategies and Games: Theory and Practice. The MIT Press, Cambridge (1999)
9. Roszkowska, E., Burns, T.R.: Fuzzy bargaining games: conditions of agreement, satisfaction, and equilibrium. Group Decis. Negot. **19**(5), 421–440 (2010)

Dynamic Route Optimization Problem Based on Variable Range Short-Term Traffic Flow Forecast

Guanghui Dai, Qianqian Shao, Yunfeng Zhang, and Siqi Zhang

Abstract A dynamic route optimization scheme based on short-term traffic flow prediction is designed. The overall idea of this paper is to use the real-time information of each section of the urban road communication network to predict the short-term traffic speed of the road in the next 5 min, in order to reflect the changes of the road traffic state of each section of the urban road traffic network. Then, the speed prediction information of each section in the urban road traffic network is converted into the estimated average travel time of each section by using the speed-time conversion formula, and then the road weight of each section in the urban road traffic network is set. Finally, the optimal route at the current time is calculated by using the improved algorithm based on the traditional Dijkstra algorithm. Before the traffic participant arrives at the destination, the scheme will cycle through the short-term prediction and route optimization of the traffic flow until the traffic participant arrives at the destination. And then achieve the balance of traffic conditions of various sections in the urban road traffic network, improve the situation of urban road traffic congestion, and avoid the emergence of bad road traffic conditions such as "navigation congestion".

Keywords Traffic flow · Dynamic route optimization · Wavelet neural network · Improved Dijkstra algorithm

1 Introduction

In recent years, with the continuous development of operational research theory, the continuous efforts of scholars have made a lot of achievements. As more and more operational research theories are applied to our social production and practical use, many research results can benefit the public. In the theoretical study of operational research, the problem of route choice has always been an important part of it. Domestic scholars have made a lot of achievements in the continuous exploration

G. Dai · Q. Shao (✉) · Y. Zhang · S. Zhang
School of Transportation Engineering, Shenyang Jianzhu University, Shenyang, China
e-mail: mr_Crowley@163.com

© The Author(s), under exclusive license to Springer Nature Singapore Pte Ltd. 2021   253
Y. Li et al. (eds.), *Advances in Simulation and Process Modelling*,
Advances in Intelligent Systems and Computing 1305,
https://doi.org/10.1007/978-981-33-4575-1_25

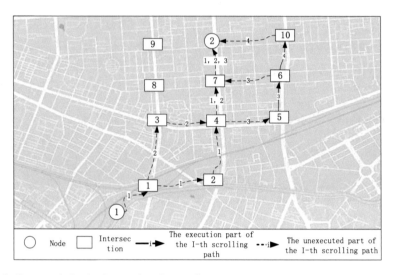

Fig. 1 Route optimization in actual road network

of the combination of theory and practice. But so far, there are not many results of combining the prediction information of urban road traffic flow in uncertain environment with the problem of dynamic route optimization of urban road traffic. There is a certain gap in our daily life and practical application, so that the urban road traffic guidance information can not completely optimize and adjust the route according to the real-time condition of urban road traffic flow (Fig. 1).

The driver wants to drive from node 1 to node 2, the original route is through intersection 1-2-4-7, to node 2, after the first adjustment, the route is adapted to pass through intersection 1-3-4-7, arrive at node 2, re-plan the follow-up route at intersection 4 according to the latest traffic information, update the route again at intersection 6, and the final driving route is 1-3-4-5-6-10-2 through intersection 1-3-4-5-6-10-2.

2 Related Works

Integrate the real-time road condition information, and accurately predict the traffic flow parameters in the future period according to the detected real-time traffic information, and use it as the basis to recommend the optimal route for travelers, so that the driver can adjust the driving route at the intersection. Before the driver reaches the end point, the driving route will be constantly adjusted according to the changes of traffic flow parameters, so it is called rolling optimization.

In 2015, Abdi and Moshii [1] compared three different traffic flow forecasting algorithm models: multilayer perceptron (MLP), radial basis function (RBF) neural network, and wavelet neural network (wavelet neural network). It is concluded that the MLP is easier to fall into the local minimum than the RBF neural network and the wavelet neural network (WNN). In the same year, Kumar et al. [2] applied the artificial neural network to the short-term prediction of traffic flow. Through the actual rural highway traffic flow data collected in the study, the effectiveness of the artificial neural network model was verified, and it was proved that the artificial neural network can achieve good results in the short-term prediction of traffic flow. In 2016, Habtemichael and Cetin [3] proposed a non-parametric data-driven short-term traffic flow prediction model algorithm based on k-nearest neighbor classification algorithm (k-nearest neighbor) and achieved good expected results. The method is simple, accurate, robust and can be used in real-time traffic control. In 2017, Laña et al. [4] discussed how to use the harmonious search optimization algorithm combined with the optimization characteristics of the current input data set to improve the parameters of the neural network. Through the discussion of the results and comparison with other adjustment methods, the potential application of this technique in multi-location proximity perceptual traffic prediction is demonstrated. In 2016, Ciarla et al. [5] introduced a multi-objective method to apply to optimal control based on the principle of optimal offline control under space and time boundaries. The goal of the proposed optimization problem is to find the best combination of segmented time without affecting the global final time, and to minimize the corresponding cost, and the optimal solution is obtained by applying the shortest route algorithm of Dijkstra.

Dynamic route optimization is a hot research topic in intelligent transportation system, which is based on real-time traffic information and parameters such as dynamic road resistance in road traffic network. According to the characteristics that the traffic behavior participants have different understanding and emphasis on the route optimization, there are also differences in the road resistance parameters on the same road section, and at the same time, but the dynamic road resistance is often the most intuitive reflection in the travel time of the whole traffic behavior. Therefore, this paper takes the traffic flow parameters of short-term traffic flow prediction as the basis, calculates the dynamic road resistance between the travel times of the road sections in the predicted state, obtains the objective function of the minimum travel time as different requirements under different weights, and studies the dynamic route optimization scheme of urban road traffic network.

Aiming at the basic needs and objectives of dynamic route optimization, and after analyzing the research status of dynamic route optimization, a scheme is proposed to combine the short-term traffic flow prediction information of urban road traffic network with dynamic route optimization algorithm. A real urban local road network example is used to verify the effectiveness of the proposed scheme. Based on the topological structure of the BP neural network model which can identify the characteristics of the complex nonlinear system, the wavelet basis function is used as the transfer function of the hidden layer to predict the traffic flow information, so as to improve the accuracy of the prediction information, so that the traffic flow prediction information can really serve the purpose of the dynamic route optimization algorithm. Finally, the travel time predicted by the wavelet neural network is used as the dynamic road section impedance. An improved algorithm based on the traditional Dijkstra algorithm is used to realize route optimization.

3 Route Optimization Model and Algorithm

The basic structure of wavelet neural network is divided into an input layer, and the input data are: t 15 min time, t 10 min time, t 5 min time, and the average speed of traffic flow at t time, so the number of nodes in the input layer is 4, one hidden layer, and the determination of the number of nodes in the hidden layer needs further analysis, an output layer, and the output data is the average speed of traffic flow at t moment 5 min, so the number of nodes in the output layer is 1.

In the initial state, the learning rate is set to 0.05, and the training algorithm adopts the gradient learning rate training algorithm.

The evolution number of training samples of wavelet neural network is set to 3000 times.

The error accuracy of wavelet neural network is 0.001.

In this paper, Morlet wavelet is used in the design of hidden layer, and its scale function does not exist and does not have orthogonality, but as a wavelet basis function commonly used in complex-valued wavelet, it can extract time information, signal floating information, and phase information.

The empirical formula for determining the number of neurons in the hidden layer is as follows:

r: the number of neural nodes in the input layer;
c: the number of neural nodes in the output layer;
y: the number of neural nodes in the hidden layer;
a: constant between [1, 10].

Morlet wavelet is adopted in designing hidden layer whose scale function does not exist and does not have orthogonality, but as wavelet function commonly used in complex wavelet, it can extract time information signal floating information and phase information. Its mathematical expressions are:

$$\psi(x) = \cos(1.75x)e^{-\frac{x^2}{2}} \tag{1}$$

The number of hidden layer neurons is determined by empirical formulas and several experiments by designers of network models. Empirical formulas for determining number of hidden layers are as follows:

$$y = \sqrt{r + c} + a \tag{2}$$

$$y = \sqrt{rc} \tag{3}$$

$$y = \log_2 r \tag{4}$$

Using classical BPR function to calculate travel time, the parameters are assumed as follows

$$t = t_i \left[1 + a_i \left(\frac{x_i}{c_i} \right)^{B_i} \right] \tag{5}$$

The classic BPR function of the US Highway Administration is suitable for long, continuous-flow highway sections, where the traffic flow composition is relatively simple, and the vehicle performance and road conditions are ideal.

Because the complexity of Dijkstra algorithm is $O(n^2)$[6], where n is the number of nodes of the road network, if it is assumed that the nodes in the traffic network are evenly distributed in the whole network plane, that is, the number of nodes in the network is proportional to the area of the network plane occupied, then in the whole process of searching the target node, the number of network nodes searched is expressed by the area swept by the search. For the classical Dijkstra algorithm, the number of network nodes searched is expressed by the area swept by the search. The area swept by the search is the area of a circle with the radius of the distance between the start and destination points, while the DKD algorithm, because of the forward and reverse search at the same time, the area swept by the search is the sum of the area of two circles whose radius is half of the distance between the start and destination points. The ratio of complexity between the two algorithms can be expressed as:

$$\lambda = 2 \times O\left(\left(\pi \left(\frac{w_{1j}}{2} \right)^2 \right)^2 \right) \Big/ O\left(\left(\pi w_{1j}^2 \right)^2 \right) \tag{6}$$

4 Simulation and Analysis

4.1 Traffic Flow Parameter Prediction

A real road network environment near a Square in Shenyang is selected as an example to verify the feasibility and effectiveness of the dynamic route optimization scheme based on wavelet neural network prediction, and the results are tested. The structure of the real road network is abstracted to facilitate simulation. The feasibility and effectiveness of the scheme are verified by simulation.

Fig. 2 Comparison between predicted traffic flow and actual traffic flow

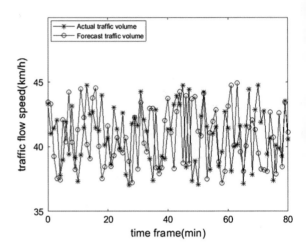

The dynamic route optimization scheme based on short-term traffic flow prediction information is based on the short-term traffic flow data predicted by wavelet neural network model, and the route is selected according to the improved algorithm based on traditional Dijkstra algorithm.

The real-time traffic flow data of a Square in Shenyang from August 13 to August 14, 2019 are used as experimental data. (The data come from the statistics of Shenyang Transportation Bureau.) (Fig. 2)

4.2 Dynamic Route Optimization Simulation

Figure 3 is a simplified diagram of the actual road network. Confirm the type of road network and set the starting point and departure time required for the departure time.

Figure 4 shows the path simplification diagram after simplifying the actual road network.

Selecting 50 travelers to optimize their travel routes, as shown in Fig. 5

It can be clearly seen that the proposed method has been optimized by an average of 8.5% compared with the original travel route, which verifies the effectiveness of the proposed method.

5 Conclusions

In this paper, the embedded wavelet neural network model is used as the short-term prediction model of traffic flow, and in view of the deficiency of BP neural network, the wavelet basis function is proposed as the transfer function of the hidden layer. The commonly used optimal route solving algorithm is briefly introduced, and then the

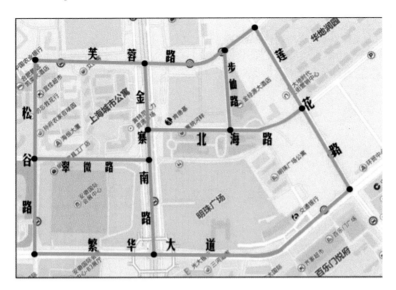

Fig. 3 Actual road network structure diagram (The data employed in this research work study was from the website of Shenyang Municipal Bureau of Communications, China, in Chinese.)

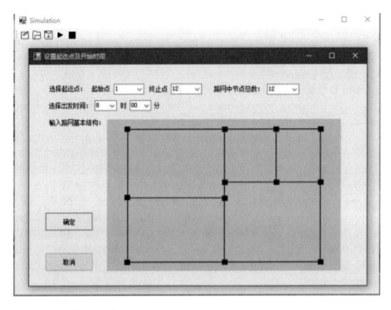

Fig. 4 Route simplification diagram

Fig. 5 Comparison between
the optimized route and the
original route

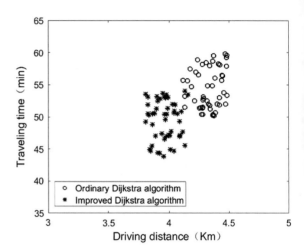

goal of route optimization is set, and then an improved bi-directional search scheme based on traditional Dijkstra algorithm is proposed. A dynamic route optimization algorithm based on short-term traffic flow prediction is designed and compared with the traditional route optimization algorithm, which shows the superiority of this scheme. Take the actual road network as the research object and make a simple road network abstraction to make it more concise and convenient to reflect the feasibility of the design scheme in this paper. After the average speed of each section of the road network is predicted by the previously designed wavelet neural network model, the improved Dijkstra algorithm is used to solve the best route. Finally, computer simulation is used to verify the feasibility of the dynamic route optimization scheme based on short-term traffic flow prediction.

The dynamic route optimization scheme based on short-term traffic flow prediction designed in this paper only involves the wavelet neural network model, the traditional Dijkstra algorithm, and some of its improved algorithms. In future research, we can choose more kinds and more advanced algorithms to solve the dynamic route optimization problem. In view of the continuous improvement of the theoretical study of fractal technology, there will be more feasible choice space for the short-term prediction method and model optimization of traffic flow, to achieve better prediction effect and higher prediction accuracy and further improve the reliability of the calculation results of the route optimization algorithm in the dynamic route optimization scheme.

Acknowledgements This research work is partially supported by the Scientific Research Project of the Educational Department of Liaoning Province 2019 of China (Project Codes: LNQN201916), Shenyang Social Science Project 2020 of China (Project Codes: SYSK2020-09-14, SYSK2020-09-32), Project Plan of China Logistics Learned Society and China Logistics and Purchasing Federation 2020 (Project Codes: 2020CSLKT3-187, 2020CSLKT3-188), Logistics Reform and Teaching Research Project 2020 of China (Project Codes: JZW2020205).

References

1. Abdi, J., Moshiri, B.: Application of temporal difference learning rules in short-term traffic flow prediction. Expert Syst. **32**(1), 49–64 (2015)
2. Kumar, K., Parida, M., Katiyar, V.K.: Short term traffic flow prediction in heterogeneous condition using artificial neural network. Transport **30**(4), 397–405 (2015)
3. Habtemichael, F.G., Cetin, M.: Short-term traffic flow rate forecasting based on identifying similar traffic patterns. Transp. Res. Part C: Emerg. Technol. **66**, 61–78 (2016)
4. Laña, I., Del Ser, J., Vélez, M., Oregi, I.: Joint feature optimization and parameter tuning for short-term traffic flow forecasting based on heuristically optimized multi-layer neural networks. In: International Conference on Harmony Search Algorithm, pp. 91–100. Springer, Singapore. (2017, February)
5. Ciarla, V., Chasse, A., Moulin, P., Ojeda, L.L.: Compute optimal travel duration in eco-driving applications. IFAC-PapersOnLine **49**(11), 519–524 (2016)
6. Deng, Y., Chen, Y., Zhang, Y., Mahadevan, S.: Fuzzy Dijkstra algorithm for shortest route problem under uncertain environment. Appl. Soft Comput. **12**(3), 1231–1237 (2012)

GAPSO-Based Traffic Signal Control in Isolated Intersection with Multiple Objectives

Yifan Chen, Feng Qiao, Lingzhong Guo, and Tao Liu

Abstract In this paper, an optimization algorithm is presented to deal with the issue of traffic signal timing in an isolated intersection in rush hour aiming at reducing traffic congestion. It selects the traffic capacity, number of stops, and delay time as the objective of the optimization, and the genetic algorithm (GA) is integrated with the particle swarm optimization (PSO) algorithm to obtain optimized traffic signal timing plan to upgrade the performance of intersection with faster convergence and higher accuracy. A numerical simulation study is conducted on MATLAB as a case study with a genetic algorithm particle swarm optimization (GAPSO), and the simulation results show that the proposed GAPSO algorithm outperforms the conventional PSO algorithms. In addition, with the presented isolated intersection multi-objective optimization model, intersection capacity is improved effectively.

Keywords Genetic algorithm · Particle swarm optimization · Multiple objective · Traffic signal timing · Isolated intersection · Rush hour

1 Introduction

As rapidly increasing number of vehicles in urban areas, traffic congestion has become one of the most serious problems in many cities around the world. In recent decades, the traffic networks where two or more roads meet or cross have played a significant role in traffic signal control and traffic system management. A proper traffic signal timing plan for an intersection will increase the traffic flow, relieve the

Y. Chen · F. Qiao (✉) · T. Liu
Faculty of Information and Control Engineering, Shenyang Jianzhu University, Shenyang 110168, China
e-mail: fengqiao@sjzu.edu.cn

L. Guo
Department of Automatic Control and Systems Engineering, University of Sheffield, Sheffield S13JD, UK

© The Author(s), under exclusive license to Springer Nature Singapore Pte Ltd. 2021 263
Y. Li et al. (eds.), *Advances in Simulation and Process Modelling*,
Advances in Intelligent Systems and Computing 1305,
https://doi.org/10.1007/978-981-33-4575-1_26

traffic jam, reduce travel delay, and minimize pollution. Therefore, it is highly desirable to develop the traffic signal timing schemes for intersections with multi-objective optimization (MOO) algorithms.

Many researchers have been devoted to studying various advanced strategies and schemes to alleviate traffic congestion in urban areas for the past few decades. The Webster model is often used in low demand traffic with fixed time control [1, 2], but these studies mainly focus on the single-objective optimization. With the actual traffic systems being more complex, these methods are not suitable for the problems aiming at optimize multiple objectives. In the past decade, a number of multi-objective evolutionary algorithms (MOEAs) for multi-objective optimization problems (MOOPs) have been investigated [3, 4]. The primary reason for using this kind of methods is that it just needs one single time to find the multiple Pareto-optimal solutions. And the genetic algorithm particle swarm optimization (GAPSO) proposed in [5] was one of the MOEAs. In order to improve the performance of PSO, algorithms in various forms have been proposed. A multi-leader strategy was used in [6] to deal with constrained multi-objective nonlinear problems, but this approach may mislead the direction of particle optimization. Multi-objective particle swarm optimization (MOPSO) is easy to fall into a local optimum, so some researchers combined MOPSO with other algorithms to improve the performance of MOPSO. MOPSO was integrated with a distribution estimation algorithm in [7]. With the further studies of urban traffic system and the development of MOEAs, in recent years, some dynamic MOO algorithms have been proposed to solve the MOOPs in traffic systems [8–12]. In these studies, the intersection capacity, average queue ratio minimum, delay time, and stop times are selected as the performance indexes in the models.

In this paper, a novel genetic algorithm particle swarm optimization (GAPSO) is applied. It chooses the intersection capacity, stop times, and delay time as the performance indexes subjected to the signal cycle, split and total intersection capacity. In order to make the saturated intersection more efficient, the top-priority optimization goal of this model is to maximize the traffic capacity of intersection during rush hour. A numerical simulation study is carried out on MATLAB to verify the effectiveness of the proposed model and algorithm, and the comparison is made to show the advantages of the proposed GAPSO over the conventional PSO in performance.

The remaining part of this paper is organized as follows. In Sect. 2, the MOO model of an isolated intersection in rush hour is established. In Sect. 3, an improved GAPSO algorithm is proposed and experimental analysis is made to show the validity of the proposed algorithm. In Sect. 4, a numerical simulation study was conducted on MATLAB for an isolated intersection in rush hour, and the simulation results are analyzed. The conclusions are drawn in Sect. 5.

2 Traffic Signal Mathematical Model for Optimization

2.1 Performance Indexes

The mathematical model of signal timing for an isolated three-phase intersection is discussed in this section together with the objectives of the research work.

In order to increase the efficiency of an intersection, three performance indexes are selected in this research work for optimization, including delay time, number of stops, and traffic capacity.

Vehicle delay time: The vehicle delay time of an isolated intersection D is expressed as follows:

$$D = \frac{\sum_i^n d_i q_i}{\sum_i^n q_i} = \frac{\sum_i^n \frac{C\left(1 - \frac{y_i(C-L)}{YC}\right)^2}{2(1-y_i)} q_i}{\sum_i^n q_i} \tag{1}$$

The basic equation is Webster's average delay equation [1].

$$d_i = \frac{C(1 - \lambda_i)^2}{2(1 - \lambda_i x_i)} + \frac{x_i^2}{2q_i(1 - x_i)} - 0.65\left(\frac{C}{q_i}\right)^{\frac{1}{3}} x_i^{(2+5\lambda_i)} \tag{2}$$

where d_i is the average delay time of the ith phase; C is the cycle length; λ_i is the split of the ith phase; x_i is the degree of saturation during the ith phase; q_i is the vehicle arrival rate of the ith phase.

Webster's delay equation consists of two parts, namely uniform delay (the first term in the equation) and random delay (two or three terms in the equation). It can be known from (2) that when the ratio of saturation tends to 1, the delay time will tend to infinity, and the result obtained is meaningless at that time. Therefore, this equation is applicable to the case where the saturation is less than 0.9. When the road conditions are oversaturated, there are many unpredictable random conditions when the vehicle travels, so the exact delay time cannot be obtained in (1). In this paper, the first term in the equation is used to obtain the delay time:

$$d_i = \frac{C(1 - \lambda_i)^2}{2(1 - \lambda_i x_i)} = \frac{C\left(1 - \frac{y_i(C-L)}{YC}\right)^2}{2(1 - y_i)} \tag{3}$$

where y_i is the traffic intensity of the ith phase; Y is the sum of the maximum flow ratios of the individual phases in a cycle; L is the lost time.

Number of stops: The number of stops H is shown as follows

$$H = \frac{\sum_i^n h_i q_i}{\sum_i^n q_i} \tag{4}$$

where h_i is the average number of stops during the ith phase, and

$$h_i = 0.9 \times \frac{C(1 - \lambda_i)}{1 - y_i} \tag{5}$$

Traffic capacity: The capacity of an isolated intersection Q is expressed as

$$Q = \sum_i^n Q_i \tag{6}$$

where Q_i is the intersection capacity of the ith phase; S_i is the intersection saturation flow of the ith phase.

$$Q_i = S_i \lambda_i = S_i \frac{y_i(C - L)}{CY} \tag{7}$$

2.2 Performance Objectives

The function of this optimization algorithm is to obtain the fluctuations of the three indexes according to the real-time traffic monitoring at the intersection, and assign the three changes to different weights according to the characteristics of the traffic flow to find the current traffic flow and make the intersection service. When the traffic volume of the intersection is high, the traffic capacity of the intersection is emphasized. When the traffic volume is low, the focus will be on the index for reducing the delay time of the vehicle and the number of stops.

The purpose of the algorithm is to improve the traffic capacity, reduce the average delay time, and the number of stops. Therefore, the multi-objective optimization model is constructed as follows:

$$\min F = \left(k_1 \frac{D}{D_0} + k_2 \frac{H}{H_0} \right) \Big/ k_3 \frac{Q}{Q_0} \tag{8}$$

where k_1 is the weight of vehicle delay time; k_2 is the weight of the number of stops; k_3 is the weight of the intersection capacity; D_0 is the initial vehicle delay time; H_0 is the initial number of stops; Q_0 is the initial intersection capacity.

In (8), the units of the three indexes are different, so each indicator is divided by their initial value and the parameters are dimensionless.

In a real situation, when the traffic volume is high, the traffic capacity is worth to be improved, and when the traffic volume is low, we might focus on reducing the delay time and the number of stops. Therefore, the traffic intensity Y is proportional to the capacity Q, while it is inversely proportional to the delay time D and the number of stops H. k_1, k_2 and k_3 are assigned as follows:

$$k_1 = \frac{1 - Y}{2}, \quad k_2 = \frac{1 - Y}{2} \text{ and } k_3 = Y \tag{9}$$

the optimization algorithm constraints are subjected to:

$$s.t. \begin{cases} ge_{\min} \leq ge_i \leq ge_{\max} \\ C_{\min} \leq C \leq C_{\max} \\ 0.6 \leq x \leq 0.9 \end{cases} \tag{10}$$

where $ge_{\min}$ and $ge_{\max}$ are the minimum and maximum of the green time during phase i ($i = 1, 2, 3$); $C_{\max}$ and $C_{\min}$ are the minimum and maximum of the signal cycle, respectively; and x is the saturability of traffic flow during each phase.

3 Multi-objective Optimization Algorithms

The classical particle swarm optimization (PSO) methods normally suggest that particles update their speed and position through individual extremum and population extremum during iteration. Therefore, the diversity of the population will gradually decrease in such an evolutionary process, and thus it is easy to fall into a local optimum. The proposed GAPSO algorithm improves the particle diversity by the crossover and mutation steps in the genetic algorithm, and solves the local optimal problem of the classical algorithm. Besides, due to the increase of particle diversity, the convergent rate of the proposed GAPSO algorithm is faster than that of the classic PSO algorithms.

3.1 Clustering Strategy

According to the fitness value calculation, the population is divided into two sub-populations, one is called non-inferior group and the other is called inferior group. When the serial number of the particle is within the threshold δ, the particle enters the non-inferior group; otherwise, it enters the inferior group. In order to have particles in the subgroup, δ changes according to the following equation.

$$\delta = t / T = 1 / (1 + e^{\text{ratio}}) \tag{11}$$

where ratio is the threshold change ratio with ratio $= e^{0.9*\mathrm{fr}} - 1$ (fr is the number of appearing empty subgroup), t is the current iteration.

The particles in the non-inferior solution set are randomly selected into the next generation, and the particles in the inferior solution set are crossed and mutated.

3.2 Genetic Crossover

The partial mapping hybridization is used to determine the parent of the crossover operation, and the parents samples are grouped in pairs, each of which repeats the following process (assuming a sample size of 10).

Generate random integers r_1 and r_2 in the interval $[1, 10]$, determine two positions, and crossover the intermediate data of the two positions, such as $r_1 = 4$ and $r_2 = 7$, respectively.

$$9\ 5\ 1\ \big|\ 3\ 7\ 4\ 2\ \big|\ 10\ 8\ 6$$
$$10\ 5\ 4\ \big|\ 6\ 3\ 8\ 7\ \big|\ 2\ 1\ 9$$

Crossing:

$$9\ 5\ 1\ \big|\ 6\ 3\ 8\ 7\ \big|\ 10\ *\ *$$
$$10\ 5\ *\ \big|\ 3\ 7\ 4\ 2\ \big|\ *\ 1\ 9$$

After the intersection, the non-repeating numbers are retained, and the conflicting numbers (with * position) are partially resolved by the method of partial mapping, and the correspondence between the intermediate segments is used for mapping. The result is:

$$9\ 5\ 1\ \big|\ 6\ 3\ 8\ 7\ \big|\ 10\ 4\ 2$$
$$10\ 5\ 8\ \big|\ 3\ 7\ 4\ 2\ \big|\ 6\ 1\ 9$$

3.3 Particle Speed Update

It is assumed that after dynamically dividing the kth population, E particles enter the non-inferior group, and $N - E$ particles enter the inferior group. Then the particle is updated.

The velocity equation of the non-inferior group is expressed as:

$$v_H^{t+1}(i\ j,) = w v_H^t(i\ j,) + c\ r\ \mathrm{pbest}_1\big(_H^t(i\ j\ x,) -_H^t(i\ j,)\big)$$
$$+ cr Y \mathrm{best}_{2\ 2}\big(_H^t - jx_H^t(i\ j,)\big)\ k = 1, 2, \ldots, U;$$

$$i = 1, 2, \ldots, E; \quad j = 1, 2, \ldots, D. \tag{12}$$

The velocity equation of the inferior group is expressed as:

$$v_L^{t+1}(i\ j,) = wv_L^t(i\ j,) + c\ r\ \text{pbest}_{1\ 1}\left({}_L^t(i\ j\ x,) - {}_L^t(i\ j,)\right)$$
$$+ c_2 r_2\left(Y\text{best}_L^t - j\ x_L^t(i\ j,)\right)i = 1, 2, \ldots, N - E \tag{13}$$

where $x_H^t(i, j)$ is the jth dimension of the current position of the particle i in the current group; $p_{\text{best}H}^t(i, j)$ is the jth dimension of the p_{best} of the particle i in the current group; $Y_{\text{best}H}^t(j)$ is the jth dimension of particle position in the current group.

3.4 Flowchart of GAPSO Algorithm

The flowchart of GAPSO algorithm has been shown in Fig. 1; after initialing the population, all the particles' fitness will be calculated. In the second step, the queue

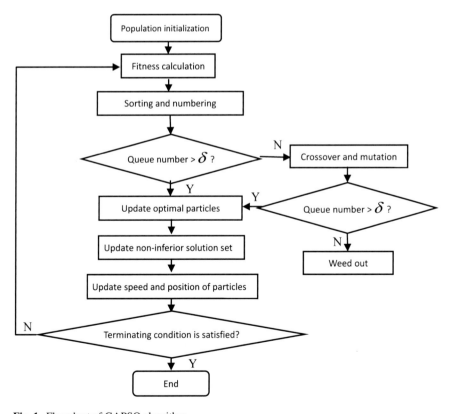

Fig. 1 Flowchart of GAPSO algorithm

number will compare with threshold δ after sorting and numbering the fitness, if the queue number is greater than δ, the optimal particles will be updated, else, those particles which less or equal to δ will be crossovered and compared with δ again, the greater group will be updated while the another group will be weeded out. The third step is to update the position and speed of the best particle after updating the non-inferior solution set. The final step is to judge whether the terminating condition is satisfied or not, if satisfied, this flow will be end, else, the flow will back to fitness calculation.

4 Experimental Analysis

4.1 Pareto Face

Mathematically, MOOP can be expressed as follows:

$$\min F(x) = \min(f_1(x), \ f_2(x), \ldots, \ f_M(x)) \tag{14}$$

Pareto-optimal set:

$$P^* = \{x \in \Omega | \neg \exists x^* \in \Omega, \ f_i(x^*) \le f_i(x), i = 1, 2, L, M\} \tag{15}$$

where $M \ge 2$ is the number of objectives; x is the feasible set of decision vectors.

Compared with the single-objective algorithm, the multi-objective search algorithm is more close to the actual problem, and the result is more valuable. What is finally obtained by the multi-objective search algorithm is not an optimal solution, but a non-inferior solution set. It is necessary to select a solution from the non-inferior solution set according to the actual problem as the final solution of the problem.

A MOOP classical account case is simulated on MATLAB R2014a by the proposed GAPSO algorithm. The non-inferior solution spatial distribution is shown in Fig. 2.

The non-inferior solution searched by the algorithm constitutes the p-plane, and the algorithm search has achieved good results.

4.2 Optimal Individual Fitness

Comparing the individual fitness evolution curves of GAPSO and classical PSO, it can be seen that the proposed algorithm is faster than the classical algorithm.

Comparing Figs. 3 and 4, the proposed algorithm approximately has the best value in the 65th generation, while the classical algorithm appears in the 125th generation.

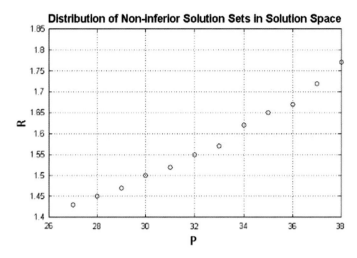

Fig. 2 Distribution of non-inferior solution

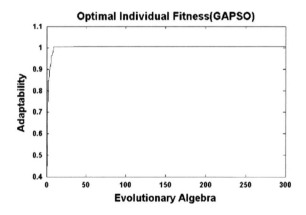

Fig. 3 Fitness of optimal individuals by GAPSO

4.3 An Example

An example was solved in the proposed GAPSO, classical PSO and NSGA-II for a common MOOP to verify its effectiveness (Table 1).

$$f(x_1, x_2) = x_1^4 + x_1 x_2 + x_2^4 - x_1^2 x_2^2$$

$$s.t. \begin{cases} -5 \le x_1 \le 5 \\ -5 \le x_2 \le 5 \end{cases} \tag{16}$$

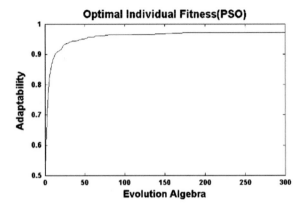

Fig. 4 Fitness of optimal individuals by PSO

Table 1 Fitness of the three algorithms

Algorithm	Parameter settings	Fitness
PSO	maxgen = 200; sizepop = 50; V_{max} = 10; V_{min} = 10; popmax = 100; popmin = −10	−1.2088 × 10⁴
GAPSO	w = 0.7298; N = 20; D = 6; eps = 10^(−6); MaxDT = 500	**−1.2261 × 10⁴**
NSGA-II	p_c = 0.9; p_m = 1/n; η_c = 20; η_m = 20	−1.2214 × 10⁴

As can be seen from this example, the GAPSO algorithm has better optimization ability than the other two algorithms.

5 Numerical Stimulation Study for an Isolated Intersection

5.1 An Example Typical Isolated Intersection

A typical isolated intersection example is selected in this paper.

In Fig. 5, it shows the diagram of an isolated intersection with three phases, and in the right side of the diagram, it describes the three phases, respectively.

The specific data of the traffic flow distribution at the early peak time of the intersection (7 am–9 am) was collected on-site and converted into the hourly average traffic volume as shown in Table 2.

Initial average delay, initial number of stops, initial saturation flow
According to the known traffic data, the equations of delay time, number of stops and traffic capacity, those can be used to find the initial average delay is 29 s, the initial number of stops is 85, and the initial traffic capacity is 4901pcu/h. And also, $D_0 = 29$, $H_0 = 85$, and $Q_0 = 4901$.

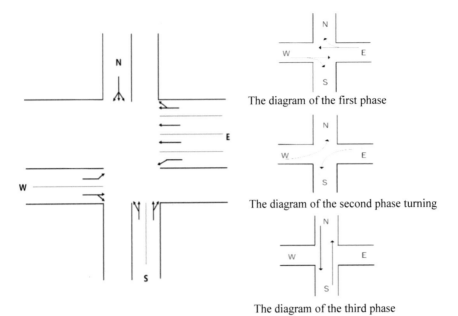

The diagram of the first phase

The diagram of the second phase turning

The diagram of the third phase

Fig. 5 Schematic diagram of an isolated intersection

Table 2 Specific data of the traffic flow

Orientation	Average traffic volume	Volume (per lane)	Total
East	Left turn lane	198	198
	1st straight lane	497	1394
	2nd straight lane	457	
	Straight/right lane	440	
West	Left turn lane	6	6
	Straight/right lane	580	580
South	Straight/left lane	304	490
	Straight/right lane	186	
North	Straight/right/left lane	111	111

Saturated flow and traffic intensity

The basic saturated flow rates of the straight, right-turn, and left turn lanes are set to 1800 pcu/h, 1550 pcu/h, and 1800 pcu/h, respectively. The equation for the saturated flow rate is as follows:

$$S = S_0 N \tag{17}$$

where S_0 is the saturated flow, N is the number of lanes.

Table 3 Traffic timing indicator table for each timing algorithm

Timing algorithm	Cycle	Green time/split			Average delay time	Average number of stops	Total traffic capacity
		I	II	III			
Now	114	57/0.5	23/0.2	25/0.22	29	85	4901
Webster	72	34/0.47	14/0.19	15/0.21	20	57	4448
MOO	152	80/0.53	29/0.19	34/0.22	36	109	5096

According to (9), the traffic intensity of the seven lane groups is calculated as: 0.1194, 0.2773, 0.0036, 0.3463, 0.1759, 0.1076, 0.0655.

Flow ratio of each phase:

$$y_1 = \max(0.2773, 0.3463) = 0.3463 \qquad y_2 = \max(0.1194, 0.0036) = 0.1194$$
$$y_3 = \max(0.1759, 0.1076, 0.0655) = 0.1759 \qquad Y = y_1 + y_2 + y_3 = 0.6416$$

MOO simulation

According to the constraints of the model, C_{min} is set to 30 and C_{max} is set to 180. Substituting the model into the GAPSO algorithm to find the best period at this time (Table 3).

According to the data in the table, compared with the existing timing scheme at the intersection, the period value obtained with the Webster method is too small, and the green signal ratio of each phase is slightly reduced, but the average delay time of the vehicle is reduced by 31%. The number of times has decreased by 33%, and the capacity has decreased by 452 pcu/h. Therefore, the Webster method is applicable when the traffic volume is small, and the effect is not good when the traffic volume is large. The multi-objective optimization method of this paper improves the traffic capacity to 5096 pcu/h, while other two indicators fell slightly. This is because the current traffic volume of the intersection is large, and it pays more attention to the timing scheme with multi-objective algorithm to improve the traffic capacity of an intersection. The delay time and the number of stop times are placed in the secondary position, and the solution of timing scheme with multi-objective algorithm is obtained. There has been a big increase in the indicator of capacity. The indicators of the stable period can be obtained with the same method.

6 Conclusions

Given that the urban intersection is always crowded in recent years and at the same time, optimizing the traffic flow by optimizing the signal timing is the best and the top-priority approach, this paper has proposed a GAPSO algorithm to optimize the traffic flow better. The PSO algorithm is employed in a numerical simulation study for optimizing a multiple objective problem of an isolated intersection for signal timing scheme. The simulation results show that with the optimization algorithm, the intersection capacity is improved effectively.

References

1. Webster, F.V., Cobbe, B.M.: Traffic signals. Road Research Technical Paper No. 56, Her Majesty's Stationery Office **4**(4), 206–207 (1966)
2. Akçelik, R.: Traffic Signals: Capacity and Timing Analysis. Research Report. Publication of Australian Road Research Board (1981)
3. Zhou, A., Qu, B.Y., Li, H., et al.: Multi-objective evolutionary algorithms: a survey of the state of the art. Swarm Evol. Comput. **1**(1), 32–49 (2011)
4. Wang, C.H., Tsai, S.W.: Multi-objective optimization using genetic algorithm: applications to imperfect preventive maintenance model. In: Proceedings of International Conference on Computer Science & Education 2011, pp. 1355–1360, Singapore (2011)
5. Wang, L., Si, G.: Optimal Location Management in Mobile Computing With Hybrid Genetic Algorithm and Particle Swarm Optimization (GA-PSO), 2010 IEEE
6. Shokrian, M., High, K.A.: Application of a multi objective multi-leader particle swarm optimization algorithm on NLP and MINLP problems. Comput. Chem. Eng. **60**(1), 57–75 (2014)
7. Cheng, T., Chen, M., Fleming, P.J., Yang, Z.L., Gan, S.J.: A novel hybrid teaching learning based multi-objective particle swarm optimization. Neurocomputing **222**(26), 11–25 (2017)
8. Cao, C.T., Xu, J.M.: Multi-object traffic signal control method for single intersection. Comput. Eng. Appl. **46**(16), 20–22 (2010)
9. Hu, H., Gao, Y., Yang, X.: Multi-objective optimization method of fixed-time signal control of isolated intersections. In: International Conference on Computational and Information Sciences 2010, pp. 1281–1284. IEEE, Chengdu, China (2010)
10. Li, Y., Yu, L.J., Tao, S.R., Chen, K.M.: Multi-objective optimization of traffic signal timing for oversaturated intersection. Math. Prob. Eng. **2013**(1683), 1–9 (2013)
11. Qiao, F., Sun, H.C., Wang, Z.Y., Fashaki, A.T., NSQGA-based optimization of traffic signal in isolated intersection with multiple objectives. Lecture Notes in Electrical Engineering, pp. 291–305 (2018)
12. Du, L., Jiao, P., Wang, H.: A Multi-objective traffic signal control model for intersection based on B-P neural networks. In: Proceedings of the 10th Asia Pacific PSO Transportation Development Conference, pp. 451–458. ASCE, Beijing, China (2014)

Multi-objective Optimization of Traffic Signal Systems on Urban Arterial Roads

Tao Liu, Feng Qiao, Lingzhong Guo, and Yifan Chen

Abstract In this paper, multi-objective optimization is used to solve the signal synchronization problem in arterial traffic roads, where a traffic dispersion module is introduced to further expand the solution space. By incorporating the models of delay time, queue length and stop times into the optimization, a first model called M1 is established. In the second model M2, the free flow speed assumption is replaced by a traffic dispersion module for better estimating the link travel time. A simulation study is then carried out on an arterial road, and the results show that the proposed strategy improves the performance of the traffic system compared to the current timing scheme and M2 has the best performance among all solutions in this paper, and the delay is reduced for about 24%.

Keywords Multi-objective optimization · Arterial urban traffic · Signal timing scheme

1 Introduction

Urban arterial roads are an important part of the urban transportation system and bear the main traffic load of an entire city. An effective signal control strategy is critical for ensuring a higher traffic capacity. Limited by urban space and economic practicability, the infrastructure load capacity of urban highways cannot permanently be kept above the increasing traffic flow. Therefore, the key to solving the problem of traffic congestion in the urban road network is to reduce the traffic congestion in the series of intersections on an arterial road [1].

T.`Liu · F. Qiao (✉) · Y. Chen
Faculty of Information and Control Engineering, Shenyang Jianzhu University, Shenyang, China
e-mail: fengqiao@sjzu.edu.cn

L. Guo
Department of Automatic Control and Systems Engineering, University of Sheffield, Sheffield S13JD, UK

© The Author(s), under exclusive license to Springer Nature Singapore Pte Ltd. 2021   277
Y. Li et al. (eds.), *Advances in Simulation and Process Modelling*,
Advances in Intelligent Systems and Computing 1305,
https://doi.org/10.1007/978-981-33-4575-1_27

Saka et al. [2] classified the intersections according to different traffic conditions and carried out signal control for each situation. A fuzzy hierarchical control method of urban road intersections was proposed by Kim [3], where he adjusted the control strategy in real time according to different traffic flow conditions and applied the genetic algorithm (GA) to the fuzzy control of intersection signals in [4] to improve the performance of the fuzzy controller. Genetic algorithm was also used to study the real-time adaptive control optimization method of traffic signals in [5]. Based on multi-intelligence framework, Khamis et al. [6] studied the adaptive multi-objective enhanced learning traffic signal optimization control method and verified its application in the experimental platform built. Aiming at obtaining wider green wave bandwidth, Qiao et al. [7] adopted particle swarm optimization (PSO) to conduct bidirectional green wave optimization for arterial traffic system.

In this paper, considering the model of delay time, queue length and stop times, a model M1 is constructed to solve the multi-objective optimization problem of arterial traffic system. Then, in order to provide more practical results, the traffic dispersion module is adopted to estimate the link travel time in the model M2. Using the non-dominated sorting genetic algorithm-II (NSGA-II) [8] and commercial optimization software, feasible solutions can be found quickly. Simulation studies on a sample road are made with VISSIM to verify the effectiveness of the proposed coordination traffic signal timing scheme. The results show that model M2 provides the best time scheme.

The rest of the paper is organized as follows. In Sect. 2, the multi-objective optimization issue is formulated as model M1 for the arterial traffic system; in Sect. 3, we formulate the extended model M2 by introducing traffic dispersion module into model M1; in Sect. 4, the simulation results are presented and analyzed. Finally, the conclusions are drawn in Sect. 5.

2 Problem Statements

According to the actual traffic situation, the optimization target of arterial system is generally to minimize the average delay or to maximize the green wave bandwidth. The minimum delay method is devoted to reducing vehicle delays at each intersection by reasonably allocating the period and offset of each intersection on the main road, while the maximum green wave zone method is devoted to increasing the number of vehicles passing through each green wave time [9]. In this paper, based on the minimum delay method, the models of line delay, queue length and stop times are incorporated into the multi-objective optimization problem.

The average delay model of the arterial traffic systems is expressed as two parts: the upward delay and the downward delay. This delay model has a good optimization effect for the traffic situation with unequal traffic flow on each road [10]. Now define D_u and D_d as the upstream delay time and the downstream delay time, respectively; denote l as queue length; and H as the stop times. The multi-objective function (1) is to minimize D_u, D_d, l and H

Fig. 1 Geometric presentation of adjacent intersection

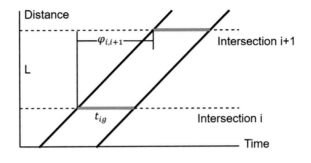

$$z = \min[D_u, D_d, l, H] \tag{1}$$

As shown in Fig. 1, let $\varphi_{i,i+1}$ be the offset between intersection i and intersection $i + 1$, $t_g(t_{ig})$ be green time (of intersection i), $t_r(t_{ir})$ be red time (of intersection i), C be the length of cycle, L be the distance between intersection i and $i + 1$, v be free flow speed. If the light is turned red when the front of traffic flow arrives at intersection $i + 1$, let t_u be the waiting time of the traffic flow, $t_{i,i+1}$ be the link travel time from intersection i to intersection $i + 1$.

$$t_{i,i+1} = \frac{L}{v} \tag{2}$$

$$t_u = \varphi_{i,i+1} - t_{i,i+1} \tag{3}$$

Let t be the evacuation time of vehicles queuing after the green light; q_u be the actual traffic flow in the upstream direction; q_{um} be the upstream saturation flow.

$$t = \frac{t_u q_u}{q_{um} - q_u} \tag{4}$$

It shows, in Fig. 2a, a schematic diagram of the delay time when the front of traffic flow is blocked by the intersection $i + 1$. In this case, we need to improve $\varphi_{i,i+1}$ to advance the green phase, so that the traffic flow will meet the green light when arriving at the intersection $i + 1$. The total delay time is represented as the area of shaded area in Fig. 2a.

Unless otherwise specified, a and a' represent the same variable in different case of vehicle delay in Fig. 2. Let d_{iu} (d'_{iu}) be the upstream delay time of the vehicle arriving at intersection i at the red (green) light. Let t_{ir} be the time of red light at the intersection i. By combining Eqs. (3) and (4), we have

$$d_{i+1,u} = \frac{q_u q_{um}(\varphi_{i,i+1} - t_{i,i+1})^2}{2(q_{um} - q_u)} \tag{5}$$

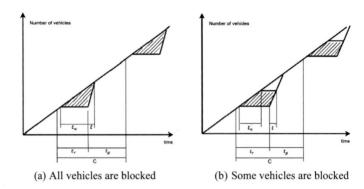

(a) All vehicles are blocked   (b) Some vehicles are blocked

Fig. 2 Delay time. **a** All vehicles are blocked, **b** some vehicles are blocked

It shows, in Fig. 2b, a schematic diagram of the delay time when the end of traffic flow is blocked by the intersection i. In this case, we need to reduce $\varphi_{i,i+1}$ to delay the green phase, so that the traffic flow can pass through the intersection in the green time. The total delay time is represented as the area of shaded area in Fig. 2b. Let t'_u be the time taken for the end of the traffic flow to pass through the intersection $i+1$.

$$t'_u = t_{i,i+1} - \varphi_{i,i+1} \tag{6}$$

The number of vehicles that fails to pass intersection $i+1$ in time in the traffic flow is $t'_u q_u$, and they need to wait for the green light in the next cycle to pass. The time required for these vehicles to pass is t'.

$$t' = \frac{t'_u q_u}{q_{um}} \tag{7}$$

By combining Eqs. (6) and (7), we have

$$d'_{i+1,u} = q_u t_{ir}\left(t_{i,i+1} - \varphi_{i,i+1}\right) - \frac{1}{2}q_u\left(t_{i,i+1} - \varphi_{i,i+1}\right)^2 + \frac{q_u^2}{2q_{um}}\left(t_{i,i+1} - \varphi_{i,i+1}\right)^2 \tag{8}$$

Let α_i be the Boolean function of intersection i in the upstream direction, n be the number of intersections.

$$D_u = \sum_{i=2}^{n}\left[\alpha_i d_{iu} + (1 - \alpha_i)d'_{iu}\right] \tag{9}$$

Similarly, for the delay time in the upward direction, let d_{id} $\left(d'_{id}\right)$ be the downstream delay time of the vehicle arriving at intersection i at the red (green) light. Let q_d be the actual traffic flow in the downstream direction; q_{dm} be the downstream saturation flow; $\varphi_{i+1,i}$ be the offset from intersection $i+1$ to intersection i; β_i be the Boolean function of intersection i in the downstream direction.

$$D_d = \sum_{i=1}^{n}\left[\beta_i d_{id} + (1 - \beta_i)d'_{id}\right] \tag{10}$$

$$d_{id} = \frac{q_d q_{dm}\left(C - \varphi_{i+1,i} - t_{i+1,i}\right)^2}{2(q_{dm} - q_d)} \tag{11}$$

$$d'_{id} = q_d t_r\left(t_{i+1,i} - C + \varphi_{i+1,i}\right) - \frac{1}{2}q_d\left(t_{i+1,i} - C + \varphi_{i+1,i}\right)^2$$
$$+ \frac{1}{2}q_d^2\frac{\left(t_{i+1,i} - C + \varphi_{i+1,i}\right)^2}{q_{dm}} \tag{12}$$

For the queue length l, let q_m be the saturation flow; x be the saturation; N be the number of vehicles arriving at the intersection in a cycle.

$$l = \frac{\exp\left(-\frac{4}{3}\sqrt{(C - t_r)q_m}\frac{1-x}{x}\right)}{2(1 - x)} + N\left(1 - \frac{C - t_r}{C}\right) \tag{13}$$

The average stop times H are expressed as the total number of stops divided by the number of vehicles arriving in a cycle, which is denoted by S and N, respectively. Let q be the actual traffic flow in all inlet; q_r be the maximum waiting traffic flow in red.

$$H = \frac{S}{N} = \frac{q\left(\frac{q_r}{q_m - q} + r\right)}{N} \tag{14}$$

To ensure that the results are reasonable, we must set boundaries for the decision variables. Let $t_{i,l}$ be the green loss time; $t_{i,b}(t_{j,b})$ be the yellow time.

$$s.t.\begin{cases} t_{i,r,min} \le t_{i,r} \le t_{i,r,max} \\ C - t_{i,r} - t_{i,l} \ge \lambda_{i,r,min}C \\ C_{min} \le C \le C_{max} \\ t_{i,r} + t_{j,r} + t_{i,b} + t_{j,b} = C \end{cases} \tag{15}$$

Combining Eqs. (1), (5), (8), (9)–(15), we obtain the first model M1. For the final optimization in Eq. (1), decision variables include the cycle (C), red light time $(t_{i,r}, t_{j,r})$, the green split (λ_i) and offset $(\varphi_{i,i+1})$. Model parameters include actual traffic flow (q), saturation flow (q_m) and travel time.

3 Incorporate Traffic Dispersion

In the model M1, the calculation method of link travel time from intersection i to intersection $i + 1$ is the ratio of the distance in between to the free flow speed of that link, as in Eq. (2). In most existing models, the free flow speed is set to a fixed value. However, in reality, the free flow speed is rarely achieved, especially when the traffic is far from being sparse. And, assuming the link travel time for all traffic that remains constant is far from being realistic. Some other models improve on this by setting the speed boundary conditions and change functions, but this introduces new variables into the model, resulting in an increase in model complexity [11].

Based on the above observations, we turn to the widely used signal timing method TRANSYT series of traffic dispersion module to estimate link travel time. The dispersion module ensembles computing the expectation of geometric variables to estimate the link travel time. The travel time can be adjusted in accordance with the upper and lower travel speeds on that link. For the upper travel speed, the free flow speed or the speed limit of that link is used. As for the lower travel speed, since there usually is no lower speed limit for urban traffic systems, the bound is set based on the simulation results from VISSIM.

Now, let t_l $\left(t_l'\right)$ be travel time from the intersection i $(i + 1)$ to intersection $i + 1$ (i) required by the vehicle with the lowest speed on line. Let t_f $\left(t_f'\right)$ be the travel time from intersection i $(i + 1)$ to intersection $i + 1$ (i) required by the vehicle with the highest speed on line. Therefore, travel time in the upward $\left(t_{i,i+1}\right)$ and downward $\left(t_{i+1,i}\right)$ directions can be calculated using the following equations.

$$t_{i+1,i} = \sum_{t=t_i'}^{t_f'} \frac{F(1-F)^{t-t_i'}}{C} t \tag{16}$$

$$t_{i+1,i} = \sum_{t=t_i'}^{t_f'} \frac{F(1-F)^{t-t_i'}}{C} t \tag{17}$$

$F = 1/(1 + \kappa t)$, and κ is an adjusting factor and is set to 0.35 [12]. As a result, the second proposed model M2 is given by replacing Eq. (2) in M1 with Eqs. (16) and (17). By comparing M1 and M2 models and substituting Eq. (16) into Eq. (2), we can solve that, when all traffic flows are assumed to travel at the speed $V = v/C - \kappa L$, model M1 is equivalent to model M2. In this case, link travel time is $t_l = t_f = LC/(v - \kappa LC)$. Mathematically, for the above to be true, the boundary condition for κ is

$$0 < \kappa < \frac{v}{lC} \tag{18}$$

As a result, the dispersion module can be reduced to the free flow model by the above transformation. Equation (2) can be considered a special case of Eq. (16), which means that model M2 has a larger solution space than model M1.

4 Simulation Study and Result Discussion

For the above proposed model, we selected three important intersections of one urban arterial road, to collect traffic data and conduct simulation experiments. Four paths with significantly larger traffic flow were selected to observe the change of delay time. It shows, in Fig. 3, the geographical topology for the test system.

During the three time periods, morning peak (7:00–9:00), off-peak (15:00–17:00) and evening peak (17:00–19:00), on-site surveys were conducted to collect traffic data. Traffic volume data suggest four critical lines that contain most of the traffic over the planning horizon. We marked these three intersections from right to left as 1, 2 and 3. Line 1 contains the traffic flow from the northbound off-ramp through the major arterial path; line 2 passes through the major arterial path in opposite directions; line 3 contains the traffic flow from east to west at intersection 3; line 4 enters the major arterial path from the east entrance of intersection 2. It is worth noting that the existing phase design is quite reasonable and does not need to be changed. It shows, in Table 1, the phase design of the current timing scheme.

For numerical simulation, the multi-objective evolutionary algorithm NSGA-II was adopted to obtain the optimal solutions. The NSGA-II is one of the most advanced multi-objective optimization algorithms based on Pareto optimal solution currently.

For performance evaluations, we adopted VISSIM to simulate and record the total delay time of each path output, in unit of minutes. Then, we compared the path performance of the three timing schemes in three different time periods. It shows, in Table 2, the current scheme and the phase lengths resulting from M1 and M2.

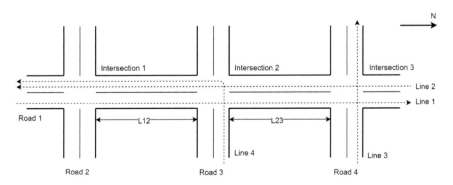

Fig. 3 Geographical topology for the test arterial line

Table 1 Design of phases of a cycle for the current timing scheme

Current scheme	Intersection 1	Intersection 2	Intersection 3
Phase 1			
Phase 2			
Phase 3			
Phase 4			

Table 2 Simulation results for three timing schemes

Scheme	Current $C = 150$	M1 $C = 160$	M2 $C = 140$
Intersection 1	(61, 46, 43)	(63, 51, 46)	(56, 48, 36)
$\varphi_{1,2}$	27	34	31
Intersection 2	(45, 46, 59)	(42, 55, 63)	(39, 48, 53)
$\varphi_{2,3}$	55	67	64
Intersection 3	(50, 28, 32, 40)	(47, 30, 36, 47)	(41, 28, 31, 40)

[*]Note that the phase lengths are arranged in the phase sequence shown in Table 1

It shows, in Fig. 4, the comparison of total delay time of three timing schemes. Obviously, the timing scheme obtained by M1 and M2 is better than current scheme, with an improvement of 21–24% in road performance.

It can be observed from Fig. 4, that even M2 has no obvious advantage over M1 in the branch (according to lines 3 and 4), from the perspective of the arterial road, M2 is a better model than M1 (according to Lines 1 and 2). That is to say, incorporation of the dispersion constraints does improve road performance, compared to free flow settings in other models. In fact, this is to be expected because the free flow speed is difficult to achieve, and using a constant travel time is an unrealistic ideal.

As can be seen from Fig. 4, the delays of line 1 in the morning peak and line 2 in the evening peak are significantly higher. This is because the traffic flow from south to north is relatively high in the morning and opposite in the afternoon. It can be seen from lines 1 and 2 that the performance of M2 is particularly improved in the

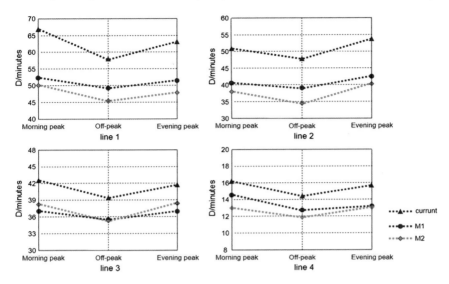

Fig. 4 Comparison of the control delay per line

off-peak, because the vehicle speed difference is larger in this period and the traffic dispersion module is more applicable. In line 3, current timing scheme adopts a fixed offset, which leads to the rise of delay, while M1 changes this situation. In line 4, subject to the left-turn phase time not changing, the delay time is not significantly improved, and the slight reduction is due to the improvement of the traffic capacity between intersection 1 and intersection 2.

5 Conclusion

In this paper, the traffic signal timing coordination on arterial road is considered as a multi-objective optimization problem, and models M1 and M2 can be used to generate signal timing schemes based on traffic flow. As a result, numerical simulation and result analysis have shown that models M1 and M2 are able to improve the performance compared with the current schemes. In addition, it can also be found that M2 is the best choice for optimizing the delay time of the main line, while M2 has no overall advantage over M1 in the lines that include branch traffic flows. Finally, the most potential development direction of traffic signal timing is to increase the scale of online control. With the rapid development of 5G communication and intelligent vehicles, the arterial systems or regional traffic signal systems can automatically change according to real-time traffic, which is the development direction in the future.

References

1. Araghi, S., Khosravi, A., Creighton, D.: A review on computational intelligence methods for controlling traffic signal timing. Expert Syst. Appl. **42**(3), 1538–1550 (2015)
2. Saka, A.A., Anandalingam, G., Garber, N.J.: Traffic signal timing at isolated intersections using simulation optimization. In: Conference on Winter Simulation, pp. 795–801 (1986)
3. Kim, J.: A fuzzy logic control simulator for adaptive traffic management. In: IEEE International Fuzzy Systems Conference, vol. 3, pp. 1519–1524 (1997)
4. Kim, J., Kim, B.M., Huh, N.C.: Genetic algorithm approach to generate rules and membership functions of fuzzy traffic controller. In: IEEE International Conference on Fuzzy Systems, vol. 1, pp. 525–528 (2001)
5. Lee, J., Abdulhai, B., Shalaby, A., Chung, E.H.: Real-time optimization for adaptive traffic signal control using genetic algorithms. J. Intell. Transp. Syst. **9**(3), 111–122 (2005)
6. Khamis, M.A., Gomaa, W.: Adaptive multi-objective reinforcement learning with hybrid exploration for traffic signal control based on cooperative multi-agent framework. Eng. Appl. Artif. Intell. **29**(3), 134–151 (2014)
7. Qiao, F., Tan, X.Y., Alexander, T.F.: Optimization of bidirectional green wave of traffic systems on urban arterial road. In: 9th International Conference on Modelling, Identification and Control (ICMIC), pp. 851–856 (2017)
8. Deb, K., Pratap, A., Agarwal, S., Meyarivan, T.A.M.T.: A fast and elitist multi-objective genetic algorithm: NSGA-II. IEEE Trans. Evol. Comput. **6**(2), 182–197 (2002)
9. Arsava, T., Xie, Y., Gartner, N.H., Mwakalonge, J.: Arterial traffic signal coordination utilizing vehicular traffic origin-destination information. In: 17th International IEEE Conference on Intelligent Transportation Systems (ITSC), pp. 2132–2137 (2014)
10. Ye, B.L., Wu, W., Mao, W.: A two-way arterial signal coordination method with queueing process considered. IEEE Trans. Intell. Transp. Syst. **16**(6), 3440–3452 (2015)
11. Hajbabaie, A., Benekohal, R.F.: A program for simultaneous network signal timing optimization and traffic assignment. IEEE Trans. Intell. Transp. Syst. **16**(5), 2573–2586 (2015)
12. Cho, H.J., Huang, T.J., Huang, C.C.: Path-based MAXBAND with green-split variables and traffic dispersion. Transportmetrica B: Transp. Dyn. **7**(1), 726–740 (2019)

Study on Traffic Volume Transferred by Bohai Strait Tunnel

Bing Wang, Fei Liu, and Yu Peng Li

Abstract The Bohai Economic Rim is a very important political and economic area, but Bohai Strait is a bottleneck and hinders this region's development. This paper studied the influence of the Bohai Strait tunnel on transferred traffic volume of Bohai Economic Rim. The influencing parameters included the network flow assignment, cargo volume, transferred traffic volume and tunnel toll. The results indicated that the Bohai Strait tunnel can undertake large proportion of traffic volume and attract vast cargo volume. Large cargo volume is transferred to the routes of direct transportation and undersea tunnel, which relieves traffic congestion of Circum-Bohai-Sea region. Cargo volume in undersea tunnel is sensitive to the tunnel toll. When tunnel toll is larger than the critical value, there is no advantage for undersea tunnel, and the tunnel toll should be taken into full consideration.

Keywords Network flow assignment · Cargo transportation · Transferred traffic volume · Tunnel toll

1 Introduction

The Bohai Sea is China's largest inland sea, surrounded by lands on three sides. It borders Liaoning to the north, Shandong to the south, Heibei and Beijing to the west. The Bohai Economic Rim constitutes the most important political, economic, cultural, international communication and large conurbations. But due to the hinder of Bohai Strait, the transportation from Liaoning to Shandong has to make a long detour through Hebei and Beijing, which leads to serious waste of transport time, transport cost and manpower. The Bohai Strait aggravates the dual tensions of transportation and energy, hinders the economic and trade exchange in the Bohai Economic Rim

B. Wang (✉) · Y. P. Li
School of Civil Engineering, Shenyang Jianzhu University, Shenyang 110168, China
e-mail: bingwer@126.com

F. Liu
Shenyang Highway Plan & Design Institute, Shenyang 110168, China

© The Author(s), under exclusive license to Springer Nature Singapore Pte Ltd. 2021   287
Y. Li et al. (eds.), *Advances in Simulation and Process Modelling*,
Advances in Intelligent Systems and Computing 1305,
https://doi.org/10.1007/978-981-33-4575-1_28

and affects the implementation of the strategy of country's overall development strategy. The Bohai Strait becomes a bottleneck, and there is an urgent need to build an undersea tunnel to short transport distance, time and effectively relieve the traffic pressure of Bohai Economic Rim.

In 1994, the former vice-premier Jiahua Zou of China affirmed the significance and necessity of Yantai-Dalian railway ferry project and Penglv-bridge-tunnel project for Bohai Strait, and Yantai-Dalian railway ferry project was listed in national development plan [1]. In 2008, the National Development and Reform Commission organized experts to discuss Bohai Strait project and formed a consensus: as for Bohai Strait project, it is advisable to carry out strategic planning at first, and then the project enters the feasibility study stage after the basic conclusions of major technology and economy are drawn. In 2009, National Development and Reform Commission organized Ministry of Communications, Ministry of Railways, Liaoning province, Shandong province, Dalian city, Yantai city, to set up a strategic planning group, and the strategic planning of Bohai Strait project was launched. In 2014, the state council issued the "Opinions of the State Council on several major policies and measures to support the revitalization of the northeast in the future (Guofa [2014] No. 28)" and put forward the requirement of "accelerating the preliminary work of Bohai Strait project". In the same year, Mengshu Wang [2], academician of Chinese Academy of Engineering, published a paper on a journal of Strategic Study of CAE. He summed up and reviewed the development of Bohai Strait cross-sea channel project, gave a comprehensive analysis of engineering solutions, investment and financing methods and provided a good reference for decision-making. In 2015, the preliminary work of Bohai Strait project was listed in Shandong and Liaoning 13th Five Year development Plan of transportation.

The Bohai Strait tunnel is a super huge project, which is larger than the Three Gorges Dam. It will take decades from the initial idea to the project implementation, and it needs careful preliminary study. Traffic volume is critical of the preliminary work of tunnel construction, and it is necessary to make a systematic research. Previous studies of Bohai Strait tunnel were focused on the landing site [3], economy [4] and construction scheme of undersea tunnel [5], etc. There were few studies on the transferred traffic volume of the Bohai Strait tunnel. This paper studied the influence of the Bohai Strait tunnel on network flow assignment, cargo volume and transferred traffic volume and the effect of tunnel toll on the network flow assignment. The aim of this paper is to provide some references for project planning and design.

2 Routs and Transport Modes

Three routes were used in this paper, including Shenyang-Yantai, Liaoyuan-Yantai and Siping-Yantai. The transportation routes included the detour by land and direct transportation, as listed in Tables 1 and 2. The multiple transport modes were selected, including highway, rail, air, sea and their intermodal transportation. The network flow

Table 1 Transportation route and distance before completion of undersea tunnel

OD pair	Transportation route			Transportation distance (km)					
Shenyang-Yantai	Direct transportation	Shenyang-Dalian-Yantai	Mode	HS	RS	HA	RA	A	
			Distance	561	562	636	637	593	
	Detour by land	Shenyang-Beijing-Yantai	Mode	H	R				
			Distance	1430	1722				
Liaoyuan-Yantai	Direct transportation	Liaoyuan-Shenyang-Dalian-Yantai	Mode	HS	RS	HA	RA		
			Distance	823	832	898	907		
	Detour by land	Liaoyuan-Shenyang-Beijing-Yantai	Mode	H	R				
			Distance	1692	1992				
Siping-Yantai	Direct transportation	Siping-Shenyang-Dalian-Yantai	Mode	HS	RS	HA	RA		
			Distance	739	750	814	825		
	Detour by land	Siping-Shenyang-Beijing-Yantai	Mode	H	R				
			Distance	1621	1796				
		Siping-fuxin-chaoyang-Beijing-Yantai	Mode	H					
			Distance	1626					

Notes H = highway transportation, R = rail transportation, A = air transportation, HS = highway sea multimodal transportation, RS = rail sea multimodal transportation, HA = highway air multimodal transportation, RA = rail air multimodal transportation

Table 2 Transportation route and distance after completion of undersea tunnel

OD pair	Transportation route			HT	RT	HS	RS	H A	RA	A
Shenyang-Yantai	Direct transportation	Shenyang-Dalian-Yantai	Mode	HT	RT	HS	RS	HA	RA	A
			Distance	557	558	561	562	636	637	593
	Detour by land	Shenyang-Beijing-Yantai	Mode	H	R					
			Distance	1430	1722					
Liaoyuan-Yantai	Direct transportation	Liaoyuan-Shenyang-Dalian-Yantai	Mode	HT	RT	HS	RS	HA	RA	
			Distance	819	828	823	832	898	907	
	Detour by land	Liaoyuan-Shenyang-Beijing-Yantai	Mode	H	R					
			Distance	1692	1992					
Siping-Yantai	Direct transportation	Siping-Shenyang-Dalian-Yantai	Mode	HT	RT	HS	RS	HA	RA	
			Distance	735	746	739	750	814	825	
	Detour by land	Siping-Shenyang-Beijing-Yantai	Mode	H	R					
			Distance	1621	1796					
		Siping-fuxin-chaoyang-Beijing-Yantai	Mode	H						
			Distance	1626						

Transportation distance (km)

Notes HT = highway tunnel multimodal transportation, RT = rail tunnel multimodal transportation assignment, cargo volume, transferred traffic volume and the influence of tunnel toll were predicted after completion of undersea tunnel in 2040 year.

3 Model and Method

3.1 Network Flow Assignment

This paper used the negative exponent network flow assignment [6] model to study the network flow assignment, and the model can be evaluated by the following expression:

$$K_i = A \cdot B_i \cdot e^{\frac{-C_0 \cdot (D_i - D_j)}{D_j}} \tag{1}$$

where K_i is the proportion of traffic volume on a certain route i to the total passenger or cargo volume, which stands for the network flow assignment. B_i is a parameter which represents a preference for choosing the route i, which is affected by some random factors. The random factors include random supply [7, 8] (weather, pavement preservation, etc.) and random demand [9] (transportation time, traffic congestion). The calculation method of B_i see Ref. [10]. C_0 is a cost parameter which is sensitive for transportation expense. D_i is transportation expense for route i. D_j is the minimum transport expense for all routes.

3.2 Cargo Volume and Transferred Traffic Volume

The base cargo volume was obtained from the OD survey data of three transport routes. The future cargo volume in 2040 year was predicted using growth rate method; see Ref. [11]. After completion of undersea tunnel, the transferred traffic volume included two types: One was the cargo volume of detour by land transferring to the route of direct transportation, and the transferred traffic volume was called T_1; the other was the total cargo volume (including detour by land and direct transportation) transferring to the route of Bohai Strait tunnel (HT and RT), and the transferred traffic volume was called T_2.

3.3 Tunnel Toll

The transportation expense has a great effect on choosing of a route for the passengers or consignors. This paper studied multiple tunnel tolls of the Bohai Strait tunnel on network flow assignment, including 1.0 time base toll, 1.5 time base toll, 2.0 time base toll, 2.5 time base toll, 3.0 time base toll, 3.5 time base toll and 4.0 time base toll. The base toll was based on the charge standard of ordinary tunnel.

4 Results

4.1 The Results of Network Flow Assignment and Cargo Volume

Before completion of undersea tunnel, the result of network flow assignment was shown in Table 3. It was seen that when there is no Bohai Strait tunnel, highway sea multimodal transportation (HS) shows the largest proportion of cargo volume, and the values K_i from high to low are 71.6% for Shenyang-Yantai route, 53.5% for Siping-Yantai route and 46.3% for Liaoyuan-Yantai route. The rail sea multimodal transportation (RS) also has larger proportion of cargo volume, and the sequence of proportion K_i is 41.8% for Liaoyuan-Yantai route, 34.5% for Siping-Yantai route and 20.1% for Shenyang-Yantai route. Rail transportation (R) takes a certain proportion in network flow assignment, and the values K_i are in order as follows: 10.9% for Liaoyuan-Yantai route, 10.6% for Siping-Yantai route and 4.6% for Shenyang-Yantai route. Other transportation modes (HA, RA, A, H) have very small proportion of traffic volume, especially for highway transportation (H) detouring through Beijing. The proportions of traffic volume of highway transportation (H) for the three routes are only 0.1%.

After completion of undersea tunnel, the result of network flow assignment was shown in Table 4. It was found that the rail undersea tunnel multimodal transportation (RT) becomes the most important transportation mode. The largest proportion of traffic volume K_i of RT is 73.9% for Liaoyuan-Yantai route, followed by 70.0% for Siping-Yantai route and 55.9% for Shenyang-Yantai route. The highway undersea tunnel multimodal transportation (HT) also is the main transportation mode, and the K_i is as following in turn: 28.7% for Shenyang-Yantai route, 15.5% for Siping-Yantai route and 11.5% for Liaoyuan-Yantai route. The highway sea multimodal transportation (HS) and rail sea multimodal transportation (RS) no longer have large proportion of traffic volume. Due to small transportation expense for shipping, there is still a certain percentage for HS and RS, but less than 10%. Other transportation modes (HA, RA, A, H, R) have very small proportion of traffic volume, and it is even 0 for mode H. So the Bohai Strait tunnel takes a great influence on network flow assignment. The main reason is that the undersea tunnel can short the transport distance and save transport expense and time.

The results of cargo volume are indicated in Tables 3 and 4. The variation tendency of cargo volume is in accord with the network flow assignment.

Table 3 Results of network flow assignment and cargo volume before completion of undersea tunnel

			Transport mode	HS	RS	HA	RA	A
Shenyang-Yantai	Direct transportation	Shenyang-Dalian-Yantai	K_i (%)	71.6	20.1	1.6	1.2	0.9
			Cargo volume (thousand ton)	6417	1798	139	105	76
	Detour by land	Shenyang-Beijing-Yantai	Transport mode	H	R			
			K_i (%)	0.1	4.6			
			Cargo volume (thousand ton)	8	415			
Liaoyuan-Yantai	Direct transportation	Liaoyuan-Shenyang-Dalian-Yantai	Transport mode	HS	RS	HA	RA	
			K_i (%)	46.3	41.8	0.5	0.4	
			Cargo volume (thousand ton)	498	449	5	4	
	Detour by land	Liaoyuan-Shenyang-Beijing-Yantai	Transport mode	H	R			
			K_i (%)	0.1	10.9			
			Cargo volume (thousand ton)	1	117			

(continued)

Table 3 (continued)

Siping-Yantai		Transport mode	HS	RS	HA	RA
Direct transportation	Siping-Shenyang-Dalian-Yantai	K_i (%)	53.5	34.5	0.6	0.5
		Cargo volume (thousand ton)	1299	838	15	13
Detour by land	Siping-Shenyang-Beijing-Yantai	Transport mode	H	R		
		K_i (%)	0.1	10.6		
		Cargo volume (thousand ton)	3	258		
	Siping-fuxin-chaoyang-Beijing-Yantai	Transport mode	H			
		K_i (%)	0.1			
		Cargo volume (thousand ton)	3			

Table 4 Results of network flow assignment and cargo volume after completion of undersea tunnel

Shenyang-Yantai				HT	RT	HS	RS	HA	RA	A
	Direct transportation	Shenyang-Dalian-Yantai	Transport mode	HT	RT	HS	RS	HA	RA	A
			K_i (%)	28.7	55.9	7.2	4.4	1.4	1.1	0.8
			Cargo volume (thousand ton)	2570	5011	647	391	127	96	69
	Detour by land	Shenyang-Beijing-Yantai	Transport mode	H	R					
			K_i (%)	0.0	0.5					
			Cargo volume (thousand ton)	0	49					
Liaoyuan-Yantai	Direct transportation	Liaoyuan-Shenyang-Dalian-Yantai	Transport mode	HT	RT	HS	RS	HA	RA	
			K_i (%)	11.5	73.9	3.2	9.0	0.4	0.4	
			Cargo volume (thousand ton)	124	795	35	97	5	4	
	Detour by land	Liaoyuan-Shenyang-Beijing-Yantai	Transport mode	H	R					
			K_i (%)	0.0	1.5					

(continued)

Table 4 (continued)

Siping-SYantai			Cargo volume (thousand ton)	0	16				
	Direct transportation	Siping-Shenyang-Dalian-Yantai	Transport mode	HT	RT	HS	RS	HA	RA
			K_i (%)	15.5	70.0	4.2	7.7	0.5	0.4
			Cargo volume (thousand ton)	377	1699	102	187	13	11
	Detour by land	Siping-Shenyang-Beijing-Yantai	Transport mode	H	R				
			K_i (%)	0.0	1.6				
			Cargo volume (thousand ton)	0	40				
		Siping-fuxin-chaoyang-Beijing-Yantai	Transport mode	H					
			K_i (%)	0.0					
			Cargo volume (thousand ton)	0					

Table 5 Results of transferred cargo volume after completion of undersea tunnel

Routes	Original volume of detour by land T_0 (thousand ton)	Total cargo volume T_T (thousand ton)	Type	T_1	T_2
Shenyang-Yantai	423	8959	transferred volume (thousand ton)	375	7581
			T_i/T_0 (%)	88.5	
			T_i/T_T (%)	4.2	84.6
Liaoyuan-Yantai	119	1075	transferred volume (thousand ton)	102	919
			T_i/T_0 (%)	86.1	
			T_i/T_T (%)	9.5	85.4
Siping-Yantai	263	2428	transferred volume (thousand ton)	224	2076
			T_i/T_0 (%)	85.0	
			T_i/T_T (%)	9.2	85.5

Notes T_i is T_1 and T_2, $i = 1, 2$

4.2 The Results of Transferred Cargo Volume

The results of transferred cargo volume including T_1 and T_2 were shown in Table 5.

After the completion of undersea tunnel, it was found that the transferred cargo volume from detour by land to the direct transportation T_1 increases significantly. The T_1 of Shenyang-Yantai route is 375 thousand ton, and the ratio of T_1 to original traffic volume of detour by land T_0 reaches up to 88.5%. The T_1/T_0 in Liaoyuan-Yantai route and Siping-Yantai route are consistent with that of Shenyang-Yantai route, also have a higher percentage, 86.1% and 85.0%, respectively. The Bohai Strait tunnel can attract the cargo volume of detour by land and reduce traffic pressure of Beijing-Tianjin-Hebei region.

The total cargo volume transferring to the route of Bohai Strait tunnel T_2 also exhibits significant growth. The transferred cargo volume T_2 of Shenyang-Yantai route is 7581 thousand ton, and the ratio of T_2 to total cargo volume T_T is up to 84.6%. The ratios of T_2/T_T in Liaoyuan-Yantai route and Siping-Yantai route are 85.4% and 85.5%, respectively. So, the Bohai Strait tunnel can attract most cargo volume in all routes and become the main transport route.

4.3 The Influence of Tunnel Toll on Network Flow Assignment

The influence of tunnel toll on network flow assignment was presented in Fig. 1. It was apparently seen that the proportion of cargo volume K_i in tunnel transportation

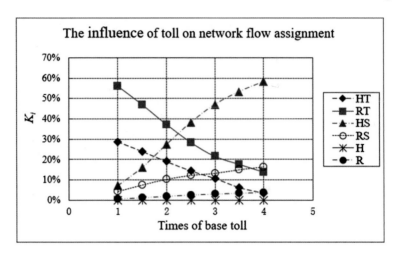

Fig. 1 Influence of tunnel toll on network flow assignment

(HT and RT) decreases with increasing tunnel toll. The proportions of cargo volume K_i in other transportation modes (HS, RS, H and R) increase with the increase of tunnel toll. Among them, the K_i of HS is growing fastest, and the next is RS. The tunnel toll has a relatively very small impact on the transport modes of H and R.

It should be noted that when the tunnel toll is 1.8 time base toll, the proportion of cargo volume K_i of HT is equal to that of HS. When tunnel toll is larger than 1.8 time base toll, the K_i of HT is smaller than that of HS, and there is no advantage for highway undersea tunnel multimodal transportation (HT). It was also found that when the tunnel toll is 2.2 time base toll, the K_i of RT is equal to that of HS. When tunnel toll is larger than 2.2 time base toll, the K_i of RT is smaller than that of HS, and there is no advantage for rail undersea tunnel multimodal transportation (RT). So the proportion of cargo volume K_i is sensitive to the tunnel toll, and it is needed to fully consider the tunnel toll after completion of undersea tunnel.

5 Conclusions

(1) Bohai Strait tunnel would take a great influence on network flow assignment. After completion of undersea tunnel, tunnel transportation (HT and RT) would undertake the largest proportion K_i and cargo volume and become the main transportation mode.

(2) After completion of undersea tunnel, the Bohai Strait tunnel could attract the large cargo volume and reduce traffic pressure of Beijing-Tianjin-Hebei region. More than 85% cargo volume of detour by land would be transferred to the direct transportation, and more than 80% total cargo volume would be transferred to the route of Bohai Strait tunnel.

(3) Tunnel toll would have a great effect on the network flow assignment. When tunnel toll would be larger than the critical value, there would be no advantage for undersea tunnel.

Acknowledgements The authors would like to acknowledge financial support provided by Liaoning Provincial Natural Science Foundation with Grant No. 20170540496 and Liaoning Provincial Key Research and Development Project 2019JH2/10100014.

References

1. Wei, L.Q., liu, X.H.: Research on Cross-Sea Channel of Bohai Strait. Economic Science Press, Beijing (2009) (in Chinese)
2. Wang, M.S.: Strategic plan of Bohai Strait cross-sea channel. Eng. Sci. **15**(12), 4–9 (2013). (in Chinese)
3. Zhao, Y., Tian, B.H., Li, P.F.: Simple discussion on compared options of landing sites and railway connection planning in Bohai Strait cross-sea channel. Strateg. Study CAE **15**(12), 39–44 (2013). (in Chinese)
4. Sun, H.Y., Lu, D.D., Sun, F.H., Feng, S.B.: Influence of the economic contacts between cities in Shandong peninsula and east Liaoning peninsula on the construction of trans-Bohai strait passageway. Sci. Geogr. Sin. **34**(2), 147–153 (2014). (in Chinese)
5. Song, K.Z.: Study on scheme comparison of bridge and tunnel for Bohai Strait cross-sea channel. Strateg. Study CAE **15**(12), 52–60 (2013). (in Chinese)
6. Wang, X.F.: Network Flow Assignment Applied in National Economic Evaluation of Planned Port Construction Project. Dalian Maritime University, Dalian (2003). (in Chinese)
7. Al-Deek, H., Emam, E.B.: New methodology for estimating reliability in transportation. J. Intell. Transp. Syst. **10**(3), 117–129 (2006)
8. Lo, H.K., Tung, Y.K.: Network with degradable links: capacity analysis and design. Transp. Res. Part B **37**(4), 345–363 (2003)
9. Clark, S.D., Watling, D.: Modeling network travel time reliability under stochastic. Transp. Res. Part B **39**(2), 119–140 (2005)
10. Wang, B., Deng, C.N., Liu, F.: The influence of Bohai Strait tunnel project on freight transportation system in Bohai Economic Rim. Appl. Mech. Mater. **587–589**, 1944–1949 (2014)
11. Wang, B., Deng, C.N., Liu, F., Zhang, J.: Discussion on contribution of Bohai Strait tunnel project to freight transportation system and economic benefit. Civ. Eng. Urban Plan., **IV**, 71–76 (2016)

Smart City, Smart Building and Smart Home

Effects of Cap Gap and Spiral-Welded Seam Composite Defects on Concrete-Filled Steel Tubes

Zhengran Lu, Chao Guo, and Guochang Li

Abstract Many old concrete-filled steel tube (CFST) arch bridges use consisted of ordinary concrete and spiral-welded steel tubes. However, low strength of spiral-welded seams (SWS) and cap gaps of concrete due to fluidity composed composite defects in CFST arch bridges. In the current work, a finite element analysis has been performed using ultrasonic scanning field tests on the bearing capacity of a serviced CFST arch bridge rib with a cap gap and reduced SWS strength exposed to a weak eccentric axial compression. The nonlinear behaviors of different components and composite defect influence on CFST performance were also studied. Composite defect influences on the bearing capacity of CFSTs were also experimentally studied. This research proposed a theoretical analysis method for the maintenance and restoration strategy of CFST arch bridges.

Keywords Concrete-filled steel tube · Bearing capacity · Gap · Composite defect · Nonlinear finite element analysis

1 Introduction

In previous decades, many imperfections have been formed in concrete core columns (CCC) and steel tubes of several concrete-filled steel tube (CFST) arch bridges in China during their initial manufacturing or service life [1]. During the last two decades, a large number of experimental and numerical research works have been performed on CFST columns exposed to axial loadings through finite element analysis (FEA). Han et al. [2] and Ellobody and Young [3] carried out elastic-to-plastic FEA on CFST columns having circular or square sections. Hassanein [4], Tao et al. [5, 6], and Ellobody and Young [7] conducted nonlinear FEA on short or long CFST columns. Hu and Su [8] and Huang et al. [9] applied ABAQUS software to analyze CFST column strength performance. Most of the above analyses focused on small-diameter CFSTs, whether straight-welded or seamless, without considering

Z. Lu · C. Guo (✉) · G. Li
School of Civil Engineering, Shenyang Jianzhu University, Shenyang, China
e-mail: guochaoglovel@126.com

© The Author(s), under exclusive license to Springer Nature Singapore Pte Ltd. 2021   303
Y. Li et al. (eds.), *Advances in Simulation and Process Modelling*,
Advances in Intelligent Systems and Computing 1305,
https://doi.org/10.1007/978-981-33-4575-1_29

the effects of concrete gap, SWS strength and segregation defects on the bearing performance of these structures. On the other hand, in existing CFSTs, especially large-diameter CFST arch bridges constructed twenty years ago, have inevitably been influenced by the mentioned defects. Liao et al. [10, 11], Han et al. [12], Huang et al. [13] have previously studied single-factor CFST rib defects. However, there few studies have been conducted on cap gap and SWS composite defects on CFSTs. In the current research, a CFST arch bridge was considered as research object and cap gaps and SWS were assumed as composite defects acting on CFST arch bridge ribs. To do so, a series of nonlinear FEA analyses were performed on CFST rib members with cap gaps and SWS composite defects under eccentric axial compression.

2 Problem Statement

This paper aimed to study the CFST arch bridge of Chang-Qing Hun River Bridge located in Shenyang City, Liaoning Province in northern China (Fig. 1).

The bridge was built in 1997 and has 120 + 140 + 120 m net span. Arch cross-sectional height and width are 2.4 and 1.8 m, respectively. Each of the ribs is consisted of 4 tubes 720 mm in diameter; spiral-welded tube (SWT) thickness is 10 mm. All K-shape, transverse, and lateral braces have similar diameter to the main rib. SWT was constructed by welding the aligned edges of curved spiral steel plates. However, the welded seam strength of SWTs was lower than that of base metal; spiral-welded seam (SWS) penetrated only about 70% [14, 15]. Today, self-consolidating concrete (SCC) is used in CFSTs because of its convenience of construction. Despite its importance, before 2001, i.e., 4 years after the construction of the above bridge was completed, no research work had been conducted on SCC application in CFST arch bridges in China [16]. As a result, ordinary concrete was employed in this bridge. Therefore, in concrete pouring process, cap gaps could be easily created at concrete-steel tube interface, particularly on the top arch rib. For the determination of real existing defects on CFST arch ribs, a 3D ultrasonic section scanning imager (MIRA A1040, ACSYS, Russia) was used to detect defects.

The results showed that the ratio of interface gap to total tube areas was in the range of 3.6–30.5% and the depth of cap gap was about 15–45 mm. Based on the

Fig. 1 Chang-Qing Hun River Bridge

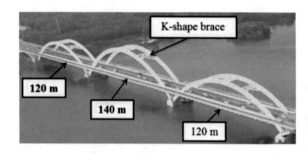

Fig. 2 Mesh of CFST
composite defects

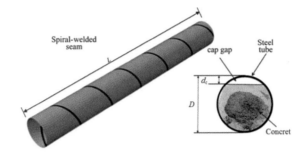

ultrasonic data, FEAs were developed for analytical investigation and estimation of cap gaps and SWS, in which section gap ratio (χ) was the most important calculation parameter:

$$\chi = \frac{d_c}{D} \tag{1}$$

where, as shown in Fig. 2, D is SWT section outer diameter and d_c is the size of cap gap and is specified as the maximum distance between CCC and SWT inner surface.

3 Methods

FEAs were performed by ABAQUS software, and SWT was simulated by a 4-node simplified integrating shell element (S4R). Also, CCC was considered as an 8-node hexahedral element (C3D8) with each node having 3 translational degrees of freedom [17, 18]. The loading eccentricity was $D/6$ and steel tube $\sigma - \varepsilon$ relationship in CFST members conformed to the trilinear material property model proposed by Tao and Wang [19]. In this case, steel tube parameters are given in Table 1.

The concrete core was drilled from serviced CFST of Chang-Qing Hun River Bridge to obtain the real strength grade of CCC using unconfined compressive strength tests. Based on the obtained results, Poisson's ratio μ_c, cylinder compressive strength f_c', and corresponding strain ε_c' values of concrete specimens 100 mm in diameter exposed to uniaxial compressive stress were found to be 0.21, 33 MPa, and 0.0033, respectively. E_c was calculated using the experimental Eq. (2) suggested in ACI318 [20] to be 2.7×10^4 MPa, in which f_c' is in MPa.

Table 1 Steel tube parameters

t mm	D mm	E_s MPa	μ_s	f_y MPa	f_u MPa	ε_y	ε_u
10	700	2.0×10^5	0.32	345	400	0.00192	0.179

μ_s Poisson's ratio, D outer diameter, t thickness

Fig. 3 Stress–strain curve

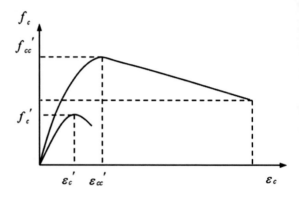

$$E_c = 4700\sqrt{f'_c} \tag{2}$$

According to the experimental results reported in [21], the following equation was developed for the prediction f_{bc}/f'_c ratio as 1.12:

$$f_{bc}/f'_c = 1.5(f'_c)^{-0.075} \tag{3}$$

where f_{bc} denotes initial equibiaxial compressive yield stress. In case, core concrete was exposed to circular confining pressures, uniaxial compressive yield strength f'_c and corresponding strain ε'_c (Fig. 3) were considerably greater than those obtained for unconfined concrete. Mander et al. [22, 23] estimated $-\varepsilon'_{cc} - \varepsilon'_c$ and $f'_{cc} - f'_c$ relations. Han et al. [24] applied Mander's model to develop equivalent $\sigma - \varepsilon$ model for the simulation of CCC plastic behavior in CFSTs, as shown in Fig. 3.

4 Results and Discussion

4.1 FEA of CFST with SWS Defect

In FEA, the strength values of SWS were 50, 70, and 100% of those of parent materials. The length L and screw pitch of seam were 3 and 0.95 m, respectively. For the evaluation of SWS strength effect on the bearing capacity of CFSTs, firstly, empty steel tube (EST) was investigated and then, CFST was evaluated, as shown in Fig. 4, which presents the comparison of $N - \varepsilon$ curves drawn for ESTs in the presence and absence of SWS defects. It was seen in Fig. 4a that, after reaching yield strength, N in ESTs with or without SWS defects appeared strain-softening until reaching a high axial strain ($\varepsilon = 0.03$).

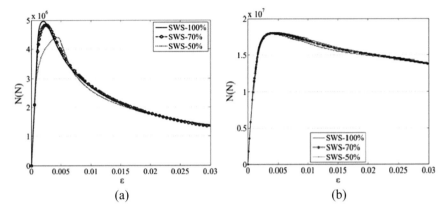

(a)   (b)

Fig. 4 Comparison of $N - \varepsilon$ relationships. **a** EST, **b** CFST

However, 30 and 50% strength loss due to SWS defects decreased the strength of EST by 2.6% and 11.2%, respectively, which led to strain-softening ability reduction. It was concluded that EST strength loss due to SWS defects was mainly because of decrease in strength contribution.

As seen in Fig. 4b, under different SWS defect conditions, CFST $N - \varepsilon$ curves were almost similar. The results showed that when inner concrete had no defects, the decrease of EST strength with SWS defects had only a slight effect on the bearing capacity of CFST. This was consistent with experiences in references [14] and [15]. Figure 5 compares typical EST deformation modes. When SWS strength was less than that of base metal, SWS exhibited local buckling on EST compression side. By decreasing the intensity of SWS, this phenomenon became more obvious.

Figure 6 compares typical CFST SWT failure modes. The results showed that CFST outer SWT strength was reduced at weld seam, but it was different from the local fold buckling failure modes of EST.

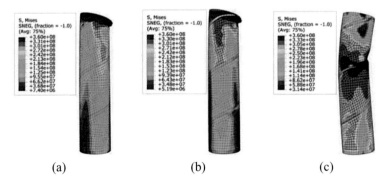

(a)   (b)   (c)

Fig. 5 EST von Mises stresses (Pa) with different SWS strengths. **a** SWS: 100%, **b** SWS: 70%, **c** SWS: 50%

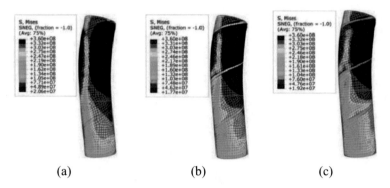

Fig. 6 Tube Mises stresses (Pa) in CFSTs with various SWS strengths. **a** Tube SWS: 100%, **b** Tube SWS: 70%, **c** Tube SWS: 50%

In addition, it was proved that CCC had strong restraint ability. If CCC had no defects, it could fully support outer steel pipe and prevent local buckling. Therefore, coinciding performance of concrete and steel tube was ensured.

4.2 FEA of CFST with Cap Gap Defect

Figure 7 compares the $N - \varepsilon$ curves of CFST without and with through-type cap gaps ($\chi = 2.1, 4.2$, and 6.3%).

As shown in Fig. 7, all CFSTs with or without cap gaps are presented typical strain-softening properties. The results complied with those of Refs. [14] and [15]. There were two yield points at $\chi = 6.3\%$ corresponding to $N - \varepsilon$ curves at peak

Fig. 7 Comparison of N versus ε relations

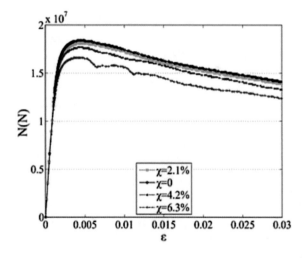

load, where ε yield stopped, and thereafter, two steps referred to strain-hardening again. The N in CFST with cap gaps was suddenly reduced after peak load; However, when $\chi = 4.2\%$ and 2.1% or when there were no cap gap defects, CFST $N - \varepsilon$ curves presented a smooth shape. Thereafter, when CCC came into contact with external SWTs, N was increased again. Figure 8 compares typical CFST failure modes. Elephant foot-shaped buckling deformation was magnified by 2 times to more clearly show deformed. This type of deformation was observed in CFST upper half. It is noteworthy that, in CFSTs with $\chi = 2.1\%$, a slight elephant foot-shaped buckling deformation was witnessed, while in those with $\chi = 4.2$ and 6.3%, many large deformations were observed. The results showed that large gaps resulted in greater spaces for local buckling deformations of CFST. CCC deformation shapes are presented in Fig. 9. It was observed that CCC had various bending deformation modes at various depths of cap gap.

With the increase of the depth of cap gap, stress and bending deformation were concentrated in compression zone and CFSTs presented local fractures and brittle failure stages.

It was noted that before the interaction of CCC and external SWTs, the greatest concrete cross-sectional deformation took place in the upper half. After that, when CCC came into contact with external SWTs, CCC cross-sectional deformation at compression region was confined by SWT. Therefore, CCC cracking position moved toward the mid-height of CFST. Concrete failure was witnessed at regions where local buckling of SWT occurred, as shown in Fig. 9.

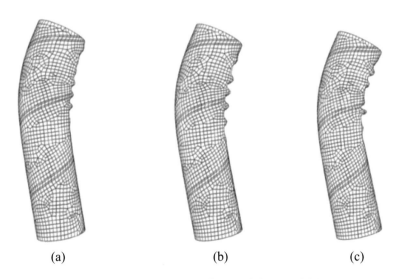

(a) (b) (c)

Fig. 8 Deformation modes of CFST. **a** $\chi = 2.1\%$, **b** $\chi = 4.2\%$, **c** $\chi = 6.3\%$

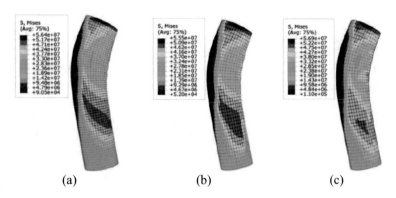

Fig. 9 Concrete Mises stresses in CFSTs with various cap gap depths. **a** $\chi = 2.1\%$, **b** $\chi = 4.2\%$, **c** $\chi = 6.3\%$

4.3 FEA of CFSTs and Cap Gap with SWS Composite Defects

Figure 10 presents the comparison of $\chi - N - \varepsilon$ curves of CFSTs with through-type cap gaps ($\chi = 2.1, 4.2,$ and 6.3%) in three dimensions.

It was observed that all CFSTs revealed typical smooth strain-softening properties in the absence of cap gap and SWS strength defects, which complied with the findings of [14] and [15]. Stress yield steps took place in the presence of cap gaps in CFST, as reported in [10] and [11]. This was the same as the results shown in Fig. 7. However, in the presence of composite defects effects at different levels of SWS strength, $N - \varepsilon$ curves of each of the components presented various rule of variation. By the decrease of the strength of SWS, yield step length and number on the $N - \varepsilon$ loss curves were increased. The results showed that $N - \varepsilon$ loss curves of CFST weld seam strength was gradually increased after peak strength. To the main reason for

Fig. 10 $\chi - N - \varepsilon$-curves

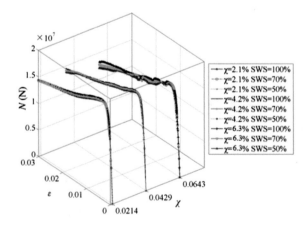

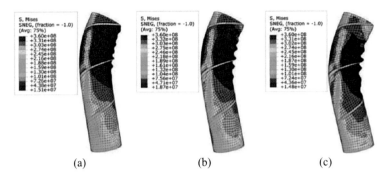

(a)   (b)   (c)

Fig. 11 Tube Mises stresses (Pa) in CFSTs with 70% SWS strength. **a** $\chi = 2.1\%$, **b** $\chi = 4.2\%$, **c** $\chi = 6.3\%$

this was the weakening of local steel tube restraint ability on concrete by SWS, which decreased CFST composite bearing capacity.

Figure 11 compares typical failure modes of CFST with 70% SWS intensity under different cap gap defects. To obtain a clearer deformed image, elephant foot-shaped buckling deformations, which were found near the upper part of CFST, were multiplied by 2.

It should be noted that, for CFSTs with $\chi = 2.1\%$, slight elephant foot-shaped buckling deformations were only observed, while for those with $\chi = 4.2\%$ and 6.3%, such deformations were more intense, as shown in Fig. 8. Unlike intact SWS, however, the shape and location of local buckling in such steel tubes were different.

In defected SWS, steel tube tensile yielding position was observed at the first welding seam, regardless of the value of χ. Along with first weld line, by increasing χ value, compressive yield also occurred on the steel tube of compressive side. As shown in Fig. 11, in case the strength of SWS was equal to 70% of the ideal value and χ was tripled from 2.1 to 6.3%, first SWS tensile yield on the upper part of CFST on tensile side was transferred to folding failure on compression side along SWS rotation direction. This showed that SWS defects were the main reason for the bending failure of CFST.

5 Conclusions

(1) Based on the ultrasonic scanning experimental measurements of Chang-Qing Hun River Bridge, a composite FEA was developed for CFST exposed to eccentric axial compressions. In comparison with the existing FEA methods, it was shown that this FEA is perfect in prediction accuracy.

(2) Through the comparison of test results, it was found that SWS defects had a weaker influence on the bearing capacity of CFST than on that of EST, especially in the absence of defect in internal concrete. However, SWS defect had a

significant influence on CFST local yield failure mode and outer SWT strength was decreased at weld seam.

(3) Considering single defect of cap gap, comparison of $N - \varepsilon$ curves showed that CFSTs with and without cap gaps had typical strain-softening characteristics. However, for CFSTs with high χ values, several yield points appeared on $N - \varepsilon$ curves due to multiple contacts between core concrete and outer SWTs.

(4) Comparing typical failure modes of CFSTs with single cap gap defect showed that cap gaps changed the pattern of CFST buckling deformation and larger gaps expanded local buckling deformation space of CFST.

(5) For CFSTs with cap gap and SWS composite defects, SWS defects weakened local steel tube restrain capacity to concrete, resulting in CFST bending failure. However, for CFSTs with the same level of SWS defects, tensile yield on the tensile side was transferred to compression side followed the rotation direction of the SWS, which indicated that welding defect degree significantly affected local failure mode.

Acknowledgements This research was supported by Liaoning Revitalization Talents Program (No. XLYC1907121), Province Natural Science Foundation of Liaoning, China (No. 20180550442), and Scientific Research Project of Liaoning Provincial Department of Education, China (No. lnjc202019, No. lnjc201904).

References

1. Zheng, J.L., Wang, J.J.: Concrete-filled steel tube arch bridges in China. Engineering **4**, 143–155 (2018)
2. Han, L.H., Yao, G.H., Tao, Z.: Performance of concrete-filled thin-walled steel tubes under pure torsion. Thin-Walled Struct. **45**(1), 24–36 (2007)
3. Ellobody, E., Young, B.: Nonlinear analysis of concrete-filled steel SHS and RHS columns. Thin-Walled Struct. **44**(8), 919–930 (2006)
4. Hassanein, M.F.: Numerical modeling of concrete-filled lean duplex slender stainless steel tubular stub columns. J. Constr. Steel Res. **66**(8), 1057–1068 (2010)
5. Tao, Z., Uy, B., Liao, F.Y.: Nonlinear analysis of concrete-filled square stainless steel stub columns under axial compression. J. Constr. Steel Res. **67**, 1719–1732 (2011)
6. Tao, Z., Uy, B., Han, L.H., Wang, Z.B.: Analysis and design of concrete-filled stiffened thin-walled steel tubular columns under axial compression. Thin-Walled Struct. **47**(12), 1544–1556 (2009)
7. Ellobody, E., Young, B.: Design and behaviour of concrete-filled cold-formed stainless steel tube columns. Eng. Struct. **28**, 716–728 (2006)
8. Hu, H.T., Su, F.C.: Nonlinear analysis of short concrete-filled double skin tube columns subjected to axial compressive forces. Marine Struct. **24**, 319–337 (2011)
9. Huang, H., Han, L.H., Tao, Z., Zhao, X.: L: Analytical behaviour of concrete-filled double skin steel tubular (CFDST) stub columns. J. Constr. Steel Res. **66**(4), 542–555 (2010)
10. Liao, F.Y., Han, L.H., He, S.H.: Behavior of CFST short column and beam with initial concrete imperfection. Experiments. J. Constr. Steel Res. **67**(12), 1922–1935 (2011)
11. Liao, F.Y., Han, L.H., Tao, Z.: Behavior of CFST stub columns with initial concrete imperfection: Analysis and calculations. Thin-Walled Struct. **70**, 57–69 (2013)

12. Han, L.H., Hou, C.C., Wang, Q.L.: Behavior of circular CFST stub columns under sustained load and chloride corrosion. J. Constr. Steel Res. **103**, 23–26 (2014)

13. Huang, Y.H., Liu, A.R., Fu, J.Y., Pi, Y.L.: Experimental investigation of the flexural behavior of CFST trusses with interfacial imperfection. J. Constr. Steel Res. **137**, 52–65 (2017)

14. Gunawardena, Y., Aslani, F.: Behaviour and design of concrete-filled mild-steel spiral welded tube short columns under eccentric axial compression loading. J. Constr. Steel Res. **151**, 146–173 (2018)

15. Gunawardena, Y., Aslani, F.: Finite element modelling of axial loaded mild-steel hollow spiral-welded steel tube short columns. In: Proceedings of 9th International Conference on Advances in Steel Structures, vol. 1, pp. 1559–1570, Hong Kong (2018)

16. Ding, Q.J.: Application of large-diameter and long-span micro-expansive pumping concrete filled steel tube arch bridge. J. Wuhan Univ. Technol.-Mater. Sci. **4**, 11–15 (2001)

17. Ramos, A.G., Jacob, J., Justo, J.F., Oliveira, J.F., Rodrigues, R., Gomes, A.M.: Cargo dynamic stability in the container loading problem -a physics simulation tool approach. Int. J. Simul. Process Model. **12**(1), 29–41 (2017)

18. Wang, H.X., Chen, S.S., Liu, Y.X., Zhang, L.J., Zhang, Z.N.: Numerical simulation and experimental validation for design improvement of packer rubber. Int. J. Simul. Process Model. **12**(5), 419–428 (2017)

19. Tao, Z., Wang, X.Q., Uy, B.: Stress-strain curves of structural and reinforcing steels after exposure to elevated temperatures. J. Mater. Civ. Eng. **25**, 1306–1315 (2013)

20. American Concrete Institute: Building Code Requirements for Structural Concrete (ACI 318-11) and Commentary. Farmington Hills, MI, USA (2011)

21. Papanikolaou, V.K., Kappos, J.: Confinement-sensitive plasticity constitutive model for concrete in triaxial compression. Int. J. Solids & Struct. **44**(21), 7021–7048 (2007)

22. Mander, J.B., Priestley, M.J.N., Park, R.: Theoretical stress-strain model for confined concrete. J. Struct. Eng. **114**, 1804–1826 (1988)

23. Mander, J.B., Priestley, M.J.N., Park, R.: Observed stress-strain behavior of confined concrete. J. Struct. Eng. **114**, 1827–1849 (1988)

24. Han, L.H., An, Y.F.: Performance of concrete-encased CFST stub columns under axial compression. J. Constr. Steel Res. **93**, 62–76 (2014)

Fuzzy Sliding Mode Control of a VAV Air-Conditioning Terminal Temperature System

Ziyang Li, Lijian Yang, Zhengtian Wu, Baoping Jiang, and Baochuan Fu

Abstract In this paper, it is developed a control strategy based on sliding mode scheme to deal with the issue of variable air volume system (VAV) with the characteristics of multiple-input and multiple-output and strong nonlinearity. Besides, fuzzy control is employed to suppress chattering caused by sliding mode control. The proposed control strategy overcomes the limitations of traditional control methods in this kind of systems.

Keywords Sliding mode control · Fuzzy control · VAV air conditioning · Chattering suppression

1 Introduction

A variable air volume (VAV) air-conditioning system belongs to an all-air-conditioning system, which can automatically adjust the air supply volume of air conditioning according to the change in air-conditioning load and indoor air parameters. Appropriate control strategies should be developed to solve the problems caused by the complexity of the control system because of the complexity of VAV air-conditioning system terminal equipment and the high requirements for the overall control of the system. Many scholars have proposed VAV air-conditioning control methods. For example, hysteresis relay feedback control is introduced into the model parameter identification of indoor temperature hysteresis characteristics, providing a new method for the identification of the characteristics of indoor temperature hysteresis [1]. For indoor fresh air demand, the total fresh air flow is dynamically corrected on the basis of the detected occupancy rate of each area and relevant

Z. Li · L. Yang · Z. Wu (✉) · B. Jiang · B. Fu
School of Electronic and Information Engineering, Suzhou University of Science and Technology, Suzhou, China
e-mail: wzht8@mail.usts.edu.cn

Suzhou Institute of Smart City, Suzhou University of Science and Technology, Suzhou, China

© The Author(s), under exclusive license to Springer Nature Singapore Pte Ltd. 2021
Y. Li et al. (eds.), *Advances in Simulation and Process Modelling*,
Advances in Intelligent Systems and Computing 1305,
https://doi.org/10.1007/978-981-33-4575-1_30

measurement values by using fresh air from the over ventilated area [2]. Furthermore, a neural fuzzy structure of a parameter self-tuning decoupled fuzzy neural PID controller is proposed [3]. The classical PID control equation with a decoupled coefficient is used as the Sugeno function to introduce the following part of the fuzzy rule, which improves the anti-interference ability of the system.

However, the design of fuzzy control is not systematic, so fuzzy control should be combined with some other control strategies in most cases to achieve the desired control effect.

In this study, the sliding mode control is applied to a VAV air-conditioning system. In the late 1950s, scholars in the former Soviet Union studied the sliding mode variable structure control. Nowadays, sliding mode control is used to solve many practical problems. Some important studies on robot control, aircraft adaptive control, power system control, and other mechanisms have been conducted. For example, in [4], a constraint design with a sliding mode strategy is proposed to improve the stability of aircraft engine control. In [5], a terminal sliding mode controller is designed to track the planned speed signal, which can timely suppress the adverse dynamic behavior of an electric vehicle after a tire blows out on a highway.

The traditional PID controller optimizes parameters dynamically in difficulty and it also has the problems of low control precision and poor stability, which cannot meet the VAV air-conditioning system with the characteristics of large lag, strong nonlinearity, and uncertainty. Due to the strong robustness of the sliding mode control to the external disturbance and modeling dynamics of the system and the advantages of order reduction, decoupling, fast response, and easy implementation. So it is suitable for VAV air-conditioning system. However, some problems occur in the sliding mode control, and the most prominent one is chattering. Therefore, this study selects the improved power index approaching law, and then selects the parameters of the sliding mode approaching law through the fuzzy control to minimize chattering.

2 VAV Air-Conditioning System Overview

The system structure diagram of a VAV air conditioner is shown in Fig. 1. As an air-conditioning system, VAV air-conditioning system is mainly composed of an air handling unit, an air supply pipeline system of the air conditioner, an indoor air supply terminal, an electric control system, and a common all-air system.

3 Mathematical Model of Air-Conditioning Room Temperature

A building where an air-conditioned room is located belongs to an office building, and the temperature of the area where the building is located is higher in summer and

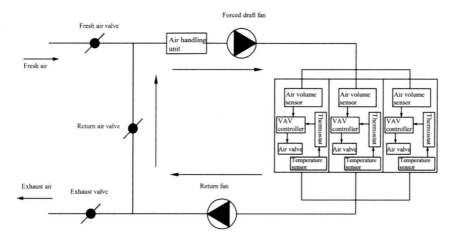

Fig. 1 VAV air-conditioning system structure diagram

daytime than in other seasons or nighttime. The air-conditioned room is affected by many factors, such as solar radiation, internal heat dissipation, equipment, lighting, and radiation of each wall. The whole room is a complex system, which cannot be described by using an accurate mathematical model. To simplify the mathematical model of room temperature, we make the following assumptions:

1. Temperature is uniformly distributed in an air-conditioned room;
2. Temperature in air-conditioned room changes evenly;
3. The heat storage of an enclosure is neglected;
4. Default adjacent rooms have the same effect on air-conditioned rooms; and
5. Indoor equipment, personnel heat load, and other parameters are unified as indoor heat load interference.

According to the law of thermodynamics, the differential equation of the heat balance of the room is as follows:

$$\rho_{air} V c_{air} \frac{dT}{dt} = G_s c_{air}(T_s - T) + k_1 F_1(T_i - T) + k_{out} F_{out}(T_{out} - T) + Q_{in} \quad (1)$$

where V is the volume of the air-conditioned room, m^3; ρ_{air} is the air density, kg/m^3; c_{air} is the specific heat capacity of air, kJ/(kg °C); T is the temperature of the air-conditioned room, °C; Q_{in} is the indoor thermal load interference; G_s is the air supply volume, kg/s; T_s is the air supply temperature, °C; k_1 is the heat transfer coefficient of inner wall of adjacent room, kW/(m^2 °C); F_1 is the heat transfer area of inner wall of adjacent room, m^2; k_{out} is the heat transfer coefficient of external wall, kW/(m^2 °C); F_{out} is the heat transfer area of external wall, m^2; T_{out} is the outdoor ambient temperature, °C; and T_i is the temperature of adjacent rooms, °C.

4 Design of the Fuzzy Sliding Mode Controller

4.1 Design of the Sliding Mode Controller

Selection and Design of the Sliding Surface and Approaching Law

An integral sliding surface can smoothen the torque, reduce the steady-state error of a system, weaken the chattering, and enhance the stability of a controller. Therefore, the integral sliding surface is defined as follows:

$$s = e + k_s \int_0^t e \, dt \tag{2}$$

where e is the deviation signal, $e = T - T_d$, T is the temperature of the air-conditioned room, and T_d is the set temperature.

To overcome the shortcomings of the exponential approaching law, we select the improved power exponential approaching law [6]:

$$\dot{s} = -\varepsilon |s|^\alpha \text{sgn}(s) - ks^\beta \tag{3}$$

where $\varepsilon > 0, 0 \le \alpha <, k \ge 0$, and β is a positive odd number. In particular, when $\alpha = 0$ and $\beta = 1$, the above formula is the exponential approaching law; when $k = 0$, the above formula is the power approaching law.

Calculation of control law

Set $G_s c_{air}(T_s - T)$ to u, the differential equation of the temperature of an air-conditioned room can be obtained as follows:

$$\rho_{air} V c_{air} \frac{dT}{dt} = u + k_1 F_1 (T_i - T) + k_{out} F_{out} (T_{out} - T) + Q_{in} \tag{4}$$

The derivation of Eq. (4) is as follows:

$$\dot{s} = \dot{e} + k_s e = \frac{1}{\rho_{air} V c_{air}} (u + k_1 F_1 (T_i - T) + k_{out} F_{out} (T_{out} - T) + Q_{in}) + k_s e \tag{5}$$

The convergence rate of the improved power index shown in the simultaneous Eq. (5) can be obtained as follows:

$$\frac{1}{\rho_{air} V c_{air}} * (u + k_1 F_1 (T_i - T) + k_{out} F_{out} (T_{out} - T) + Q_{in}) + k_s e$$
$$= -\varepsilon |s|^\alpha \text{sgn}(s) - ks^\beta \tag{6}$$

Therefore, the sliding mode control law based on the improved power exponential approach law can be obtained as

$$u = \rho_{air} V c_{air} \left(-\varepsilon |s|^{\alpha} \text{sgn}(s) - ks^{\beta} - k_s e \right) - k_1 F_1 (T_i - T)$$
$$- k_{out} F_{out} (T_{out} - T) - Q_{in} \tag{7}$$

To prove the stability of the system, we define the Lyapunov candidate function as

$$v = \frac{1}{2} s^2 \tag{8}$$

$$\dot{V}(s) = \dot{s}s = (\dot{e} + k_s e)s$$
$$= \left(\frac{1}{\rho_{air} V c_{air}} \times (u + k_1 F_1 (T_i - T) + k_{out} F_{out} (T_{out} - T) + Q_{in}) + k_s e \right) s \tag{9}$$

Integrating Eq. (9) into the above equations, we can obtain $\dot{V}(s) = -\varepsilon s |s|^{\alpha} \text{sgn}(s) - ks^{\beta+1}$. According to the limitation of the approaching law parameters, we can determine that the value range of each parameter is $\varepsilon > 0, 0 \leq \alpha < 1$, $k \geq 0$, and β is positive odd. Therefore, $\dot{V}(s) < 0$, and the system is asymptotically stable.

4.2 Design of the Fuzzy Switching Surface

Fuzziness of Approaching Law Parameters

In this study, the improved power index approaching law is selected, which can reduce chattering to a certain extent, but in order to better improve the robustness of the air-conditioning system in the operation process, and ensure that the system can quickly enter the sliding mode state and suppress chattering at the same time, a one-dimensional fuzzy controller is designed in this paper. The comparison between MATLAB simulation and sliding mode controller without fuzzy controller is made.

The input signal of the fuzzy adaptive sliding mode controller is the temperature error e, and the output signal is the approach law parameters ε and k. The fuzzy subsets of input variables are {negative big, negative middle, negative small, invariant, positive small, positive middle, positive big}, that is, {NB, NM, NS, ZO, PS, PM, PB}. The fuzzy subsets of output variables are {NB, NS, ZO, PS, PB}. The input quantization is in the $(-1, 16)$ region, the output ε quantization is in the $(0.001, 0.01)$ region, and the k quantization is in the $(0, 30)$ region.

Table 1 Fuzzy control rule table of parameter ε and k

e	NB	NM	NS	ZO	PS	PM	PB
ε	NB	ZO	PS	PB	PB	PB	PB
k	NB	ZO	PS	PB	PB	PB	PB

Membership Function and Fuzzy Rules

The membership function is selected as a triangle membership function, and the range width of each value of the fuzzy subset is equal. According to the actual control effect, the fuzzy control rules of the approaching law parameters ε and k are shown in Table 1.

According to the above fuzzy rules, when the deviation signal e is large, the fuzzy control system selects larger ε and k to ensure that the system can approach the sliding surface at a faster speed. When e is small, the fuzzy control system selects smaller ε and k to reduce the approach speed and weaken system chattering [7].

5   Simulation Analysis

5.1   Design of a Simulation Model

The air-conditioned room is 6 m in length, 5 m in wide, and 4 m in height. The initial temperature in the room is 30 °C, and the air supply temperature is 16 °C. We assume that the temperature change curve between 7:00 and 19:00 in the building area and the temperature change curve of the next room in the air-conditioned room are as follows (Fig. 2):

Considering that the air-conditioned room belongs to the office building, we can make the following assumptions for the heat load caused by the change in indoor personnel and the operation of computers and other equipment (Fig. 3).

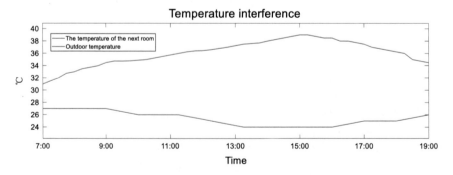

Fig. 2 Outdoor temperature and temperature change of next room

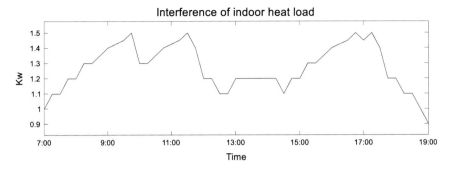

Fig. 3 Indoor thermal load interference

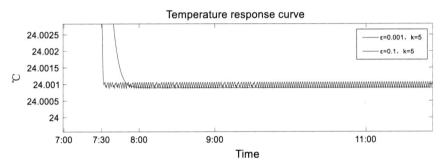

Fig. 4 Temperature response curve with $\varepsilon = 0.1$ and $k = 5$

5.2 Influence of Approaching Law Parameters on Control Effect

When ε is set to 0.1 and K is set to 5, we can find that chattering is obvious after the system reaches stability although the response speed is fast. When ε is set to 0.001 and K is set to 5, the transition time increases although chattering is reduced (Fig. 4).

5.3 Simulation Model and Results of Fuzzy Sliding Mode Controller

When the fuzzy controller designed in Sect. 4.2 is used, the temperature response curve from 7:00 to 19:00 is presented in Fig. 5. The system is robust when the external variables are greatly disturbed.

The temperature response curve of the fuzzy sliding mode control is compared with the curve under the conditions of $\varepsilon = 0.1$ and $k = 5$. After the introduction of the fuzzy control, the system not only has a shorter transition time than the case of $\varepsilon =$

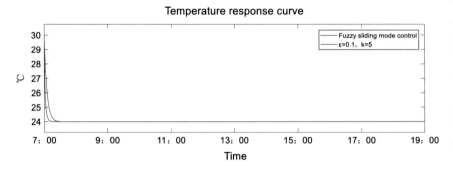

Fig. 5 Temperature response curve comparison chart

0.1 and $k = 5$ but also has a smaller steady-state static difference. Fuzzy control can suppress chattering and improve the dynamic and steady performance of the system (Fig. 6).

The air supply volume of an air-conditioned room changes with the change in indoor and outdoor heat load. The air supply volume of the system at 7:00–19:00 is as follows (Fig. 7).

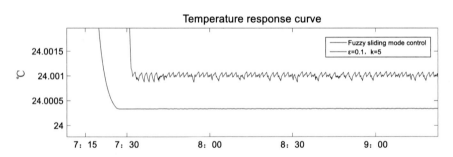

Fig. 6 Local temperature response curve after fuzzy control

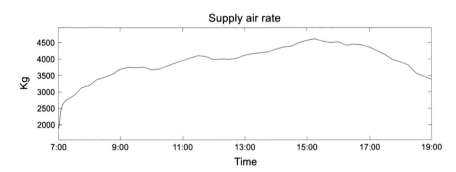

Fig. 7 Change in the air supply volume of the system

6 Conclusion

Applying fuzzy sliding mode control to the VAV air-conditioning system can overcome the limitations of some traditional control methods and improve the overall stability of the system. This study provides a new idea for the control method of VAV air conditioning. An excellent control scheme is of great significance to the popularization of VAV air conditioning. In this study, the temperature model of an air-conditioned room is established on the basis of the characteristics of VAV air conditioning, a sliding mode control is used to regulate air volume, and temperature is successfully controlled. Moreover, according to fuzzy theory, a fuzzy controller is designed to soften the sliding mode control law, effectively suppress the chattering of the system, and improve the dynamic performance of the system.

Acknowledgements This work was partly supported by the NSFC (Grant Nos. 61803279, 11871366, and 61672371), the Qing Lan Project of Jiangsu, the Open Project Funding from Anhui Province Key Laboratory of Intelligent Building and Building Energy Saving, Anhui Jianzhu University (Grant No. IBBE2018KX02ZD), the Natural Science Foundation of the Jiangsu Higher Education Institutions of China (Grant No. 18KJB460026), the Suzhou Science and Technology Foundation (Grant No. SYG201813), and Jiangsu Province Graduate Practice Innovation Program (Grant No. SJCX19_0844).

References

1. Li, X.M., Lin, S.D., Zhang, J.L., Zhao, T.Y.: Model parameter identification of indoor temperature lag characteristic based on hysteresis relay feedback control in VAV systems. J. Build. Eng. **25**, id: 100839 (2019)
2. Xu, X.H., Wang, S.W., Sun, Z.W., Xiao, F.: A model-based optimal ventilation control strategy of multi-zone VAV air-conditioning systems. Appl. Therm. Eng. **29**(1), 91–104 (2008)
3. Ganchev, I., Taneva, A., Kutryanski, K., Petrov, M.: Decoupling fuzzy-neural temperature and humidity control in HVAC systems. IFAC PapersOnLine **52**(25), 299–304 (2019)
4. Yang, S.B., Wang, X., Wang, H.N., Li, Y.G.: Sliding mode control with system constraints for aircraft engines. ISA Trans. **98**, 1–10 (2020)
5. Yang, L., Yue, M., Liu, Y.C., Guo, L.: RBFNN based terminal sliding mode adaptive control for electric ground vehicles after tire blowout on expressway. Appl. Soft Comput. J. **92**, id: 106304 (2020)
6. Tian, H.S., Xie, S.S., Miao, Z.G., Wang, L., Ren, L.T.: Fuzzy sliding mode controller for aeroengine based on an improved power exponential reaching law. Fire Control Command Control **41**(02), 108–112 (2016) (in Chinese)
7. Attia, A.H., Rezeka, S.F., Saleh, A.M.: Fuzzy logic control of air-conditioning system in residential buildings. Alexandria Eng. J. **54**(3), 395–403 (2015)

Study on the Treatment of Cutting Fluid Wastewater Chromaticity by Multiphase Fenton System

Nana Wu, Yang He, Qiang Liu, Youchen Tan, Licheng Zhang, Na Huang, and Yulin Gan

Abstract The composition of the cutting fluid wastewater is complex, the chromaticity is high, the liquid is grayish white, and the visibility is poor. In this research work, it used a multiphase Fenton system composed of catalysts CuO/CeO_2 and H_2O_2 to treat the organic wastewater chromaticity, and various influencing factors were studied. The optimal conditions of the experiment would be: The dosage of the catalyst was 1.6 g/L, the dosage of H_2O_2 was 0.5 mol/L, the reaction time was 180 min, the initial pH was 8, and the reaction temperature was 25 °C. At this time, the removal efficiency of chromaticity could reach 70%.

Keywords Multiphase Fenton system · Cutting fluid wastewater · Chromaticity

1 Introduction

With the vigorous development of the mechanical processing industry, cutting fluids with cooling, cleaning, lubricating and anti-rust properties are more and more widely used in the field of metal processing [1, 2]. The replacement of old cutting fluid and its subsequent cleaning process will produce a certain amount of cutting fluid wastewater. It mainly contains base oil, synthetic lubricant, antioxidant, defoamer and other substances. It is generally in a highly emulsified state, with complex ingredients, and is a highly concentrated organic wastewater that is difficult to degrade [3].

N. Wu (✉) · Y. He · Q. Liu · Y. Tan · Y. Gan
School of Municipal and Environmental Engineering, Shenyang Jianzhu University, Shenyang, China
e-mail: nanawu0816@xmu.edu.cn

Q. Liu
e-mail: 728859476@qq.com

L. Zhang
Urban Planning and Architectural Design Institute, Shenyang Jianzhu University, Shenyang, China

N. Huang
College of Architecture and Urban Planning, Shenyang Jianzhu University, Shenyang, China

© The Author(s), under exclusive license to Springer Nature Singapore Pte Ltd. 2021   325
Y. Li et al. (eds.), *Advances in Simulation and Process Modelling*,
Advances in Intelligent Systems and Computing 1305,
https://doi.org/10.1007/978-981-33-4575-1_31

Fig. 1 Schematic diagram
of Fenton reaction

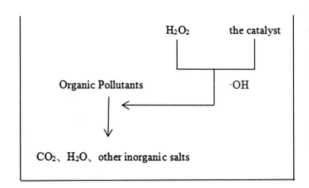

The multiphase Fenton method is a popular advanced oxidation method studied in recent years. The principle of this system is that organic pollutants react with hydroxyl radicals ($\cdot$OH) and are oxidized into CO_2, H_2O and other inorganic salts [4]. The generation of hydroxyl radicals ($\cdot$OH) comes from the catalyst in the reaction system. The catalysts of the heterogeneous Fenton reaction system are divided into solid-phase oxides or supported catalysts [5]. In the system, H_2O_2 is catalyzed on the surface of the catalyst to produce $\cdot$OH, which uses the extremely high oxidizing property of hydroxyl radicals to degrade organic pollutants in water [6]. The following is a schematic diagram of the Fenton reaction (Fig. 1).

The study found that the multiphase Fenton technology has the effect of removing the chromaticity of organic wastewater. Wang et al. [7] found that Fenton oxidation method can treat flax wastewater with a COD concentration of 1747 mg/L and a chromaticity of 200 times. The experimental results show that the COD removal efficiency is 57%, and the chromaticity removal efficiency reaches more than 90% after the reaction time is 1 h. Zhang et al. [8] synthesized the Cu/Al-silica black heterogeneous Fenton catalyst using silica as a carrier and constructed a heterogeneous Fenton catalytic oxidation system for laboratory high concentration organic wastewater treatment experiments. The results showed that when the reaction temperature was 25 °C, the pH of the reaction system was 5, the catalyst dosage was 6 g/L, and the hydrogen peroxide dosage was 55 mmol/L. After 120 min of reaction, the chromaticity removal efficiency was 80%. After five repetitions, the catalytic system still maintained an ideal treatment effect.

2 Methods

The steps of this experiment would be: first take 200 ml of experimental water into the beaker, add the catalyst, turn on the peristaltic pump and add H_2O_2 into the beaker at a uniform rate. After a period of time, use a pipette to suck the supernatant and measure and record the color value of the solution.

The catalyst used in the research work is a supported catalyst, and the catalyst CuO/CeO_2 supported by CeO_2 is prepared by the deposition precipitation method.

In this research work, the ET720 microcomputer platinum-cobalt chromaticity tester produced by Lovibond was used to determine the chromaticity of the water sample. By comparing with the blank water sample, the absorbance of the sample at a specific wavelength is converted into the concentration value of the parameter to be measured, and the value can be directly read on the liquid crystal display to obtain the water sample chromaticity. The inoLab pH 730 desktop precision pH meter produced by WTW in Germany was used to determine the pH value of the water sample. The instrument needs to be calibrated before use.

3 Results and Discussion

3.1 Effect of Reaction Time on Removal Efficiency of Chromaticity by Multiphase Fenton System

In this experiment, it studied the effects at six different reaction time instants of 60 min, 120 min, 180 min, 240 min, 300 min and 360 min on the chromaticity removal efficiency.

From Fig. 2, as the reaction time increases, the removal efficiency of chromaticity gradually increases first and then tends to be gentle. When the reaction time reached 180 min, the chromaticity removal efficiency reached 70%. This is because the catalyst CuO/CeO_2 catalyzes the ·OH produced by H_2O_2 to rapidly oxidize and degrade organic pollutants. With the increase of time, more and more ·OH will be produced, resulting in an increase in the removal efficiency. However, the total amount of

Fig. 2 Effect of reaction time on removal efficiency of chromaticity

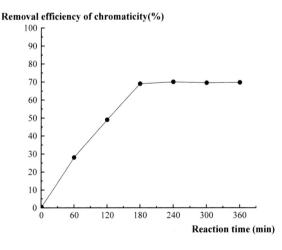

organic pollutants is limited. After 180 min, even if there is still ·OH, the chromaticity removal does not change and remains stable. So, 180 min would be the best reaction time.

3.2 Effect of Catalyst Dosage on Removal Efficiency of Chromaticity by Multiphase Fenton System

The catalyst dosage can affect not only the reactivity of the reactants, but also the mass transfer rate during the oxidative degradation process. Moreover, the catalyst CuO/CeO_2 itself has a certain adsorption effect on the reactants and the removal substances [9]. Therefore, this experiment set seven different gradients of 0.4 g/L, 0.8 g/L, 1.2 g/L, 1.6 g/L, 2.0 g/L, 2.4 g/L and 2.8 g/L to investigate the effect of catalyst dosage on the degradation of chromaticity.

From Fig. 3, with the increase of the catalyst dosage, the chromaticity removal efficiency changes little, and the removal rate was the highest when the dosage was 1.6 g/L. However, the minimum and maximum are 68% and 70%, respectively, and the difference is only 2%. This may be because the minimum amount of catalyst added is much higher than the amount of catalyst required to remove chromaticity. Thus, 1.6 g/L was chosen as the best catalyst dosage.

Fig. 3 Effect of catalyst dosage on removal efficiency of chromaticity

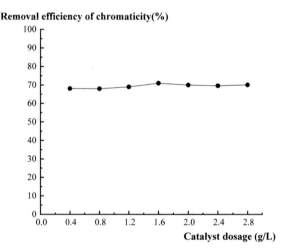

Fig. 4 Effect of H_2O_2 dosage on removal efficiency of chromaticity

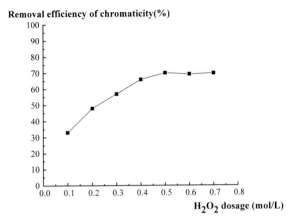

3.3 Effect of H_2O_2 Dosage on Removal Efficiency of Chromaticity by Multiphase Fenton System

The dosage of H_2O_2 will directly affect the production rate and production of ·OH, which are the main functional groups of oxidative degradation [10]. In the experiment, it studied seven different H_2O_2 dosages of 0.1 mol/L, 0.2 mol/L, 0.3 mol/L, 0.4 mol/L, 0.5 mol/L, 0.6 mol/L and 0.7 mol/L on the degradation of chromaticity.

According to Fig. 4, as the dosage of H_2O_2 increases, the chromaticity removal efficiency increases first and then becomes gentle. The removal efficiency reached the maximum when the dosage of H_2O_2 was 0.5 mol/L, which could be about 70%. This is because, as the amount of H_2O_2 added increases, the amount of ·OH generated during the effective time will increase. At the same time, the adsorption of the catalyst CuO/CeO_2 and H_2O_2 itself has strong oxidizing, which will also accelerate the oxidative degradation reaction, increasing in the removal efficiency of chromaticity. However, when the H_2O_2 dosage is greater than 0.5 mol/L, the ·OH content required to remove chromaticity is relatively small, resulting in the ineffective decomposition of H_2O_2 [11]. Therefore, the chromaticity removal efficiency does not change much after H_2O_2 excessive. The optimal H_2O_2 dosage used in this experiment was 0.5 mol/L.

3.4 Effect of Reaction Temperature on Removal Efficiency of Chromaticity by Multiphase Fenton System

During the chemical reaction, temperature is an important factor that affects the reaction rate [8]. In the experiment, it studied five different reaction temperatures of 25 °C, 35 °C, 45 °C, 55 °C and 65 °C on the degradation of chromaticity.

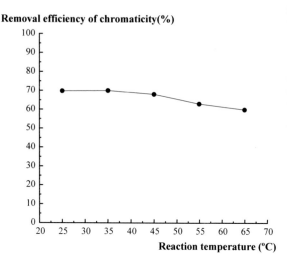

According to Fig. 5, initially, with the increase in temperature, the removal efficiency of chromaticity did not change much, the temperature increased again, and the removal efficiency decreased. This is because increasing the temperature will promote the reaction and facilitate the progress of the oxidative degradation reaction of ·OH. However, if the temperature is too high, the H_2O_2 in the solution will become unstable and decompose [12], resulting in a decrease in the ·OH content, so the removal efficiency of chromaticity decreases after the temperature is 35 °C. This experiment does not adjust the reaction temperature, and normal temperature is enough.

3.5 Effect of Initial pH on Removal Efficiency of Chromaticity by Multiphase Fenton System

The pH value of the Fenton reaction system is one of the important influencing factors. The pH value not only affects the existence form of copper ions in the solution but also affects the activity of the reaction catalyst [8]. In this experiment, it studied eight initial pH values and the effect of the initial pH value on the degradation of chromaticity.

From Fig. 6, the removal efficiency of chromaticity is significantly better under alkaline conditions than under acidic conditions. The removal efficiency is maintained at around 70% under alkaline conditions, and around 65% under acidic conditions. This is because the alkaline reaction can effectively avoid the side reaction of CuO under acidic conditions [13]. Even if part of the ·OH is wasted, the removal of chromaticity can also rely on the adsorption of the catalyst CuO/CeO₂ to reduce. The effect of reduced ·OH content, which in turn shows a higher chromaticity removal

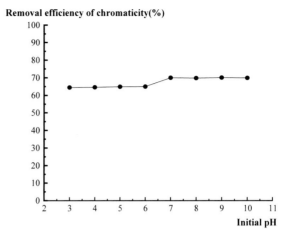

Fig. 6 Effect of initial pH on removal efficiency of chromaticity

efficiency under alkaline conditions. Since the pH was 7, the removal rate started to change. To ensure accuracy, the optimal pH was 8.

4 Conclusion

In this research work, the catalyst CuO/CeO_2 and H_2O_2 constitutes a multiphase Fenton system to study the effect of experiments on the removal of chromaticity of cutting fluid wastewater. In addition, it studied the effects of reaction time, catalyst dosage, H_2O_2 dosage, reaction temperature and initial pH in the experiment and determined the best experimental conditions for the heterogeneous Fenton system. At that time, the chromaticity removal efficiency could reach 70%. Through research work, it is once again confirmed that the multiphase Fenton system has a certain effect on removing the chromaticity of organic wastewater.

Acknowledgements This project was supported by the National Natural Science Foundation of China (51778374). The authors thank the anonymous reviewers for their valuable comments and suggestions.

References

1. Han, L.H., Zhang, X.L., Zhang, X.D.: Treatment of waste emulsified cutting fluid with inorganic and organic composite flocculants. J. Hebei Inst. Technol. (Natural Science Edition). **24**(1), 115–119 (2002) (in Chinese)
2. Zhang, Q., Zhang, Z. W., Wang, X.: Research of application of ceramic membrane in waste metal cutting fluid treatment. Contemp. Chem. Ind. **47**(4), 707–710 (2018) (in Chinese)

3. Zhou, N.L., Wang, Z.Q., Xu, X.D.: Study on treatment of metal working fluids wastewater by Fenton/UV. Environ. Sci. Technol. **22**(6), 6–7 (2009) (in Chinese)
4. Liu, X.C., Zhou, Y.Y., Zhang, J.C.: Insight into electro-Fenton and photo-Fenton for the degradation of antibiotics: mechanism study and research gaps. Chem. Eng. J. **347**, 379–397 (2018)
5. Liu, B.C., Gao, J.B., Cao, B.S.: Ultrasound-assisted Fenton oxidation degradation of oilfield fracturing flowback fluid. Oilfield Chem. **37**(2), 358–361 (2020) (in Chinese)
6. Zhang, X.Y., Ding, Y.B., Tang, H.Q.: Degradation of bisphenol A by hydrogen peroxide activated with $CuFeO_2$ microparticles as a heterogeneous Fenton-like catalyst: efficiency, stability and mechanism. Chem. Eng. J. **236**, 251–262 (2014)
7. Wang, L., Du, M.A., Li, X.: Advanced treatment of wastewater from flax production by Fenton-oxidation process. Ind. Water Wastewater **39**(1), 52–54 (2008) (in Chinese)
8. Zhang, J., Ding, M.H., Pan, K.: Experiment on high concentration organic wastewater in multiphase Fenton catalytic oxidation laboratory. Res. Explor. Lab. **38**(12), 57–60 (2019) (in Chinese)
9. Lin, K., Zhou, M., Zhou, H.J., Zhang, F., Zhong, Z.X., Xing, W.H.: Controlled synthesis of Cu_2O microcrystals in membrane dispersion reactor and comparative activity in heterogeneous Fenton application. Powder Technol. **36**(2), 847–854 (2019)
10. Zhang, J., Wang, Z.B., Zhu, Y.P., Ji, Z.M.: Fenton reagent-microelectrolysis pretreatment test of nitrobenzene wastewater. J. Yangzhou Univ. (Natural Science Edition) **9**(2), 74–78 (2006) (in Chinese)
11. Wu, C., Zhu, J.J., Cui, Z.W.: Synergetic degradation of phenol waste water by microwave-assisted Fenton reagent. Chem. World **59**(6), 334–340 (2018) (in Chinese)
12. Zhou, L.Y., Guo, W.: CuO-H_2O_2 heterogeneous catalytic oxidation of dye wastewater. Ind. Water Treat. **33**(06), 61–64 (2013) (in Chinese)
13. Yang, W.L., Gang, Z.X., Wu, J.: Advanced treatment of secondary biochemical effluent by CuO-Fe_2O_3/γ-Al_2O_3/H_2O_2/O_3. Chem. Ind. Eng. Prog. **37**(06), 2399–2405 (2018) (in Chinese)

Optimization for Finite Element Model of a Steel Ring Restrainer with Sectional Defect

Qiang Zhang, Yue Ma, Jin Feng, and Zhanfei Wang

Abstract Performances of the steel ring restrainer (SRR) as a typical steel structural component can be impaired due to a presence of initial geometrical defects or metallic corrosion. In this paper, finite element (FE) models of the SRRs with a sectional defect were optimized, and then a convenient and effective method was proposed to establish FE models. By comparing the force–displacement, stress distribution, stress triaxialities and analysis duration, it is found that analytical results of optimized models are good agreement with analytical results of models with solid elements, and meanwhile, optimized models present the higher calculation efficiency. The monotonic loading tests of SRRs were conducted to further verify the effectiveness of the optimized models. The optimized method of the FE model was proposed, and it can provide a reference for the research of related topics. The experimental results also show that the proposed method can obtain accurate analytical results in a relatively short time.

Keywords Finite element model · Steel ring restrainer · Sectional defects

1 Introduction

In current earthquakes, restrainers have shown good restraint effects in both the longitudinal and transverse directions, so that they can effectively reduce occurrences of the bridge unseating [1]. In China, due to the late start of the research on the mechanical performance of the restrainers, the restrainers are just as an attachment. Therefore, it is still necessary to further explore the mechanical performance of the

Q. Zhang (✉) · J. Feng · Z. Wang (✉)
Shenyang Jianzhu University, Shenyang 110168, China
e-mail: zhangqiang@stu.sjzu.edu.cn

Z. Wang
e-mail: ZFWang@sjzu.edu.cn

Y. Ma
Applied Technology College, Dalian Ocean University, Wafangdian 116300, China

333

restrainers. Many scholars have studied a lot of research on the mechanical properties of rigid stoppers, cables and SRRs, and their research methods are mainly theoretical calculations, experiments and numerical simulations [2–4].

For structural components of complex configuration and forces reception, theoretical calculations have difficulty to accurately analyze their mechanical properties under various external conditions and design parameters. On the other hand, it requires extremely high-standard test condition to obtain convincing results by tests. As the era of evolution of computer technology, the applicability of finite element analysis software has been improved. Numerical simulation methods can obtain more accurate results, meanwhile consume less manpower and material and have become one of the most commonly used analytical methods by most scientific researchers. The modeling method is chosen to establish the FE model, such as element type, meshing method and model simplification measures, and has a great impact not only on the accuracy of the analytical results, but also on operation costs and calculation time. Therefore, choosing an appropriate modeling method is a critical step.

The previous study showed that the section size of the device is the main factor affecting its restraint ability [3, 4]. The effect of the overall section size was assessed instead of the local section size. Therefore, it is necessary to further study the mechanical properties of the SRR with sectional defects. This paper optimizes the finite element model of the SRR with sectional defects, proposes a method for establishing an optimized FE model with high efficiency and high accuracy and compares it with the experimental results and simulation results of the solid model. The comparison of the results verifies the effectiveness of the optimization model. Researchers and engineers can use this model to obtain more accurate analysis results in a short time.

2 SRRs with Sectional Defects

The sectional defects of the SRR can be divided into two main categories. One of them is the error during processing in the factory or the initial defect of the steel plate, and the other is caused by the influence of the external environment on the device during the engineering application, such as the corrosion in the atmospheric environment and the defect due to impact. The cross-sectional defect rate, which is the ratio of the difference between the defective cross-sectional area and design cross-sectional area to the design cross-sectional area, is often used to describe the degree of defects [5]. In this paper, 15% and 30% are used as the cross-sectional defect rate, and the design cross-sectional area is 120 mm^2. Figure 1 depicts a typical SRR with sectional defects. The R is the radius of the steel ring, and D is the diameter of the guide pulley. In this paper, D and R are 40 mm and 120 mm, respectively. The A-A section represents the cross section of the defect on the steel ring.

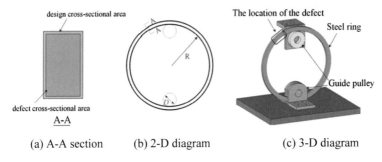

(a) A-A section　　　　(b) 2-D diagram　　　　(c) 3-D diagram

Fig. 1 SRR with sectional defects

3 Finite Element Model and Optimization Method

3.1 Solid Element

For the numerical simulation of three-dimensional geometric members, choosing 8-node linear brick stress/displacement elements (C3D8R) is intuitive and reliable [6] (Fig. 2).

The C3D8R was selected to model the steel ring, and the guide pulleys were modeled by the analytical rigid body. The degree of freedom in the x-axis direction of the upper guide pulley is free, and the unidirectional displacement is employed along the x-axis direction. The lower guide wheel adopts fixed constraint. In the sectional defect area, transition area and other areas, the fine mesh, medium mesh and coarse mesh are used, respectively. The steel yield strength, ultimate strength and Young's modulus are 294 MPa, 424 MPa and 216 GPa, respectively.

Fig. 2 C3D8R model

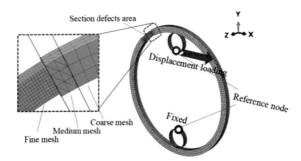

Fig. 3 CPS4R model and
two optimization models

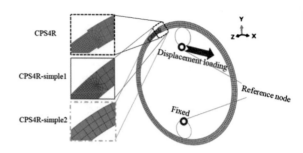

3.2 Optimization Models

Combining the previous study, 4-node bilinear plane stress elements (CPS4R) are
used to simulation the steel ring with sectional defects. The model is shown in Fig. 3.

In Fig. 3, the CPS4R-simple1 and CPS4R-simple2 are the two optimization
models that are mainly compared and analyzed in this paper. The CPS4R model:
The steel ring adopts CPS4R. The boundary conditions and loading pattern are the
same as the C3D8R model. The method is almost the same as the C3D8R model. The
CPS4R-simple1 optimization model: Compared with the CPS4R model, this model
ignores defects in section height but applies such defects to the thickness direction
based on the same cross-sectional defect rate. The CPS4R-simple2 optimization
model: Compared with CPS4R-simple1, a unified meshing method is adopted.

4 Numerical Simulation Study

4.1 Force–Displacement Relationship

The force–displacement is very important to the engineering design of the SRR.
Therefore, the effect of modeling methods with different sectional defect rate on
the force–displacement was investigated. It can be seen from Fig. 4 that the force–
displacement almost completely overlaps before the ultimate state.

4.2 Stress Distribution

The stress distribution of four types of FE models when the cross-sectional defect
rate is 15% are shown in Table 1. In the table, $\delta_1 = 155$ mm, $\delta_2 = 240$ mm, δ_3
$= 255$ mm, the selected displacements are respectively in the stable stiffness stage,
the stiffness gradient stage and the ductile stretching stage [5]. These displacements
were selected to compare the stress distribution at the different types of FE models.

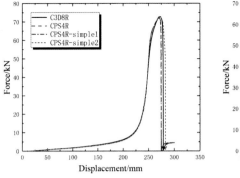

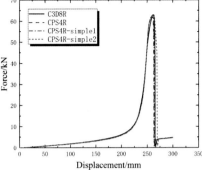

(a) The cross-sectional defect rate:15%   (b) The cross-sectional defect rate:30%

Fig. 4 Force–displacement

Table 1 Stress distribution of four model types

Model type	$\delta_1 = 155$ mm	$\delta_2 = 240$ mm	$\delta_3 = 255$ mm
C3D8R			
CPS4R			
CPS4R-simple1			
CPS4R-simple2			

It is shown, in Table 1, that there are minor differences on the stress distribution of four types of FE models.

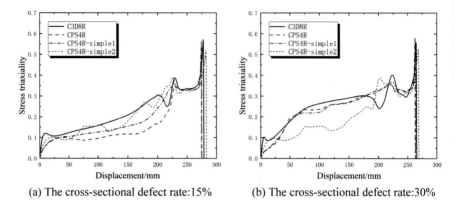

(a) The cross-sectional defect rate:15%   (b) The cross-sectional defect rate:30%

Fig. 5 Stress triaxiality

4.3 Stress Triaxiality

The stress triaxiality can reflect the stress state of the element and affect the fracture failure process [7]. The stress triaxiality of the first fractured element in the finite element model is extracted, and the element is deleted when the stress triaxiality is approximately equal to 2/3. When the cross-sectional defect rate is 15%, the stress triaxiality of CPS4R-simple2 and C3D8R models is relatively close. When the cross-sectional defect rate is 30%, stress triaxialities of CPS4R, CPS4R-simple2 and C3D8R are relatively close. Therefore, for the C3D8R model and the CPS4R model, the stress triaxiality is different when the defect section area is small.

4.4 Analysis Time

The time required for the finite element analysis often depends on the computer configuration and the complexity of the model. For the computer selected by the author, taking a finite element model with a 15% cross-sectional defect rate as an example, it took 55 min to analyze a C3D8R model, it took 6 min to analyze a CPS4R model, and it took 4 min to analyze a CPS4R-simple1 model, while the analysis of a CPS4R-simple2 is only 3 min. Therefore, the calculation efficiency of the CPS4R-simple2 is higher for this paper (Fig. 5).

5 Optimization Model Validation

In order to verify the effectiveness of the proposed optimized model, it designed, in this paper, four specimens of the SRR with different parameters to monotonic loading tests. The dimensions of the specimens are shown in Table 2. In which, the A, A_{min} and α are design cross-sectional area, defect cross-sectional area and cross-sectional defect rate, respectively. In the test, a MTS is used to drive the upper guide pulley to produce a horizontal displacement until the specimen is broken and the lower guide pulley is fixed.

By comparing the final fracture mode of the specimen, it is shown that the CPS4R-simple2 model can be employed to simulate the failure mode of SRRs (Fig. 6).

It illustrated, in Fig. 7, the comparison of force–displacement of the experiment and numerical simulation. The maximum error of the ultimate bearing capacity and ultimate displacement of the experiment results and finite element results are 4.5% and 3.7%, respectively. The force–displacement of the specimen and the FE model are basically same, which proves the validity of the CPS4R-simple2 model.

Table 2 Geometrical dimensions of specimens

Name	R/mm	D/mm	A/mm	A_{min}/mm	α/%
SRR-SD1	120	40	120	108	10
SRR-SD2	145	40	120	99.6	17
SRR-SD3	170	90	120	98.4	18
SRR-SD4	170	140	120	102	15

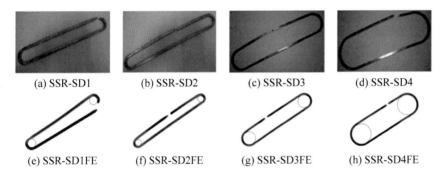

(a) SSR-SD1   (b) SSR-SD2   (c) SSR-SD3   (d) SSR-SD4

(e) SSR-SD1FE   (f) SSR-SD2FE   (g) SSR-SD3FE   (h) SSR-SD4FE

Fig. 6 Comparison of failure modes between experiment results and analytical results

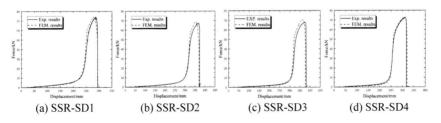

(a) SSR-SD1   (b) SSR-SD2   (c) SSR-SD3   (d) SSR-SD4

Fig. 7 Comparison of results from numerical simulation and experiment

6 Conclusions

1. This paper proposes a method for establishing an optimized FE model of the SRR with sectional defects. The validity of the model is proved by comparison with the C3D8R model and test results, and the calculation efficiency is higher. The CPS4R-simple2 model is the best choice for the optimized model of this paper.
2. For the FE analysis of the SRR with sectional defects, the force–displacement and the stress distribution of the analysis results of the C3D8R model and the CPS4R model are almost identical. However, the stress triaxiality is slightly different when the defect section area is small.
3. For the FE model of the SRR with sectional defects using CPS4R, the mesh has little effect on the force and displacement results.

References

1. Chen, W.F., Duan, L.: Bridge Engineering Handbook—Seismic Design, 2nd edn. CRC Press Taylor & Francis Group, Boca Raton, FL, USA (2014)
2. Xing X.K., Wang Y.F., Lei Z.L., Li Q., Dang H., Liu X.: Research progress of the unseating prevention devices for highway bridge. Earthq. Resistant Eng. Retrofitting **41**(04), 140–145+114 (2019) (in Chinese)
3. Wang Z.F., Sun J.B., Cheng H.B., Ge H.B.: Study on mechanical properties of a steel ring restrainer with buffer capacity. Bridge Constr. **48**(06), 18–23 (2018) (in Chinese)
4. Sun, J.B., Wang, Z.F., Xue, D.W., Ge, H.B.: Concept and behavior of a steel ring restrainer with variable stiffness and buffer capacity. J. Earthq. Tsunami (2020)
5. Xu, S.H., Wang, H., Li, A.B., Wang, Y.D., Su, L.: Effects of corrosion on surface characterization and mechanical properties of butt-welded joints. J. Constr. Steel Res. **126**, 50–62 (2016)
6. SIMULIA: ABAQUS standard manual, version 6.14. The Dassault Systèmes, Realistic Simulation, Providence, RI, USA (2013)
7. Kanvinde, A.M., Deierlein, G.G.: The void growth model and the stress modified critical strain model to predict ductile fracture in structural steels. J. Struct. Eng. **132**(12), 1907–1918 (2006)

An Epidemic Prevention Robot System Based on RoboMaster Technology

Tao Li, Lei Cheng, Huanlin Li, Yanjie Wu, and Guang Li

Abstract Based on the RoboMaster AI Challenge, a solution is proposed to use robots to replace epidemic prevention personnel to perform tasks. The epidemic prevention robots are designed for schools, office buildings and other public places to complete the monitoring of the body temperature of the entering personnel and disinfect the environment, etc. The robot designed in this study can go through deep learning to allow the body temperature of the entering person to be more accurately measured. The study introduces the working principle of the intelligent anti-epidemic disinfection robot and mainly explains the control of the gimbal to spray disinfectant and the autonomous path planning of the robot. The simulation experiment results show that these anti-epidemic robots can effectively complete their tasks. This has laid a good foundation for the next step of research on the epidemic prevention robot system.

Keywords Epidemic prevention robot · COVID-19 · TensorFlow · ROS · MATLAB

1 Introduction

The history of the epidemic is a terrible disasters history. Three thousand years ago, the smallpox virus brought huge disasters to Indians and even the world. The spanish flu that originated in the United States in 1918 spread globally, causing more than 21 million deaths [1]. Covid-19 is currently raging globally. At the time of writing, 16,000,000 people worldwide have been infected with Covid-19, and 600,000 people have died. The epidemic will also hit the global economy. During the H1N1 outbreak, the US GDP growth rate was negative; during the SARS outbreak, China's GDP growth rate declined; during the Covid-19 outbreak, the epidemic

T. Li (✉) · L. Cheng · H. Li · Y. Wu · G. Li
School of Information Science and Engineering, Wuhan University of Science and Technology, Wuhan, China
e-mail: 501320331@qq.com

© The Author(s), under exclusive license to Springer Nature Singapore Pte Ltd. 2021
Y. Li et al. (eds.), *Advances in Simulation and Process Modelling*,
Advances in Intelligent Systems and Computing 1305,
https://doi.org/10.1007/978-981-33-4575-1_33

hit small- and medium-sized enterprises a lot, and the contribution of small- and medium-sized enterprises to China's GDP was approximately 60%, and about 80% in terms of employment, which almost affects China's economic and financial stability [2]. Therefore, how to prevent and control the epidemic is a matter of great concern to governments.

In recent decades, with the development of science and technology, robots have been widely used in many fields. However, after the outbreak of Covid-19, there are very few robots that can participate in the prevention and control of the epidemic, so people are basically fighting the epidemic in person. On the battlefield, this also caused many policemen, delivery workers and couriers to contract the epidemic. Fortunately, shortly after the outbreak, researchers actively developed various anti-epidemic robots and put them into the battlefield. All kinds of robots appeared. A 5G-powered patrol robot manufactured by Guangzhou Gosuncn Robot Company is used for temperature monitoring and mask detection in airports and shopping malls [3]. In another example, the mobile Aimbot robot [4] and humanoid Cruzr robot from Shenzhen-based UBTECH Robotics is being used in school and hospital. They can measure temperature and confirm whether the masks are worn correctly. About robots for disinfection, in Danish, using high intensity UV-C light with a wavelength of 254 nm, the UVD Robots manufactured by Blue Ocean Robotics are used to kill bacteria and viruses [5]. In Singapore, PBA Group which is a local company has produced a disinfection robot called Sunburst UV Bots [6]. They use lider to complete autonomous navigation and use UV lamp to disinfect. In addition to the UV disinfection, there is also the method of spraying disinfectant. In March 2020, the VHP Robots were deployed in Hong Kong's railway station, using hydrogen peroxide to disinfect the compartments and stations [7]. In Shanghai, the Keenon Robotics can disinfect by using UV lamp and hydrogen peroxide, but they need remote-controlled from a phone or tablet. The use of robots has not only reduced the direct contact between staff, but also greatly reduced the work burden of the epidemic fighting personnel, making great contributions to the fight against the epidemic. The development and application of intelligent technology will play an increasingly important role in epidemic prevention and control. Based on RoboMaster technology, this article proposes a design scheme of an epidemic prevention robot, which completes functions such as body temperature monitoring, spraying disinfection and autonomous navigation.

The paper consists of the following parts. The second section gives the design ideas of the robot system. The third section describes the various robots and what we do simulation work. The fourth section presents the conclusions of our research and the outlook for the future anti-epidemic robots.

2   System Structure

Based on the current situation of the epidemic situation, we propose a design scheme for an epidemic prevention robot system based on the RoboMaster AI robot. Two types of robots are planned to be designed. The first type is a temperature detection

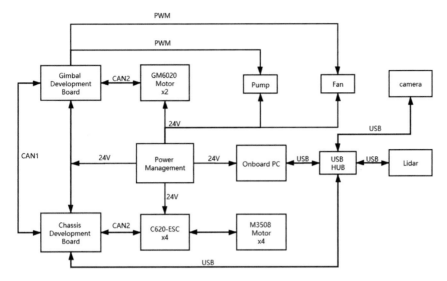

Fig. 1 Structure of the system design

robot, which is mainly responsible for monitoring the temperature of moving personnel at designated locations, finding people with abnormal body temperature in time and controlling the further spread of the epidemic; the second type is a disinfection robot, which is mainly responsible for disinfecting the area to reduce the risk of cross-infection.

For the design of a single robot, the robot can use a unified model to save costs. Temperature detection robots mainly use robot vision technology, so they are equipped with infrared cameras and RGB cameras. The disinfection robot will not be equipped with a camera and will only use lidar for composition and positioning. The structure of the system design is shown in Fig. 1.

3 Function Design and Practice

The following of this section describes the basic parts of the system. The first part uses an infrared camera and an RGB camera, adopts the KCF tracking algorithm and trains a recognition model through TensorFlow to realize the functions of pedestrian tracking and body temperature detection. In the second part, the spraying model is established according to the automatic disinfection function of the robot, which lays the foundation for the intelligent disinfection of the robot. In the third part, it introduces the research of robot autonomously determining the motion trajectory on ROS [8] by using AMCL algorithm and A* algorithm.

3.1 Thermal Imaging Visual Tracking

The hardware of the vision solution consists of an infrared imaging sensor and an RGB monocular camera. The infrared imaging sensor is used to obtain the thermal radiation video stream. From it, the approximate object contour boundary can be obtained. It can be used to track the moving objects when the environment is not much complex, but it is quite difficult to track and recognize an object when the light is bright or the environment changes greatly. About this question, the RGB camera can improve it. The infrared image and the RGB image can be complemented to get temperature and light texture information at the same time.

Tracking objects can be divided into two situations: bright light with the environment complicate and dim light with the environment ordinary. In the first case, tracking people with abnormal body temperature is best achieved through RGB vision. For the second case, when the light is insufficient and the environment changes little, it is better to use infrared images for analysis, identification and tracking.

Implementation methods of identification and tracking. Recognition function is to train offline .pb models by TensorFlow, including face recognition model and pedestrian recognition model. The main function is to identify people with abnormal body temperature. The edge information, texture information, color information and centroid inflection point information can all be extracted in the acquired image. In addition, other features of the face and body area can also be obtained. Using this method can make tracking relatively simple, and the system has good robustness. In order to improve the tracking ability under high-speed motion and ensure that the target is not easily lost, the commonly used and mature KCF tracking algorithm is adopted. The tracking effect is shown in Fig. 2.

For temperature detection, firstly, pedestrians and their faces are detected in RGB images. In the infrared image, choose the area corresponding to the RGB image of the detected pedestrians forehead. And then obtain the temperature information from

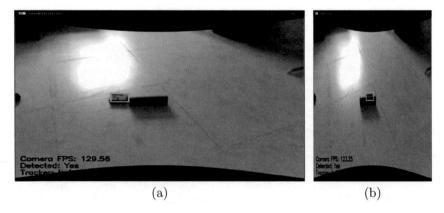

(a)   (b)

Fig. 2 KCF tracking algorithm renderings. **a** shows the start of tracking with KCF, and **b** shows the tracking after a few frames

Fig. 3 Tracking test under infrared image and RGB image

intercepted infrared image, and average the local value as the body temperature of the detected person. If the body temperature is found to be greater than 37.3 °C, the person will be added to the key tracking target, the KCF algorithm will be used for tracking and real-time feedback of position information, and the pedestrian's position information will be read when the pedestrian needs to be further checked. Using this information enables the robot to conduct autonomous navigation and tracking. The tracking test effect with the infrared image and the RGB image is shown[1] in Fig. 3.

Disadvantages and improvements: Due to the instability of the infrared image, the measured temperature will jump at ±0.2 °C. Therefore, the critical point of 37.3 °C should be suspicious. For this situation, a more effective treatment should be used method. For the same person under test, taking 10 times as the detection cycle, if the temperature measured more than 6 times is the same value, then this value is taken as the measured temperature value. In the subsequent cycles, when the error between the measured temperature value and the current value exceeds 0.3 °C, the currently confirmed temperature value will be updated, otherwise it will remain unchanged.

3.2 Automatic Disinfection Function Design

The automatic disinfection function is one of the most important functions that a disinfection robot needs to have. According to the recommendations of WHO experts, 70% medical ethanol can be used repeatedly special equipment for disinfection, such as a thermometer. 0.5% of sodium hypochlorite to disinfect the surfaces of objects frequently touched at home or in medical institutions [9]. Sodium hypochlorite is the main component of 84 disinfectant. Therefore, 84 disinfectant is used as a disinfection weapon for epidemic prevention robots.

This section will introduce the realization of automatic disinfection tasks in two parts. The first part will introduce the hardware design, and the second part will introduce the spraying model design based on MATLAB.

Hardware Design The RoboMaster development board type C (STM32f407IG) designed by DJI is used as the gimbal control board and chassis control board. The M3508 motors are used to drive the robots to move, and the GM6020 motors are used

---

[1] Informed consent was obtained from the participant included in the study.

to control the rotation of the gimbal and the spraying direction of the disinfectant. The disinfectant is pumped up from the medicine box by a pump and atomized into small droplets by a fan and a nozzle.

Simulated Disinfectant Spraying Model After the sprayer atomizes, the disinfectant will form a group of small droplets. We assume that each small droplet is spherical, and the small droplets can form a cone-shaped droplet group. Based on this, we carry out modeling analysis (Figs. 4 and 5).

The movement of droplets in the air should consider the influence of air resistance. According to the knowledge of aerodynamics and Newton's second law [10], some differential equations can be listed:

$$\begin{cases} m\ddot{x} = -k\dot{x}^2 \\ m\ddot{y} = -k\dot{y}^2 \\ m\ddot{z} = -mg - k\dot{z}^2 \ (t > t_1) \\ m\ddot{z} = -mg + k\dot{z}^2 \ (t \le t_1) \end{cases} \tag{1}$$

where t represents the movement time of the droplet; t_1 represents the rise time.

The simulation result Fig. 6 is obtained by solving Eq. (1) and performing MATLAB simulation. When the spray speed is constant, the greater the angle of the spray, the larger the disinfection range. When the angle of the spray is constant, the greater the spray speed, the farther the range. And ejecting droplets of the initial velocity

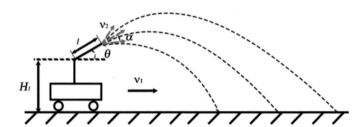

Fig. 4 Schematic diagram of spraying disinfectant

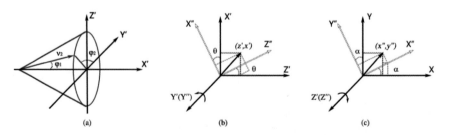

Fig. 5 Coordinate establishment and transformation. In **a** φ_1 represents the spray opening angle; φ_2 represents the radial angle of injection. In **b** represents the relative coordinate system; In **c** represents the absolute coordinate system

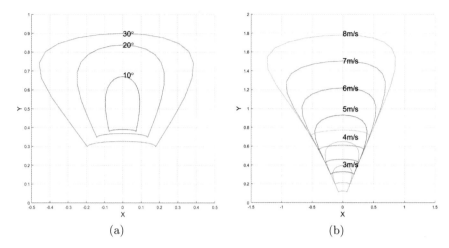

Fig. 6 Relationship between the spray range of the disinfectant and the spray opening angle and the initial spray velocity. In **a**, the initial spray velocity is 5m/s, and the spray opening angle is 10°, 20°, 30°. In **b**, the spray opening angle is 30°, and the initial spray velocity changes in 3–8m/s

is determined by the speed of the fan, the spray opening angle is determined by the pump power. Therefore, according to the relationship between simulation and calculation, the robot can choose the power of the pump and the speed of the fan according to the width and depth of the terrain of the disinfection area to achieve the optimal disinfection efficiency.

3.3 AMCL Is a Probabilistic Localization System for 2D Mobile Robots

Adaptive Monte Carlo Localization(AMCL) is used in the localization section. AMCL is a probabilistic localization system for 2D mobile robots. It achieves adaptive (or KLD sampling) Monte Carlo localization. Based on the known map, particle filters are used to track the robot's attitude. The basic idea of traditional Monte Carlo localization refers to particle filtering and Bayesian filtering. From the viewpoint of Bayesian theory, state estimation problem is to calculate the credibility of the current state according to the previous series of existing data recursively, which is mainly divided into two steps: prediction and update. The particles are sampled from the motion model, and the current confidence is used as the start point. The measure model is selected to determine the particle weight distribution. The initial confidence is obtained from those particles which are randomly generated by the prior distribution (set the sampling points), and the same weight factor is assigned to each particle. KLD is introduced by AMCL method, KLD method takes the sample collection after being weighted not the re-sampling as input. AMCL method produces the sample set until reaches its approximation error of statistics. ROS has prepared the relevant

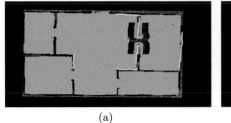
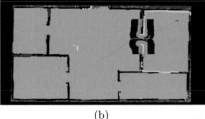

(a)   (b)

Fig. 7 Localization simulation results. **a** shows the motion planning of the robot from the starting point, and **b** shows the simulation when the robot returns

function package which can easily running by subscribing to the relevant topics and setting some parameters.

In the planning section, A* algorithm [11] is quoted to get the effective routine which is also the shortest. The whole area would be simplified to the area contained with square grids, turning the map to the two-dimensional array. The number of the grids, which is calculated by the program, is the routine needed. The shortest routine would be searched from the start point to the surrounding area. Each start point is determined by the grid which has the minimum sum of the distances from A and B. In the ROS platform, a virtual scene is constructed and motion planning is carried out (Fig. 7).

4   Conclusions and Future Work

This article introduced a body temperature monitoring robot that could perform deep learning and a disinfection robot that could navigate autonomously. Most of the robots on the market now complete the epidemic prevention work independently. If the robot fails, the entire anti-epidemic system may be paralyzed. In the future work, we plan to group multiple robots into a robot system. Divide the task to be completed into multiple parts, arrange specific tasks for the robots in the system, and cooperate to complete the tasks through a collaborative method, so that the anti-epidemic work can run stably.

References

1. Liu, W.M.: The origin of the 1918 epidemic and Its Global Spread. J. Glob. Hist. Rev. **000**(001), 296–306 (2011) (in Chinese)
2. Wang, G.H.: A lesson from hstory: impact of the epidemic on financial stability and risk control measures. J. Chin. Banker **03**, 48–50 (2020). (in Chinese)
3. China combines edge patrol robots and 5G to detect Covid-19 cases. https://www.mwee.com/news/china-combines-edge-patrol-robots-and-5g-detect-covid-19-cases (2020/03/17)

4. Robot Man: UBTECH AIMBOT Covid-19 Robot (2020/06/03)
5. Bogue, R.: Robots in a contagious world. J. Ind. Rob. (2020). https://doi.org/10.1108/IR-05-2020-0101
6. Sia, M.: Sunburst UV Bot Autonomus UV Disinfecting Robot (2020/05/27)
7. MTR Corp deploys VHP Robot for deep cleaning of trains. https://www.railway-technology.com/news/mtr-corp-vhp-robot-cleaning-trains2020/03/12
8. Quigley M.: ROS: an open-source robot operating system. In: International Conference on Robotics and Automation, vol. 3, no. 3.2, p. 5. (2009)
9. Tan, C.X.: WHO on Novel Coronavirus (2019-NCOV) infection prevention and control. J. Chin. J. Infect. Control **19**(03), 286–287 (2020)
10. Quan, L.Z., Wang, J.S., Xi, D.J.: Establishment and verification of the aerodynamics model of targeted herbicide robot spraying liquid. J. Trans. Chin. Soc. Agric. Eng. **33**(15), 72–80 (2017) (in Chinese)
11. Huang, J.H., Wu, J.H.: Path planning of indoor service robot based on improved A* algorithm. J. Technol. Market **27**(03), 62–63 (2020) (in Chinese)

Research on the Fire Resistance of Grout Sleeve Splicing Joint for Precast Stadium

Xudong Yang, Fan Gu, Liqiang Liu, and Qinghe Li

Abstract Based on the sequential coupling method and the temperature rise curve according to ISO-834, the ABAQUS numerical models of grout sleeve splicing joint was established. At a different fire exposure time, the temperature field and ultimate tensile strength of splicing joint were numerically analyzed. Meanwhile, by means of the stress distribution analysis of grouting material, the failure mechanism of splicing joint was put forward. Numerical simulation result shows that under the action of ultimate tensile load, the Mises stress of rebar decreases gradually along axial direction from uploaded end to non-loaded end, the Mises stress of sleeve increases gradually along axial direction from sleeve end to middle cross section, and the stress distribution of grouting material is relatively uniform. Furthermore, with the increase of fire exposure time, the stress state of grouting material changes from trip-direction compressive state to trip-direction tensile state gradually, which leads to the ultimate tensile load of grout sleeve splicing joint decreases.

Keywords Grout sleeve splicing joint · Fire resistance · Stress distribution · Sequential coupling method · ABAQUS

1 Introduction

Nowadays, prefabricated technology has been applied in large shopping malls, office buildings, school buildings, stadiums, multi-storey residential buildings, and so on. For prefabricated stadiums, there are two main development trends, light-steel structures and precast reinforced concrete structures. The former is mainly applied as

X. Yang
Sports Department, Shenyang Jianzhu University, Shenyang, China

F. Gu (✉) · L. Liu · Q. Li
School of Civil Engineering, Shenyang Jianzhu University, Shenyang, China
e-mail: guzhaozheng@yeah.net

Advances in Intelligent Systems and Computing 1305,
https://doi.org/10.1007/978-981-33-4575-1_34

temporary stadiums and gymnasiums such as assembled swimming pools and assembled canopy stadiums, and the latter is mainly applied as large-scale precast reinforced concrete stadium structures. In precast stadium structures, the prefabricated parts manufactured in factory are mainly connected by grouted splice connector as a whole and effectively realize the design requirement of assembly equivalent to cast-in-place.

Grouted splice connector has the advantages of reliable connection performance, easy installation, and scholars have carried out sufficient experimental researches and numerical simulation researches on the mechanical properties of grout sleeve splicing joint and precast reinforced concrete structure with grouted splice connector.

For grout sleeve splicing joint, Ling et al. [1] carried out an experimental study on the connection performance of grout sleeve splicing joint under incremental tensile load and derived the equations to evaluate structural performance. Zheng et al. [2] took an experimental research on the effect of compressive strength and expansion rate of grouting material on the bond behavior of splicing joint and derived the determination of bond performance by experimentally measuring the exterior surface transverse strain of grout sleeve in curing stage. Kuang et al. [3] and Abdel et al. [4] carried out experimental study and numerical study on the mechanical behavior of grout sleeve splicing joint under the action of high-stress reversed tension and compression load and large-deformation reversed tension and compression load and put forward the effect of grout content on the bond capacity between rebar and grouting material and together with the tensile capacity of joint. Gu et al. [5–8] conducted numerical research on the failure mechanism of grout sleeve splicing joint under the action of tension load and cyclic load and discussed the influence of grouting material strength, sleeve rib space on the bond behavior of joint. Furthermore, Ahmad et al. [9] proposed a new type of sleeve splice connector consisted of steel bars with tapered nuts confined in grouted steel pipes and presented its structural performance under increasing axial tension load. Eliya and George [10] proposed a grout sleeve that accommodates current production tolerances in addition to be economical and easy to produce and conducted the experimental investigation on its connection performance.

For precast reinforced concrete structure with grouted splice connector, Ali et al. [11] carried out experimental research on the adhesive performance of grout sleeve splicing joint in beam components and discussed the effect of interlocking mechanism on bond strength of sleeve. Ameli et al. [12] took a test of half-scale bridge column-to-cap beam assemblies to investigate their response under cyclic quasi-static load and revealed the seismic ability of precast concrete joints constructed with grouted splice connectors in cap beam. Nerio and Fabio [13] carried out the cyclic test on a precast reinforced concrete column-to-foundation grouted duct connection with a full-scale specimen under the action of cyclic bending combined with axial compression. In addition, Ramin et al. [14] proposed a special precast concrete wall-to-wall connection system by using two steel U-shaped channels which can resist multidirectional imposed loads and reduces vibration effects and took numerical study on its energy dissipation, stress, deformation, and concrete damage in the plastic range. Arthi and

Jaya [15] conducted comparative study on structural behavior of precast shear wall-diaphragm connection and the monolithic connection under seismic loading and indicated that the damage parameter and the interaction between structural members play a crucial role in the modeling of precast connections.

With the rapid development of prefabricated construction, the fire resistance of grout sleeve splicing joint has aroused more and more people's concern. At present, the research on its mechanical performance and damage mechanism under elevated temperature is slightly insufficient. In this paper, based on sequential coupling method and the temperature rise curve according to ISO-834, the numerical simulation research on the connection performance and damage mechanism of grout sleeve splicing joint under elevated temperature was carried out, in order to provide theoretical reference to the performance assessment of grout sleeve splicing joint after fire.

2 Numerical Model of Grout Sleeve Splicing Joint

According to the actual structure of grout sleeve for connecting $\Phi 16$ mm rebars, in accordance with ASTM A1034/A1034M-10a [16], ACI 318-14 [17] and JGJ 355-2015 [18], the finite element numerical model of grout sleeve splicing joint was built by ABAQUS. For reducing the number of elements, 1/2 model in axial direction was selected, with xz plane-symmetry displacement boundary conditions on the right surface of numerical model, where is the middle cross section of grout sleeve splicing joint in axial direction. Meanwhile, according to the axisymmetric characteristics of splicing joint geometry, the generatrix model was adopted to establish.

In the process of heat transfer analysis, DCAX4 element, namely linear axisymmetric heat transfer four-node quadrilateral element, was used for representing rebar, grouting material, and sleeve. In the process of elevated temperature connection performance analysis, CAX4 element, namely four nodes axisymmetric solid element without distortion, was used for representing grouting material. Because of the regular shape of numerical model, the tetrahedral free meshing technology was adopted, and the local grid refinement was applied on grouting material. Mesh generation of the generatrix of 1/2 model of grout sleeve splicing joint is shown in Fig. 1.

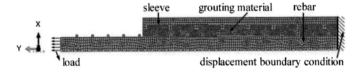

Fig. 1 Numerical model of grout sleeve splicing joint

3 Temperature Field Analysis

3.1 Temperature Rise Curve and Thermal Parameters

In the process of temperature field analysis of structure fire resistance, two kinds of temperature rise curve according to ISO-834 [19] and ASTM E119-20 [20] are usually adopted. The heating up effect by these two temperature rise curves is the same in general, and the former temperature rise curve by ISO-834 was adopted in this paper.

In the process of heat transfer analysis, thermal parameters of rebar, ductile cast iron sleeve, and grouting material were evaluated according to the formula in literature [21–23], as shown in Tables 1, 2, and 3.

3.2 Temperature Field Analysis

The heat transfer approaches between objects can be divided into heat conduction, heat convection, and heat radiation. The heat conduction depends on thermal conductivity and specific heat volume, as mentioned in Sect. 3.1. The heat convection depends on the convective heat transfer coefficient of fire surface. Considering the concrete protective layer outside the sleeve, the value of the convective heat transfer coefficient of splicing joint was taken as 25 W/(m^2 °C) [24]. The thermal radiation depends on the comprehensive radiation coefficient of fire surface, the value was taken as 0.5 [25]. The temperature nephogram of splicing joint at difference fire exposure time could be calculated, as shown in Fig. 2. By extracting the temperature data of the middle cross section of splicing joint along axial direction, that is, the bottom section of half structure numerical model, the relationship between temperature and fire exposure time could be obtained, as shown in Fig. 3. It can be seen that there is a certain temperature difference between rebar, grouting material, and sleeve in the initial period of fire. With the increase of fire exposure time, the temperature difference decreases gradually. When the fire exposure time reaches 60 min, the temperature of steel rebar reaches 800 °C and has almost lost its carrying capacity.

Table 1 Thermal parameters of steel rebar

Temperature/(°C)	Thermal conductivity/(W · m^{-1} · °C^{-1})	Linear expansivity/(°C^{-1})	Specific heat volume/(J · kg^{-1} · °C^{-1})	Density/(kg · m^{-3})
20	47.6	1.21×10^{-5}	477	7800
100	45.8	1.24×10^{-5}	497	
300	41.4	1.32×10^{-5}	568	
500	37.0	1.40×10^{-5}	669	
700	32.6	1.48×10^{-5}	800	

Table 2 Thermal parameters of ductile cast iron sleeve

Temperature/(°C)	Thermal conductivity/(W · m^{-1} · °C^{-1})	Linear expansivity/(°C^{-1})	Specific heat volume/(J · kg^{-1} · °C^{-1})	Density/(kg · m^{-3})
20	29.9	1.40×10^{-8}	495	7300
100	29.8		511	
300	29.2		565	
500	28.8		665	
700	27.8		924	

Table 3 Thermal parameters of grouting material

Temperature/(°C)	Thermal conductivity/(W · m^{-1} · °C^{-1})	Linear expansivity/(°C^{-1})	Specific heat volume/(J · kg^{-1} · °C^{-1})	Density/(kg · m^{-3})
20	1.36	1.83×10^{-5}	913	2300
100	1.36	1.01×10^{-5}	962	
300	1.34	0.28×10^{-5}	1064	
500	1.10	0.10×10^{-5}	1156	
700	0.85	0.04×10^{-5}	1170	

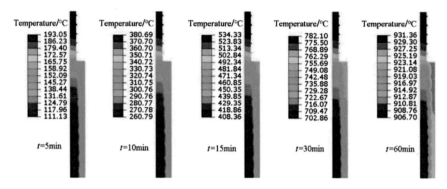

Fig. 2 Temperature nephogram of splicing joint at a different fire exposure time

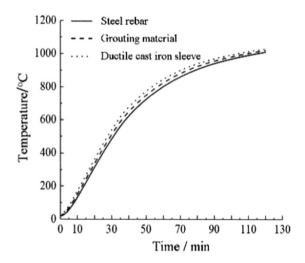

Fig. 3 Relationship between the temperature of splicing joint and fire exposure time

4 Connection Performance Analysis at Elevated Temperature

4.1 Mechanical Properties of Materials at Elevated Temperature

In the process of establishing numerical model for study the mechanical performance of grout sleeve splicing joint at elevated temperature, the concrete damaged plasticity model provided by ABAQUS was adopted for representing the constitutive relationship of grouting material at normal temperature, which means that tension crack and

Table 4 Mechanical parameters of steel rebar at elevated temperature

Temperature/(°C)	Young's modulus/(GPa)	Yield strength/(MPa)	Poisson's ratio
20	200	400	0.3
100	200	400	
200	200	400	
300	151	308	
400	120	244	
500	80	180	

crushed destruction are the failure mechanism of grouting material, and the elastic-plastic constitutive model of high-strength concrete with reference to literature [26] was adopted for representing the constitutive relationship of grouting material at elevated temperature. Meanwhile, the constitutive relationship of HRB400 rebar and ductile iron sleeve at elevated temperature with reference to literature [27, 28] were adopted, and the mechanical parameters of steel rebar, ductile cast iron sleeve, and grouting material at elevated temperature could be gotten, as shown in Tables 4, 5, and 6.

Table 5 Mechanical parameters of ductile cast iron sleeve at elevated temperature

Temperature/(°C)	Young's modulus/(GPa)	Yield strength/(MPa)	Poisson's ratio
20	169	370	0.3
100	165	370	
200	161	362	
300	158	298	
400	152	233	
500	146	190	

Table 6 Mechanical parameters of grouting material at elevated temperature

Temperature/(°C)	Young's modulus/(GPa)	Yield strength/(MPa)	Poisson's ratio
20	38.00	100	0.2
100	35.77	98	
200	25.70	91	
300	17.30	76	
400	10.68	59	
500	5.71	43	

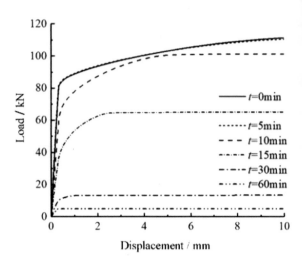

Fig. 4 Relationship between tensile load and axial displacement of splicing joint

4.2 Connection Performance at Elevated Temperature

The sequential coupling method was adopted in this paper. Firstly, the temperature field of splicing joint was simulated, and the result file *.odb was applied to the splicing joint model as predefined load. Secondly, the tensile load was exerted on the steel rebar with displacement loading method, and the load-displacement curves of splicing joint at a different fire exposure time can be obtained, as shown in Fig. 4. Compared with the ultimate tensile load of splicing joint at normal temperature, when the fire exposure time reaches 15 min, the ultimate tensile load drops to about 50%. When the fire exposure time reaches 30 min, the ultimate tensile load is only about 20% of that at normal temperature. According to Fig. 3, when the fire exposure time reaches 20, 25, and 35 min, the temperature of splicing joint reaches about 300 °C, 400 °C, and 500 °C, respectively. At a different fire exposure time, that is at a different temperature, under the action of corresponding ultimate tensile load, the stress nephogram of grout sleeve splicing joint could be obtained, as shown in Fig. 5.

Figure 5a and b show that the Mises stress of rebar decreases gradually along axial direction from uploaded end to non-loaded end, and the Mises stress of sleeve increases gradually along axial direction from sleeve end to middle cross section. Figure 5c and d shows that the stress distribution of grouting material is relatively uniform with little numerical variation, and the maximum value of stress usually occurs in the area in contact with the rebar ribs and sleeve ribs. In addition, when the fire exposure time is less than 20 min, the maximum principal stress of grouting material is compressive stress, implying that the grouting material is in trip-direction compressive state, and a relatively obvious oblique compression zone in grouting material is formed between the sleeve rib and the corresponding rebar rib. When the fire exposure time reaches 25 min, the grouting material located between the

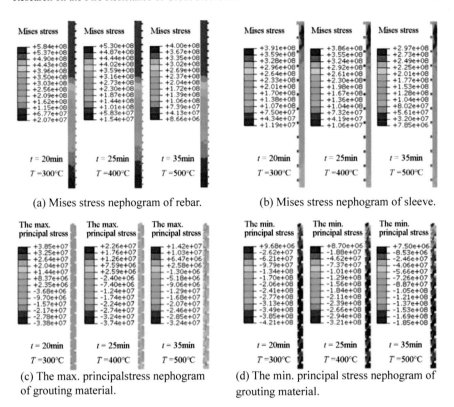

(a) Mises stress nephogram of rebar.   (b) Mises stress nephogram of sleeve.

(c) The max. principalstress nephogram of grouting material.

(d) The min. principal stress nephogram of grouting material.

Fig. 5 Stress nephogram of grout sleeve splicing joint under the action of corresponding ultimate tensile load at a different fire exposure time

sleeve rib and the corresponding rebar rib, its maximum principal stress changes to tensile stress, and there is still a very obvious oblique compression zone between the sleeve rib and the corresponding rebar rib. When the fire exposure time is greater than 35 min, the grouting material in the area of central sleeve along axial direction, its minimum principal stress gradually changes to tensile stress, which means that grouting material in trip-direction tensile state has almost lost its adhesive capacity, and the ultimate tensile strength of grout sleeve splicing joint decreases accordingly.

5 Conclusions

Based on the sequential coupling method, the connection performance of grout sleeve splicing joint at elevated temperature was studied, and numerical simulation result shows relevant conclusions as follows.

1. Under the action of ultimate axial tensile load, the Mises stress of rebar decreases gradually along axial direction from uploaded end to non-loaded end, the Mises stress of sleeve increases gradually along axial direction from sleeve end to middle cross section, and the stress distribution of grouting material is relatively uniform.

2. With the increase of fire exposure time, the stress state of grouting material changes from trip-direction compressive state to trip-direction tensile state gradually, and the oblique compression zone in grouting material also disappears accordingly, which leads to the ultimate tensile strength of grout sleeve splicing joint decreases.

Acknowledgements This research work is supported by Scientific Research Project of Liaoning Science and Technology Department (Project Codes: 2015020600) and the Sub-project III under National Key Technology Research and Development Program of the Ministry of Science and Technology of China (Project Codes: 2018YFD1100403-03).

References

1. Ling, J.H., Ahmad, B.A.R., Izni, S.I.: Feasibility study of grouted splice connector under tensile load. Constr. Build. Mater. **50**(15), 530–539 (2014)
2. Zheng, Y.F., Guo, Z.X., Zhang, X.: Effect of grout properties on bond behavior of grouted pipe splice. KSCE J. Civ. Eng. **22**(1), 2951–2960 (2018)
3. Kuang, Z.P., Zheng, G.Y., Jiao X.T.: Experimental study on effect of mechanical behavior of grout sleeve splicing for reinforced bars due to lack of grout. J. Tongji Univ. (Nat. Sci.). **47**(7), 946–956 (2019) (in Chinese)
4. Abdel J.M., Beale R.G., et al.: Cyclic loading applied to sleeve couplers for tube and fitting scaffolds. ACMSM25, Lect. Notes in Civil Eng. **37**(1), 819–830 (2019)
5. Gu, F., Wu, T.G., Zhao, W.J., Dui, X.: Numerical study on the static response of grouting-sleeve reinforcement-connection component under axial tension load. Appl. Mech. Mater. **353**(1), 3312–3315 (2013)
6. Gu, F., Xian, A.K., Zhao, W.J., Wei, W.: Influence of grouting-material compressive-strength on the mechanical properties of grouting-sleeve reinforcement-connection component. Adv. Eng. Res. **40**(1), 87–90 (2015)
7. Gu, F., Wei, W., Zhao, W.J., Xian, A.K.: Study on the mechanical properties of grouting-sleeve reinforcement-connection component under the action of cyclic tension load. Adv. Eng. Res. **40**(1), 104–108 (2015)
8. Gu, F., Song, D.K., Ming, Y., Zhang, Y.Y.: Research on fire resistance of steel sleeve grouting connector. J. Shenyang Jianzhu Univ. (Nat. Sci.). **35**(3), 420–427 (2019) (in Chinese)
9. Ahmad, B.A.R., Loh, H.Y., Izni, S.I.: Performance of grouted splice sleeves with tapered bars under axial tension. Appl. Mech. Mater. **789**(1), 1176–1180 (2015)
10. Eliya, H., George, M.: Non-proprietary bar splice sleeve for precast concrete construction. Eng. Struct. **83**(1), 154–162 (2015)
11. Ali, A.S., Ahmad, B.A.R., Mohd, Z.B.J., Johnson, A., Sayadi, A.: The relationship between interlocking mechanism and bond strength in elastic and inelastic segment of splice sleeve. Constr. Build. Mater. **55**(31), 227–237 (2014)
12. Ameli, M.J., Parks, J.E., Brown, D.N., Pantelides, C.: Seismic evaluation of grouted splice sleeve connections for reinforced precast concrete column-to-cap beam joints in accelerated bridge construction. PCI J. **60**(2), 80–103 (2015)

13. Nerio, T., Fabio, M.: Cyclic test on a precast reinforced concrete column-to-foundation grouted duct connection. Bull. Earthq. Eng. **18**(1), 1657–1691 (2020)
14. Ramin, V., Farzad, H., Hafez, T., Mohd, S.J., Farah, N.A.: Development of a new connection for precast concrete walls subjected to cyclic loading. Earthquake Eng. Eng. Vib. **16**(1), 97–117 (2017)
15. Arthi, S., Jaya, KP.: Seismic performance of precast shear wall-slab connection under cyclic loading: experimental test versus numerical analysis. Earthq. Eng. Eng. Vib. **19**(1), 739–757 (2020)
16. ASTM Designation, ASTM A1034/A1034M-10a: Standard test methods for testing mechanical splices for steel reinforcing bars. Am. Soc. Test. Mater. (2015)
17. ACI Standard, ACI 318-14: Building code requirements for structural concrete and commentary. Am. Concr. Inst. (2014)
18. People's Republic of China Industry Standard, JGJ 355-2015: technical specification for grout sleeve splicing of rebars. Ministry of Housing and Urban-Rural Development of the People's Republic of China (2015) (in Chinese)
19. International Standard, ISO 834: Fire resistance tests-elements of building construction. Int. Organ. Stand. (2014)
20. ASTM Designation, ASTM E119-20: Standard test methods for fire tests of building construction and materials. Am. Soc. Test. Mater. (2020)
21. Song X.Y.: The analysis and research on fire resistance performance of reinforced concrete columns. Master's Dissertation, Changsha: Hunan University (2006) (in Chinese)
22. Lie, T.T., Barbaros, C.: Method to calculate the fire resistance of circular reinforced concrete columns. ACI Mater. J. **88**(1), 84–91 (1991)
23. Li, Y.Q., Ma, D.Z., Xu, J.: Fireproof Design Calculation and Structural Treatment of Building Structures. China Architecture & Building Press, Beijing (1991). (in Chinese)
24. Ehab, E., Colin, G.B.: Modelling of unbonded post-tensioned concrete slabs under fire conditions. Fire Saf. J. **44**(2), 159–167 (2009)
25. Yaman, S.S.A., Riadh, A.M., Ian, B.: Experimental and numerical study of the behaviour of heat-damaged RC circular columns confined with CFRP fabric. Compos. Struct. **133**(1), 679–690 (2015)
26. Guo, Z.H., Shi, X.D.: Experiment and Calculation of Reinforced Concrete at Elevated Temperatures. Tsinghua University Press, Beijing (2011)
27. Hu, H.T., Dong, Y.L.: Experimental research on strength and deformation of high-strength concrete at elevated temperature. China Civ. Eng. J. **35**(6), 44–47 (2002) (in Chinese)
28. Zhou W.J., Liu Y.J., Wang X.: Numerical simulation of high temperature tensile performance of sleeve grouting steel bar. J. Water Resour. Archi. Eng. **14**(5), 170–176 (2016) (in Chinese)

Mechanical Property of Grout Sleeve Splicing Joint Under Reversed Cyclic Loading for Precast Stadium

Xudong Yang, Fan Gu, Qinghe Li, and Liqiang Liu

Abstract The seismic performance of grout sleeve splicing joint was investigated in this paper. Under the action of reversed cyclic loading, the numerical model of grout sleeve splicing joint was developed and was evaluated in terms of the variation of stress distribution, failure process, hysteresis loop, and energy dissipation. Numerical simulation result shows that the seismic performance of grout sleeve splicing joint mainly depends on the stress state of grouting material. In the first two loading cycles, the mechanical interaction with uneven distribution is the main load transmission mechanism from rebar to sleeve, and the fracture damage in grouting material near the end of sleeve is the main mode of energy dissipation. Between the second and the seventh loading cycles, through the stress redistribution, the grouting material in the area far from the end of sleeve gradually undertakes the energy dissipation function by means of fracture damage, and the obvious oblique compressive zone in grouting material is formed, which is the force transmission path from rebar to sleeve. Meanwhile, it decreases the elastic modulus of entire grouting material to a stable value.

Keywords Grout sleeve splicing joint · Reversed cyclic loading · Seismic performance · Energy dissipation · Numerical simulation

1 Introduction

Prefabricated structure adopts the construction mode of factorization of prefabricated parts, mechanization of field construction, scientific organization and management, which can improve the labor production efficiency and guarantee the quality of parts. In addition, prefabricated structure has the characteristics of energy saving,

X. Yang
Sports Department, Shenyang Jianzhu University, Shenyang, China

F. Gu (✉) · Q. Li · L. Liu
School of Civil Engineering, Shenyang Jianzhu University, Shenyang, China
e-mail: guzhaozheng@yeah.net

© The Author(s), under exclusive license to Springer Nature Singapore Pte Ltd. 2021
Y. Li et al. (eds.), *Advances in Simulation and Process Modelling*,
Advances in Intelligent Systems and Computing 1305,
https://doi.org/10.1007/978-981-33-4575-1_35

consumption reduction, environmental protection, realizing the sustainable development of economic benefit, environmental benefit, and social benefit. At present, prefabricated technology has been applied in office buildings, large shopping malls, stadiums, multi-story residential buildings, and so on. For prefabricated stadiums, there are two main development trends which are light-steel structures and precast reinforced concrete structures. The former is mainly applied in temporary gymnasiums such as assembled swimming pool and assembled canopy stadium, as shown in Fig. 1; the latter is mainly applied in large-scale precast stadium, as shown in Fig. 2.

In precast reinforced concrete stadium structure, the prefabricated parts are mainly assembled by grout sleeve to form as a whole and jointly resist external load. The quality of prefabricated parts manufactured in factory is easy to be guaranteed, and the splicing joint of prefabricated parts is slightly weak comparatively. Therefore, the failure of grout sleeve splicing joint is one of the main failure forms of precast reinforced concrete stadium structures, and scholars have carried out experimental researches and numerical simulation researches on the mechanical properties of grout sleeve splicing joint. Ali et al. [1] carried out an experimental research on the adhesive performance of grout sleeve splicing joint in beam components and discussed the effect of interlocking mechanism on bond strength of sleeve. Ahmad et al. [2] and Ling et al. [3] took experimental researches on the connection behavior of grout sleeve splicing joint under the action of increasing tensile load and put forward the

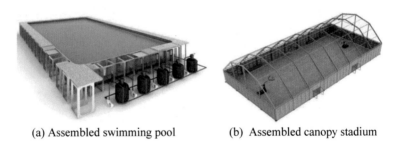

(a) Assembled swimming pool   (b) Assembled canopy stadium

Fig. 1 Light-steel structures

(a) Citi field, New York   (b) Lucas oil stadium

Fig. 2 Precast reinforced concrete structures of large-scale stadium

influence of sleeve types, sleeve length, and rebar embedment length on the tensile strength of grout sleeve splicing joint. Zheng et al. [4] took an experimental research on the effect of compressive strength and expansion rate of grouting material on the bond behavior of splicing joint. Gu et al. [5–7] conducted numerical researches on the failure mechanism of grout sleeve splicing joint under the action of axial tension load and discussed the influence of grouting material strength, sleeve rib space on the bond behavior of grout sleeve splicing joint. On the other hand, Kuang et al. [8] and Abdel et al. [9] carried out experimental studies and numerical studies on the mechanical behavior of grout sleeve splicing joint under the action of high-stress reversed tension and compression load and large-deformation reversed tension and compression load and put forward the effect of grout content on the tensile capacity of joint. Nerio and Fabio [10] carried out the cyclic test on a precast reinforced concrete column-to-foundation grouted duct connection with a full-scale specimen under the action of cyclic bending combined with axial compression. Ameli et al. [11] and Yan et al. [12] took experimental researches on the column-to-beam assemblies to investigate their response under cyclic quasi-static load and revealed the seismic ability of precast concrete joints constructed with grouted splice connectors in cap beam. Arthi and Jaya [13] conducted comparative study on the structural behavior of precast shear wall-diaphragm connection and the monolithic connection under seismic load.

At present, scholars have carried out sufficient research on the static performance of grout sleeve splicing joint under the action of unidirectional tensile load, but the research on its mechanical performance and failure mechanism under reversed cyclic loading is insufficient comparatively, which reflects its bond performance in precast reinforced concrete structure in the case of minor earthquakes. In this paper, the numerical research on the mechanical performance of grout sleeve splicing joint under the action of reversed cyclic loading was carried out, for providing theoretical reference to the optimal design of grout sleeve splicing joint.

2 Numerical Model of Grout Sleeve Splicing Joint

According to the actual structure of grout sleeve for connecting Φ25 mm rebars, the finite element numerical model of grout sleeve splicing joint was built by ABAQUS. For reducing the number of elements, 1/2 model in axial direction was selected, with xz plane symmetry displacement boundary conditions on the right surface of numerical model, where is the middle cross section in axial direction of grout sleeve splicing joint. In accordance with ASTM A1034/A1034M-10a [14] and JGJ 355-2015 [15], the high-stress reversed cyclic loading with the rebar stress amplitude of $[-0.5f_{yk}, 0.9f_{yk}]$ was exerted on the end of rebars, where f_{yk} is the standard value of the yield strength of rebar, and the loading cycle times is 20. Furthermore, according to the axisymmetric characteristics of the splicing joint geometry, the generatrix model was adopted to establish, and the mesh generation of generatrix numerical model is as shown in Fig. 3. In the process of building numerical model, the concrete

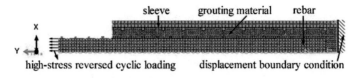

Fig. 3 Numerical model of grout sleeve splicing joint

damaged plasticity model [16] was adopted for representing the constitutive relation of grouting material, which means that tension crack and crushed destruction are the failure mechanism of grouting material. Meanwhile, the constitutive data of grouting material [17, 18], the constitutive data of HRB400 rebar and ductile iron sleeve with reference to the literatures [19–21] were adopted as their stress-strain relations, respectively.

3 Simulation and Analysis

3.1 Stress Distribution Analysis of Grouting Material

High-stress reversed cyclic loading was exerted on the end of rebars, which caused the rebar axial stress amplitude to be $[-0.5f_{yk}, 0.9f_{yk}]$ in accordance with JGJ 355-2015 and ASTM A1034/A1034M-10a mentioned in Sect. 2 that means the axial stress of rebar varies between [−200 MPa, 360 MPa]. The numerical simulation result shows that the mises stress of sleeve varies between [−75 MPa, 190 MPa] during cyclic loading, which implies that both the rebar and the sleeve are in elastic deformation stage, and no damage or stiffness degradation occurs. Therefore, we should focus on the stress distribution of grouting material, which plays a crucial role in the adhesive performance of grout sleeve splicing joint under the action of high-stress reversed cyclic loading. To explore the stress distribution of grouting material in the process of reversed cyclic loading, the minimum principal stress nephogram of grouting material was obtained when the load reaches the tensile amplitude, as shown in Fig. 4a–d corresponds to loading cycle times of 1, 5, 15, 20, respectively. It should be noted that the grouting material is in triple compressive state; therefore, the minimum principal stress of grouting material is concerned.

Figure 4 shows that with the increase of cycle times, the maximum value and its corresponding occurrence position of the minimum principal stress of grouting material are constantly varying, as shown in Fig. 5. As mentioned above, the grouting material is in triple compressive state, and its minimum principal stress is negative; therefore, the maximum value here refers to the maximum value of the absolute value of the minimum principal stress of grouting material. When the cycle times is 1, the maximum value of the minimum principal stress of grouting material is about −122.7 MPa, as shown in Figs. 4a and 5a. Its corresponding occurrence position is

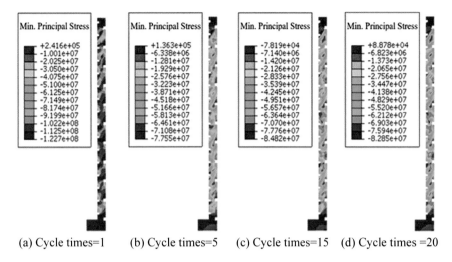

(a) Cycle times=1   (b) Cycle times=5   (c) Cycle times=15   (d) Cycle times =20

Fig. 4 Min. principal stress nephogram of grouting material at difference cycle times

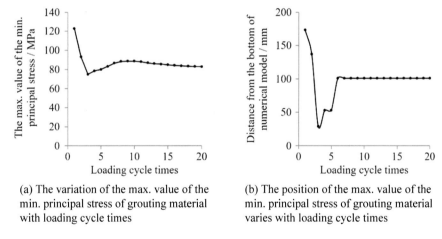

(a) The variation of the max. value of the min. principal stress of grouting material with loading cycle times

(b) The position of the max. value of the min. principal stress of grouting material varies with loading cycle times

Fig. 5 Variation of the min. principal stress of grouting material with loading cycle times

located at 173 mm away from the bottom of numerical model, as shown in Fig. 5b. As the numerical model adopts the 1/2 model in axial direction, this position corresponds to the region that contacts the third rib of the embedded reinforcement inside the sleeve from the loading end. At this time, the mechanical interaction between grouting material and rebar ribs and sleeve ribs with uneven distribution is the main mode of load transmission mechanism from rebar to sleeve, and the oblique force transfer zone in grouting material has not formed. With the increase of loading cycle times, the grouting material near the end of sleeve is constantly damaged, and stress redistribution occurs consequently. In the first 6 loading cycles, the maximum value and its corresponding occurrence position of the minimum principal stress of grouting material vary dramatically. After the sixth loading cycles, the occurrence position for the maximum value of the minimum principal stress of grouting material is stability fixed in the central regions of numerical model that is the regions of 1/4 sleeve length away from sleeve end along axial direction, where the minimum principal stress of grouting material eventually stabilizes within the range of [−85 MPa, −75 MPa], as shown in Fig. 5. Meanwhile, the obvious oblique compressive zone in grouting material is formed, which is the force transmission path from rebar to sleeve, as shown in Fig. 4d.

3.2 Hysteresis Loop of Grouting Material

According to Fig. 5, the occurrence positions of the maximum value of the minimum principal stress of grouting material when loading cycle times are 1 and 3 were selected as the analysis object that is the positions of 173 and 29 mm away from the bottom of numerical model, which corresponds to the area that contacts the third and the thirteenth ribs of the embedded rebar inside sleeve from the loading end, respectively. From Fig. 5, it can be seen that these two positions correspond to the locations of the extremum of the principal stress of grouting material. By extracting numerical simulation data at above two positions, the minimum principal stress—the minimum principal strain hysteresis loop of grouting material at the most unfavorable position can be obtained, as shown in Fig. 6. Figure 6a shows that in the early stage of reversed cyclic loading, in the area near the end of sleeve, the mechanical interaction between grouting material and rebar ribs and sleeve ribs is

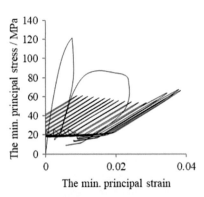

(a) The stress-strain hysteresis loop of grouting material at the 3rd rib of embedded rebar from the end of sleeve

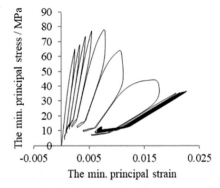

(b) The stress-strain hysteresis loop of grouting material at the 13th rib of embedded rebar from the end of sleeve

Fig. 6 Minimum principal stress—the minimum principal strain hysteresis loop of grouting material at the most unfavorable position

the main mode of load transmission mechanism from rebar to sleeve, and the fracture damage in grouting material in the area near the end of sleeve is the main mode of energy dissipation. After two loading cycles, the elastic modulus of grouting material in the area near the end of sleeve rapidly decreases to a stable value, and no longer has energy dissipation capacity. Between the second and the seventh loading cycles, through the stress redistribution of grouting material, the grouting material in the area far from the end of sleeve gradually undertakes the energy dissipation function with full hysteresis loop, as shown in Fig. 6b. With the development of fracture damage, the elastic modulus of entire grouting material decreases to a stable value with no capacity of energy dissipation, and the final residual strain of entire grouting material is about 0.005.

According to the variation role of stress distribution of grouting material with loading cycle times, non-uniform sleeve rib spacing can be considered in the design of sleeve. Along the axial direction from the sleeve end into the center of sleeve, the sleeve rib spacing should be designed from sparse to dense gradually, which could reduce the fracture damage level of grouting material in the region close to sleeve end in the early stage of reversed cyclic loading. On the other hand, the sleeve rib spacing in the area near the end of sleeve should not be too sparse. The moderate sleeve rib spacing can provide better constrains on grouting material in triple compressive state and make grouting material play its role well in load transmission. Furthermore, in order to eliminate the stress concentration of the grouting material in contact with the sleeve rib, the sleeve rib should be designed in the form of chamfering, which guarantees the more stable workability of grout sleeve splicing joint and improves the seismic resistance performance of precast reinforced concrete structure.

4 Conclusions

In this paper, the numerical model of grout sleeve splicing joint was established. Under the action of high-stress reversed cyclic loading exerted at the end of rebars, the stress distribution variation of grouting material with loading cycle times was gotten, and numerical simulation result shows the conclusions as follows.

1. Under the action of reversed cyclic loading exerted on the end of rebars, with the rebar axial stress amplitude of $[-0.5f_{yk}, 0.9f_{yk}]$, both rebar and sleeve are in the elastic deformation stage, and no damage or stiffness degradation occurs. With the increase of loading cycle times, the maximum value and its corresponding occurrence position of the minimum principal stress of grouting material are constantly varying.

2. In the early stage of reversed cyclic loading, in the area near the end of sleeve, the mechanical interaction with uneven distribution between grouting material and rebar ribs and sleeve ribs is the main mode of load transmission mechanism from rebar to sleeve, and the fracture damage in grouting material near the end of sleeve is the main mode of energy dissipation. With the increase of loading

cycle times, through the stress redistribution of grouting material, the grouting material in the area far from the end of sleeve gradually undertakes the energy dissipation function by fracture damage, and the obvious oblique compressive zone in grouting material is formed, which is the force transmission path from rebar to sleeve. Meanwhile, it decreases the elastic modulus of entire grouting material to a stable value.

3. According to the stress distribution and its variation role of grouting material during reversed cyclic loading, some aspects should be considered in the sleeve design process. The sleeve rib should be designed in the form of chamfering to eliminate the stress concentration of the grouting material in contact with the sleeve rib. The moderate sleeve rib spacing should be concerned to provide better constrains on grouting material in triple compressive state and make grouting material play its role well in load transmission. From the sleeve end into the center of sleeve along the axial direction, the sleeve rib spacing should be designed from sparse to dense gradually to reduce the fracture damage level of grouting material in the region close to sleeve end in the early stage of reversed cyclic loading.

Acknowledgements This research work is supported by Scientific Research Project of Liaoning Science and Technology Department (Project Codes: 2015020600) and the Sub-project III under National Key Technology Research and Development Program of the Ministry of Science and Technology of China (Project Codes: 2018YFD1100403-03).

References

1. Ali, A.S., Ahmad, B.A.R., Mohd, Z.B.J., Johnson, A., Sayadi, A.: The relationship between interlocking mechanism and bond strength in elastic and inelastic segment of splice sleeve. Constr. Build. Mater. **55**(31), 227–237 (2014)
2. Ahmad, B.A.R., Ling, J.H., Zuhairi, A.H., et al.: Development of splice connections for precast concrete structures. Adv. Mater. Res. **980**(1), 132–136 (2014)
3. Ling, J.H., Ahmad, B.A.R., Izni, S.I.: Feasibility study of grouted splice connector under tensile load. Constr. Build. Mater. **50**(15), 530–539 (2014)
4. Zheng, Y.F., Guo, Z.X., Zhang, X.: Effect of grout properties on bond behavior of grouted pipe splice. KSCE J. Civ. Eng. **22**(1), 2951–2960 (2018)
5. Gu, F., Xian, A.K., Zhao, W.J., Wei, W.: Influence of grouting-material compressive-strength on the mechanical properties of grouting-sleeve reinforcement-connection component. Adv. Eng. Res. **40**(1), 87–90 (2015)
6. Gu, F., Zhang, P., Zhao, W.J., Zhang, D.: Numerical study on the static response of grouting-sleeve reinforcement-connection component with different sleeve-rib space. Appl. Mech. Mater. **578**(1), 882–885 (2014)
7. Gu F., Song, D.K., Ming Y., Zhang, Y.Y.: Research on fire resistance of steel sleeve grouting connector. J. Shenyang Jianzhu Univ (Natural Science) **35**(3), 420–427 (2019) (in Chinese)
8. Kuang, Z.P., Zheng, G.Y., Jiao, X.T.: Experimental study on effect of mechanical behavior of grout sleeve splicing for reinforced bars due to lack of grout. J. Tongji Univ. (Nat. Sci.) **47**(7), 946–956 (2019) (in Chinese)
9. Abdel, J.M., Beale, R.G., Shatarat, N.K.: Cyclic loading applied to sleeve couplers for tube and fitting scaffolds. ACMSM25. Lect. Notes Civ. Eng. **37**, 819–830 (2019)

10. Nerio, T., Fabio, M.: Cyclic test on a precast reinforced concrete column-to-foundation grouted duct connection. Bull. Earthq. Eng. **18**(1), 1657–1691 (2020)
11. Ameli, M.J., Parks, J.E., Brown, D.N., Pantelides, C.: Seismic evaluation of grouted splice sleeve connections for reinforced precast concrete column-to-cap beam joints in accelerated bridge construction. PCI J. **60**(2), 80–103 (2015)
12. Yan, Q.S., Chen, T.Y., Xie, Z.Y.: Seismic experimental study on a precast concrete beam-column connection with grout sleeves. Eng. Struct. **155**(15), 330–344 (2018)
13. Arthi, S., Jaya, K.P.: Seismic performance of precast shear wall-slab connection under cyclic loading: experimental test vs. numerical analysis. Earthq. Eng. Eng. Vib. **19**(1), 739–757 (2020)
14. ASTM Designation, ASTM A1034/A1034M-10a: Standard test methods for testing mechanical splices for steel reinforcing bars. Am. Soc. Test. Mater. (2015)
15. People's Republic of China Industry Standard, JGJ 355–2015: Technical specification for grout sleeve splicing of rebars. Ministry of Housing and Urban-Rural Development of the People's Republic of China (2015) (in Chinese)
16. ABAQUS: ABAQUS analysis user's manual. US, ABAQUS Inc. (2003)
17. Nobuyuki I., Satoshi H., et al.: Research and development of super-high strength precast reinforced concrete column. Toda Constr. Tech. Res. Rep. **31** (2002) (in Japanese)
18. Hiroshi T.: Experimental study of ultra-high strength precast reinforced concrete columns. Concr. Inst. Proc. **24**(2), 727–732 (2002) (in Japanese)
19. Yu, K.L., Guang, M.S., Hao, F.X.: High strain low cycle fatigue and anti-seismic behavior of HRB400 QST reinforced steel bars. Adv. Mater. Res. **250**(1), 1128–1133 (2011)
20. Wang, Y., Zheng, W.Z.: Some cognition to HRB400 steel. Low Temp. Arch. Technol. **24**(2), 7–9 (2002) (in Chinese)
21. Liu, J.H., Li, G.L., He X.Y., Zhao, X.B., Liu, G.S.: Correlation between matrix and tensile behavior of ductile cast iron. Foundry Tech. **30**(3), 329–332 (2009) (in Chinese)

Self-test of Athletic Ability for the Elderly Using Inertia Motion Capture Device

Jun Sun, Donghua Li, and Lianjie Lv

Abstract In order to detect gait by the elderly themselves at home and apply gait rules on pension industry, we designed experiments and used inertia motion capture device, Perception Neuron Pro, to capture motion data. In this paper, it tested the subject by Functional Ambulation Performance criterion, CALC data, and BVH were used data to increase the data extraction speed. Thirdly, we designed the process of extracting gait parameters, and it investigated the law of lower limb movement and the correlation between step length and lower limb movement. In this way, the elderly can know their physical condition and changes exactly. Accordingly, the elderly can take appropriate countermeasures timely.

Keywords Motion capture · Gait parameter · Self-test

1 Introduction

In 2020, the elderly population in China has reached 248 million, makes up almost 17.2% of total population. This proportion will increase to 27.4% in 2050 [1]. As the declining physical strength and energy, it is in danger of causing stroke, hemiplegia, or even mortality to the elderly. Meanwhile, due to the sub-replacement fertility and increasing life expectancy, the elderly issue will become a major social problem in China [2]. Therefore, establishing a system is urgent for the elderly to do self-checking and self-training.

J. Sun (✉) · D. Li · L. Lv
School of Mechanical Engineering, Shenyang Jianzhu University, Shenyang, China
e-mail: sy_sunjun@163.com

In recent years, motion capture technology has been wildly applied in the pension industry. Eichler et al. [3] applied multiple depth cameras on evaluating movement of stroke patients. Hyunkyu et al. [4] combined Kinect and vibration feedback glove and set up a fun and enjoyable dance rehabilitation system. Rueangsirara [5] achieved the recognition of abnormal gait by two Kinects. Guimaraes et al. [6] used two shoe-mounted inertial sensors to achieve real-time step detection. Miyawaki et al. [7] evaluated if the homemade walking aid device is useful for the elderly with Vicon motion analysis device. Optical motion capture devices, like Vicon and Optitrack, have high requirements for light and environmental condition, and the reflective markers have to be attached on subject's body. These are not convenient for domestic use. Markless optical motion device is economical and practical, but one Kinect can only record one side of patient body.

The inertia sensor is not only portable and user-friendly, but also has good environmental adaptability and accuracy. Using inertial motion capture equipment, the elderly can get gait data at home. Once the gait data is abnormal, it will show that there is something wrong with body. The elderly can pay more attention to their behavior and seek medical aid in time if necessary. More than this, these data can be provided to professional physicians as objective physical condition to diagnose the recovery of patient, who got hemiplegia or injured in sports. Surgeons can provide a personalized rehabilitation program for each patient, which can improve the efficiency of rehabilitation and avoid secondary damage.

2   Perception Neuron Pro

Perception Neuron Pro motion capture suit is equipped with 17 sensors. Each sensor is a 9-axis inertial sensor, which is integrated with triaxial gyroscope, accelerometer, and electronic compass. Before capturing motion, we only need to wear them in position as shown in Fig. 1a. After that, the supporting software, Axis Neuron Pro, can calibrate the sensors and show human movement in real time as shown in Fig. 1b.

3   Experiment of Human Walk Capture

In order to get the lower limb motion data, human gait parameters, and the correlation between step length and the lower limb joint motion in sagittal plane, we took a series of experiments. Firstly, functional amplitude performance (FAP) test was conducted on a male subject which is 175 cm height and 75 kg weight to ensure the athletic ability is at average level [8]. According to the score criteria of FAP test, the subject's functional amplitude performance score is 98 out of 100, which is nearly perfect. Then, we asked the subject to walk in steps of 20 cm, 30 cm, 40 cm, 50 cm, and 60 cm, respectively, and capture the walking.

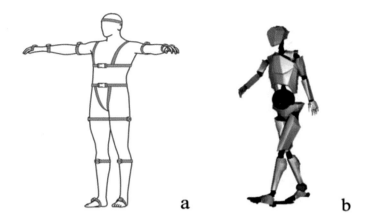

Fig. 1 Perception Neuron Pro suit

4 Extracting and Analysis of Motion Data

4.1 Extraction of Human Gait Parameters

Axis Neuron Pro has a numbered index for each joint as shown below in Fig. 2. We can extract each joint data, such as position, angle, velocity, and acceleration, by the index from data file. We extracted and analyzed the velocity and position of left ankle in sagittal plane and got a method of extracting gait parameters. Figure 3 shows the position and velocity trajectory of the ankle joint. We found four feature points within one cycle. The leg starts to swing at T_a. Swing foot is at its height at T_b. The leg stops swing at T_c. The next swing period starts at T_d. According to the flowchart as shown in Fig. 2, T_a, T_b, T_c, and T_d can be extracted. Step length is the displacement of the ankle which travels from T_d to T_a. The position of the ankle at

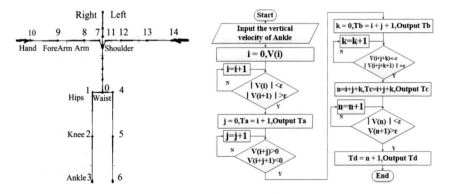

Fig. 2 Human model and flowchart of extracting feature moments

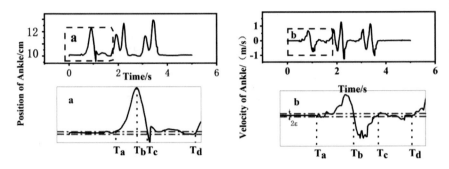

Fig. 3 Position and velocity trajectory of ankle joint

T_b is the step height. The time between T_a and T_c is swing period. The time between T_c and T_d is supporting period. Time between T_a and T_d is the whole gait period. Using the data at these four moments, we can get the human walking parameters, includes swing period, support period, gait period, step length, and step height.

We found that the step length and velocity of first step are significantly different from the normal steps. The reason was found by analyzing trajectories on Figs. 8, 9, and 10. At the beginning of first step, the extension of hip and knee joint are not as intense as normal steps. The ankle joint does not flex to make forefoot pedal on the ground. The speed is low without power provided by pedaling. It shows that the body takes a conservative step strategy at first step.

4.2 Motion of Lower Limb Joints and the Center of Mass

Motion of Center of Mass Motion trajectory of the center of mass is shown in Fig. 4. At position A, the center of mass is at its height. At this position, the center of mass is located directly above the support feet. At position B, the center of mass fell to its lowest. The projection of center of mass in the vertical direction is in the middle of the two supporting feet. We get the following conclusions about position A and B. At position A, the velocity is the smallest in the forward direction, while the lateral velocity is zero. The acceleration in the forward direction is the smallest and the acceleration in the lateral direction is zero. At position B, the velocity is the largest in the forward direction but is zero in the lateral direction. The acceleration is zero in the forward but the largest in the lateral direction.

This is because at position A, one leg of the subject stood upright for supporting, the other one prepared to swing. The center of mass is at the highest point. The center of mass only bears gravity and the reaction force from ground. At position B, the subject had stepped forward and prepared to retract the last supporting foot. The resultant force is zero.

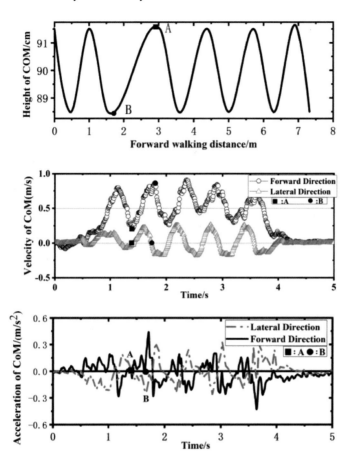

Fig. 4 Position, velocity, and acceleration trajectory of CoM

Motion of Lower Limb Joints—Walk is driven by the joints of the lower limbs. Therefore, we did analysis on the angular trajectories of the hip joint, knee joint, and ankle joint. These trajectories can help the elderly understand their own walking habits better and provide reference walking patterns. Sports anatomy divided the movement plane into coronal plane, horizontal plane, and sagittal plane, as shown in Fig. 5. The rotation of joints in the sagittal plane is defined as flexion and extension. We analyzed the angular trajectories of the extension and flexion of each lower limb joint in the sagittal plane.

In the angle trajectory of the hips, as shown in Fig. 6, the angle of extension is much larger than the flexion. At the beginning of step, hip joint of swing leg extends forward actively. As the swing leg moves forward, the support leg flexes passively, resulting that the angle of flexion is much smaller than the extension. The knee joint can only flex due to joint movement restrictions. Due to this, the flexion angle of knee in Fig. 6 is always above zero. The flexion of knee joint makes the shank of

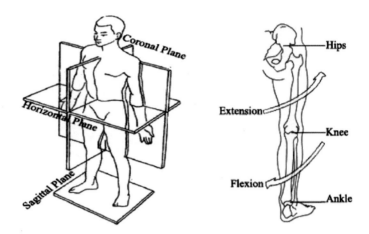

Fig. 5 Anatomical terms in human movement

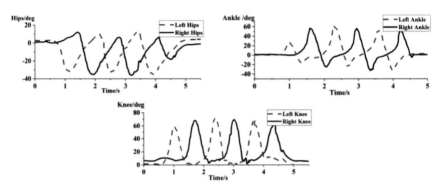

Fig. 6 Angle trajectories of lower limbs joints

the swing leg be perpendicular to the ground before landing, which increases the contact area to improve landing stability. At the beginning of the step, in order to provide power for walking, the ankle joint flexes to make the heel off the ground so that the sole of the foot can pedal on the ground to provide power. After the swing foot leaving the ground, the ankle joint begins to stretch to prepare for the heel to land first. Because the knee joint adjusts the position of the lower leg in advance to make the leg as vertical as possible to the ground, the ankle joint does not need a lot of flexion to make the sole of the foot parallel to the ground during the process of landing. That is why the flexion angle of ankle joint is larger than that of extension in Fig. 6.

Correlation between Step Length and Angle of Lower Limb Joints—In order to explore the influence of step length on the motion of lower limb joint, we carried out five sets of walking capture experiments with different step length and analyzed how the motion of center of mass and the rotation of lower limb joints are affected

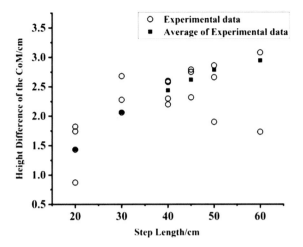

Fig. 7 Relationship between step length and height difference of the COM

by step length. As shown in Fig. 7, the displacement of center of mass in the sagittal plane increases with the increasing of step length. After second-order polynomial fitting, the formula of height difference trajectory is as follow:

$$H = -0.10276 + 0.09114 * L + 0.00067 * L^2 \tag{1}$$

where H is the height difference of the center of mass and L is the target step length.

The correlation coefficient between the step length and the height difference of the centroid is 0.9743. This shows that as the step length increasing, the height difference of the center of mass increases too. Once disturbances occur, falling are easy to occur when height difference get big. In order to prevent falling, the elderly should not take a big step at risk but keep a small step length in daily walking.

We got the angle trajectories of each lower limb joint in different step length. Observing Figs. 8a, 9, and 10a, the maximal angle of extension increased with the increasing of step length. Figures 8b and 10b show that the minimal angle of flexion decreased with the increasing of step length. In order to understand the regularity quantitatively, the correlation coefficient of the step length with joint angle (the maximum of extension and the minimum of flexion) was calculated. Then, the curve which shows the relationship between step length and extreme joint angle was fitted by the second-order polynomial. The correlation coefficient and the curve formula are shown in Table 1.

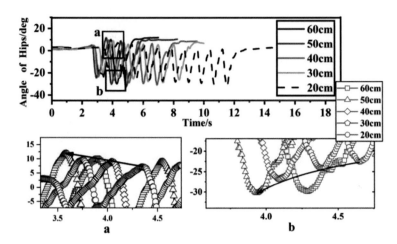

Fig. 8 Angle trajectories of hips in a different step length

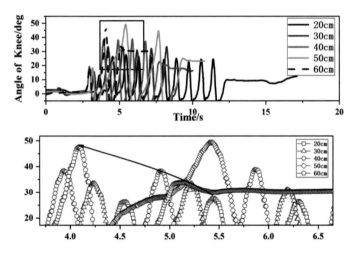

Fig. 9 Angle trajectories of ankle in a different step length

5 Conclusions

The inertial motion capture device can help the elderly proceed self-test of athletic ability at home. Gait parameters and the trajectory of lower limb joints cannot only make the elderly understand their daily behavior quantitatively, but also can be used as criterion of physical condition. For the elderly, they can realize what kind of exercise is beyond their athletic ability and prevent injury from not doing these. For surgeons, they can get stroke or hemiplegia patients' physical condition by gait parameters and the trajectory of lower limb joints to take appropriate countermeasures. For further work, more behaviors are desired to be captured and analyzed. Integration

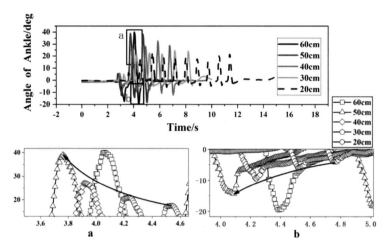

Fig. 10 Angle trajectories of ankle in a different step length

Table 1 Correlation coefficient between step length and amplitude of joint rotation angle and fitting curve about the relationship between step length and amplitude

Subject	Correlation coefficient	Fitted curve
Hips flexion	0.9871	$\Phi_{HipsFl} = 3.28 + 0.182 * L + 0.0026 * L^2$
Hips extension	−0.9828	$\Phi_{HipsEx} = -22.75 + 0.1165 * L - 0.0533 * L^2$
Knee flexion	0.9941	$\Phi_{KneeFl} = 13.09 + 0.5083 * L + 0.0042 * L^2$
Ankle flexion	0.9649	$\Phi_{AnkleFl} = 24.48 - 0.7672 * L + 0.0211 * L^2$
Ankle extension	−0.9389	$\Phi_{AnkleEx} = -11.69 + 0.5442 * L - 0.0119 * L^2$

with more medical knowledge, the self-test and self-diagnose will be more scientific and authoritative.

References

1. McLaughlin, S.J., Chen, Y., Tham, S.S.X., Zhang, J., Li, L.W.: Healthy aging in china: benchmarks and socio-structural correlates. Res Aging **42**(1), 23–33 (2020)
2. Ko, P.C., Yeung, W.-J.J.: Childhood conditions and productive aging in China. Soc. Sci. Med. 229, 60–69 (2019)
3. Eichler N., Hel-Or, H., Shmishoni, I., Itah, D., Gross, B., Raz, S.: Non-invasive motion analysis for stroke rehabilitation using off the shelf 3D sensors. In: 2018 International Joint Conference on Neural Networks (IJCNN), pp. 1–8. IEEE, Rio de Janeiro, Brazil (2018)
4. Park, H., Lee, J., Bae, J.: Development of a dance rehabilitation system using kinect and a vibration feedback glove. In: 2015 15th International Conference on Control, Automation and Systems (ICCAS), pp. 1878–1880. IEEE, Busan, South Korea (2015)

5. Rueangsirarak, W., Uttama, S., Kaewkaen, K., Hubert Shum, P.H.: Identifying abnormal gait in older people during multiple-tasks assessment with audio-visual cues. In: 2018 15th International Conference on Electrical Engineering/Electronics, Computer, Telecommunications and Information Technology (ECTI-CON), pp. 780–783. IEEE, Chiang Rai, Thailand (2018)
6. Guimarães, V., Sousa, I., Correia, M.V.: Detection and classification of multidirectional steps for motor-cognitive training in older adults using shoe-mounted inertial sensors. In: 2019 41st Annual International Conference of the IEEE Engineering in Medicine and Biology Society (EMBC), pp. 6926–6929. IEEE, Berlin, Germany (2019)
7. Miyawaki, K., Saito, R., Saito, A., Kobayashi, Y., Kizawa, S., Obinata, G.: Development and evaluation of an electric walking machine. In: 2018 International Symposium on Micro-NanoMechatronics and Human Science (MHS), pp. 1–7. IEEE, Nagoya, Japan (2018)
8. Nelson, A.J.: functional ambulation profile. Phys. Ther. **54**(10), 1059–1065 (1974)

A Method Based on CNN + SVM for Classifying Abnormal Audio Indoors

Jian Liu, Shuyan Ning, Sanmu Wang, Jiarui Yi, and Mingrui Zhao

Abstract In this paper, a novel classification algorithm combined convolutional neural network with support vector machine (CNN + SVM) is proposed for classifying abnormal audio indoors to solve the emerging problems, for which video surveillance may have obstacles, blind spots and therefore cannot protect the privacy under the scenarios. First of all, in the experiments, the quality of the audio signal is improved by pre-emphasis, framing, and windowing methods. Secondly, to obtain sufficient audio information, Mel frequency cepstral coefficient is selected as a parameter for feature extraction. Lastly, multilayer perceptron (MLP), convolutional neural network (CNN), support vector machine (SVM), and CNN + SVM are used to classify eight types of audio signals according to the complexities of the indoor environment. The result of the proposed experiments indicates that the CNN + SVM combination algorithm exhibits a higher accuracy rate for the classification of audio compared to that of the traditional single classification algorithm. It outperforms other methods for indoor abnormal audio classification in terms of applicability.

Keywords Abnormal sound · Convolutional neural network · Support vector machine · Classification

J. Liu (✉) · S. Ning (✉) · M. Zhao
Faculty of Information and Control Engineering, Shenyang Jianzhu University, Shenyang, China
e-mail: jeanliu10@163.com

S. Ning
e-mail: 2244763409@qq.com

S. Wang
Shenyang Sanfeng Electricity Co., Ltd., Shenyang, China

J. Yi
Department of Electrical and Computer Engineering, University of Texas at El Paso, El Paso, TX, USA

1   Introduction

Most recently, with the improvement of people's living standards, people have paid increasing attention to the safety for their living environment. Many people have installed the video surveillance system for elder people, infants, and so on to prevent accidents. Traditional video surveillance has some problems such as blind spots, object occlusion, and insufficient light [1, 2]. Besides, it cannot protect personal privacy. As the development of digital audio technology, the acquisition and transmission technology of audio signals has become increasingly advanced. Many researchers dedicate to study the understanding of environmental sounds by machines. Feature extraction and classification of audio signals attract attentions from various research fields in auditory scene analysis at present [3]. Chen classifies the environmental sounds under the city scenario [4]. Xin and Chen focused on four types of audio files, which are explosions, gunshots, alarm sounds, and human voices (calls for help) for feature extraction and classification [5]. Some other researchers classify music types according to different genres [6].

Abnormal sounds can effectively reveal and characterize abnormal situations, so audio monitoring can be set up indoors for abnormal audio detection. At present, there are many choices for audio recognition and classification. The traditional ones include Gaussian mixture model (GMM), Naive Bayes (NB), support vector machine (SVM), etc. However, these methods have low accuracy for the audio classification of complex scenes. In recent years, deep learning algorithms such as deep neural networks (DNN) and convolutional neural networks (CNN) have become a popular research trend [7, 8]. Therefore, a novel classification algorithm combined convolutional neural network with support vector machine (CNN + SVM) model is proposed for improving the accuracy of the indoor audio classification and abnormal situations detection in time based on deep learning in this paper.

2   CNN + SVM

The CNN + SVM is a combined classification algorithm by using the convolutional layer and pooling layer in the CNN to further extract and filter the audio feature parameters and inputting them as new features into the SVM for classification before the full connection. The combined model is shown in Fig. 1. The part that should be fully connected is sent to the SVM as a new feature vector.

The procedure of the CNN + SVM combination algorithm is stated as follows:

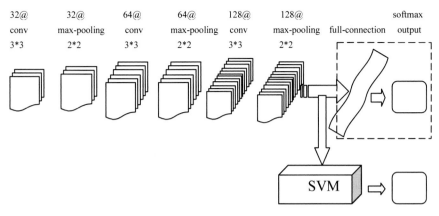

32@	32@	64@	64@	128@	128@		softmax
conv	max-pooling	conv	max-pooling	conv	max-pooling	full-connection	output
3*3	2*2	3*3	2*2	3*3	2*2		

Fig. 1 Combined classification model

Algorithm: CNN + SVM

Step 1: Sort out the processed audio feature data;
Step 2: Establish a CNN model;
Step 3: Send some audio data into CNN for training;
Step 4: Save the trained CNN model;
Step 5: Send the other audio data into the CNN model for
Step 7: Send the new feature vectors into SVM for training;
Step 8: Use the test set data to obtain classifi-cation results
          and calculate the accuracy in SVM.

2.1 CNN

CNN is the most widely used deep learning model by far. The hidden layers of CNN include a convolutional layer, a pooling layer, and a fully connected layer. The convolutional layer can perform further feature extraction on the extracted audio feature parameters. The pooling function uses the overall characteristics of the adjacent data at a certain location as the output of the CNN at that location [9]. Through the pooling layer, we can reduce the data dimension and keep the overall characteristics of the data unchanged. Therefore, the calculation process of the pooling layer is also called down-sampling. The data is continuously convolution and pooling. Then finally enter the fully connected layer to transform the multi-dimensional vector into a one-dimensional vector and uses the softmax loss function for classification output when passing through the last fully connected layer.

Loss function is also known as cost function. The process of deep learning training is the process of minimizing the loss function. The value of the loss function represents the degree of agreement between the predicted value of the CNN model and the label value and also represents the accuracy of the actual prediction process of the model [10]. The gradient of the loss function during the training process must be large enough and predictable. The softmax loss function is:

$$L_i = -\log\left(\frac{e^{f_{y_i}}}{\sum_j e^j}\right) \tag{1}$$

In Eq. (1), it is supposed that we have an array y, y_i represents the ith element in y, and the value in the log is the ratio of the exponent of the element ($e^{f_{y_i}}$) to the exponent of the sum of the elements ($\sum_j e^j$). It represents the softmax value of the correct classification of this group of data. The larger the proportion of it, the smaller the loss and the system meets the requirements.

2.2 SVM

Support vector machine (SVM) algorithm is to find a hyperplane to maximize the margin of the two types of data and to well divide the classified data. The formula for confirming the hyperplane is shown in Eq. (2). w^* is the normal vector, which determines the direction of the hyperplane, and b^* is the displacement, which can determine the distance between the hyperplane and the origin.

$$f(x) = \text{sgn}(w^*x + b^*) = \text{sgn}\left\{\sum_{i=1}^{n} a_i^* y_i * (x_i x) + b^*\right\} \tag{2}$$

The support vectors are the vectors closest to the classification line or the classification hyperplane. They are generated by training and play a decisive role in classification decision-making. Therefore, these key support vectors can be directly relied on for sample classification and eliminating other useless information. At the same time, the computational complexity of SVM will be greatly reduced, avoiding a series of problems caused by excessive sample dimensions. For multi-classification problems, SVM can be used to construct multiple classifiers. The construction methods are mostly indirect, namely one-versus-rest (OVR SVMs) and one-versus-one (OVO SVMs). OVO constructs an SVM between any two types of samples, so (n(n-1))/2 SVMs need to be designed for n types of samples [11]. OVR is used in this paper. During OVR training, samples of a certain category are classified as positive training samples, and the remaining samples are classified into another category as negative training samples, and so on so those n categories of samples are constructed N SVMs [12].

3 Audio Processing and Feature Extraction

3.1 Audio Processing

To find the spectrum with the same signal-to-noise ratio in high and low frequencies, pre-emphasis is used to increase the high-frequency part and highlight the high-frequency resonance peak. Generally, pre-emphasis is performed through a high-pass filter, and the equation is shown below, where a is the pre-emphasis coefficient, usually $0.9 < a < 1.0$.

$$H(z) = 1 - az^{-1} \tag{3}$$

Since the audio signal is not stable to make the input signal smooth and continuous when the Fourier transform is performed, frame processing is adopted. The framing usually adopts overlapping framing to avoid excessive changes between two adjacent frames. The overlap between the previous frame and the next frame of audio data is called frameshift. The length of each frame of audio data is called the frame length. According to the short-term stability of the voice signal, the frame length is generally taken as 10–30 Ms. The ratio between frame length and frameshift is generally 2–50.

To highlight the voice waveform near n, the windowing method is used to multiply the corresponding element $s'()$ in each frame by the window function $w()$. which is:

$$s(n) = s'(n) * \omega(n) \tag{4}$$

The Hamming window protects high-frequency information better and preserves the details of audio signal data [13]. Therefore, it is adopted the Hamming window. The Hamming window function formula is shown in (5). Where N is the window length, and the width of the window function is the frame length.

$$w(n) = \begin{cases} 0.54 - 0.46\cos[2\pi n/(N-1)], 0 \le n \le N-1 \\ 0, n = \text{other} \end{cases} \tag{5}$$

3.2 Feature Extraction

Choosing appropriate feature parameters is the key to ensuring the accuracy of classification. At present, in feature extraction, the commonly used features include linear prediction coefficient (LPC), linear prediction cepstrum coefficient (LPCC), and Mel frequency cepstrum coefficient (MFCC). MFCC has better robustness and noise resistance [14]. The operation process is shown in Fig. 2. MFCC is a parameter made according to the human auditory system. The actual unit of audio data is Hz. The critical frequency bandwidth increases with the increase of frequency, which is

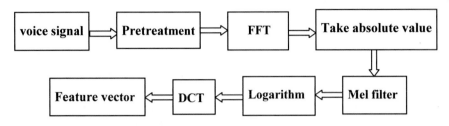

Fig. 2 MFCC principle process

consistent with the increase of Mel frequency. The speech frequency can be divided into a series of triangular filter sequences, and the Mel filter bank generally uses triangular bandpass filters.

The Mel frequency roughly corresponds to the logarithmic distribution of the actual frequency. The corresponding relationship between the Mel frequency Mel(f) and the actual frequency f is shown in the following:

$$Mel(f) = 2595 \lg(1 + f/700) \tag{6}$$

4 Experiment and Analysis

Considering some special sounds that may appear in the indoor environment, Six groups of signals are filtered out as abnormal audio classification objects in AudioSet, they are crying, screaming, dog barking, painful groaning, sneezing, and breaking glass. At the same time, the sounds of speaking and clocks are selected as normal audio classification objects. The waveform diagrams of the eight audio frequencies are shown in Fig. 3. Each group has 400 samples, 70% of which are used as training samples, and 30% are used as test samples. The original audio is in the Ogg format. To support the feature extraction algorithm, it is converted to wave format by using a fast audio converter.

After MFCC feature extraction, the audio time-domain waveform signal is converted into a spectrogram of frequency-domain features, as shown in Fig. 4. The spectrogram can more clearly show the feature information of the audio signal and is fully prepared for the subsequent classification of the classifier. The brighter area in the figure is the area with more obvious features.

In this experiment, MLP, CNN, SVM, and CNN + SVM are used to classify audio features to compare the performance of classification. First, multilayer perceptron (MLP) is used for classification. As a classification function, MLP is to map multiple input data sets X to category output O [15]. MLP contains an input layer and an output layer, with multiple hidden layers in the middle, and the neurons between the layers

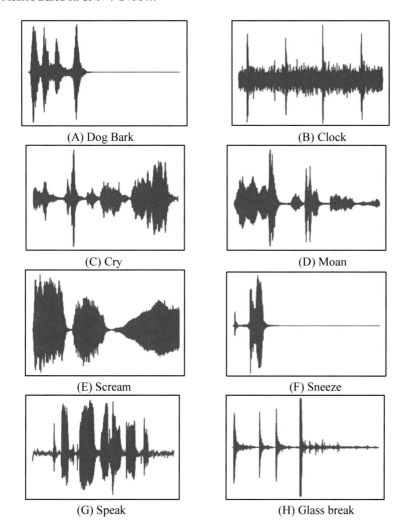

(A) Dog Bark (B) Clock

(C) Cry (D) Moan

(E) Scream (F) Sneeze

(G) Speak (H) Glass break

Fig. 3 Audio waveform graph

are fully connected. The output function f of the hidden layer adopts the sigmoid function. The output of the output layer is softmax logistic regression.

The structure of CNN is shown in Table 1, a combination of a three-layer convolution layer and a maximum pooling layer max-pooling, as well as a fully connected layer and an output layer. The convolution kernels of the convolution layer are all 3 * 3 in size, and the core size of the pooling layer is 2 * 2. To prevent over-fitting, the dropout parameter is 0.5.

SVM is constructed using the OVR method to realize eight categories. First, set crying, screaming, dog barking, painful moaning, sneezing, breaking glass, speaking, and clocks when training eight groups of training samples as A, B, C, D, E, F, G,

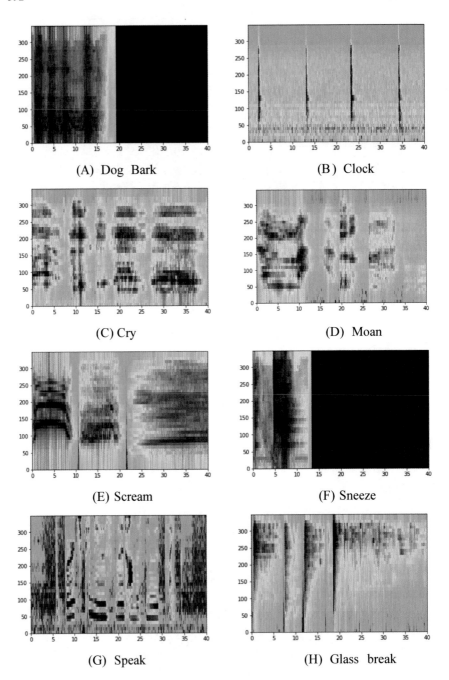

Fig. 4 Spectrogram of audio features during the experiment

Table 1 CNN structure

Neural network layer	Nuclear size	Number of feature maps
Input layer		
Convolutional layer	3 × 3	32
Maximum pooling layer	2 × 2	32
Convolutional layer	3 × 3	64
Maximum pooling layer	2 × 2	64
Convolutional layer	3 × 3	128
Maximum pooling layer	2 × 2	128
Fully connected layer		

Table 2 Comparison of accuracy of different classification methods

Classification	Accuracy (%)
MLP	65.42
CNN	72.83
SVM	83.21
CNN + SVM	91.24

H. A is a set of positive sets, the rest is a set of negative sets, and so on to get eight sets of training results. Then, test through these sets of training results and take the largest value of each test result as the final classification result.

The experimental results of using MLP, CNN, SVM, and CNN + SVM are shown in Table 2.

It can be seen from the experimental results that the accuracy of CNN is higher than that of MLP. During the experiment, due to too many parameters in MLP, it is easy to cause over-fitting and also poor training. However, due to the presence of convolutional layers on CNN, it greatly reduces the parameters and improves accuracy. CNN has a lower accuracy rate than SVM, mainly because of insufficient data in the experimental data set. The effect of using the CNN + SVM combined classification model is better than the single model and its accuracy rate is 8% higher than that of the SVM alone. It can be seen that the use of the CNN + SVM combined classification model has good applicability for abnormal audio classification.

5 Conclusions

In this paper, an abnormal sound detection research is performed for indoor monitoring security issues. In terms of feature extraction, MFCC is used as an audio feature parameter, which exhibits outstanding stability and robustness. In terms of the audio features classification, CNN + SVM combined model is proposed, which has a classification accuracy of 91.24% for the eight sound categories selected in the experiment. According to experiments, the combined classification model CNN + SVM has a higher accuracy rate than that of the single classification model and can better identify and classify abnormal sounds. However, due to the insufficient amount of data in the experimental data set, the deep learning algorithm does not fully exert its advantages. In future research work, it is expected that the data set will be expanded, more research will be focused on the training and learning number, model optimization, and classification accuracy.

Acknowledgements This research work is partially supported by the Natural Science Foundation of Liaoning Province, China (Project Codes: 201602616), Jiangsu Province Elevator Intelligent Safety Key Construction Laboratory Project, China (Project Codes: JSKLESS201707).

References

1. Li, L., Li, Y.: Design and Implementation of Sound Source Localization in Intelligent Video Surveillance System. Comput. Simul. **24**(9), 378–381 + 401 (2013) (in Chinese)
2. Liu, G., Zhang, Z.Z.: An intelligent video surveillance system with auditory function. Telev. Technol. **27**(1), 164–167 (2014) (in Chinese)
3. Mesaros, A., Heittola T.V.: Scene classification, an overview of case 2017 challenge entries. In: 2018 16th International Workshop on Acoustic Signal Enhancement (IWAENC) (2018)
4. Chen, B.: Research on Urban sound classification model based on deep neural network. J. Zhejiang Univ. Technol. **47**(2), 199–203 (2019) (in Chinese)
5. Xin, X., Chen, S.D.: Audio classification algorithm using latent probabilistic semantic model and K-nearest neighbor classifier. J. Huaqiao Univ. (Nat. Sci Ed.) **21**(3), 196–200 (2016) (in Chinese)
6. Fu, W., Yang, Y.: Audio classification method based on convolutional neural network and random forest. Comput. Appl. **38**(S2), 58–62 (2018) (in Chinese)
7. Lu, L.: A SVM-based audio event detection system. Int. Conf. Electr. Control Eng. **27**(9), 67–78 (2010)
8. Parascandolo, G., Huttunen, H., Virtanen, T.: Recurrent neural networks for polyphonic sound event detection in real life recordings. IEEE Trans. Acoustics, Speech Sig. Proc. (ICASSP) 6440–6444 (2016)
9. Mesaros, A., Heittola, T., Virtanen, T.: TUT database for acoustic scene classification and sound event detection. Sig. Proc. Conf. (EUSIPCO). European (2016)
10. Mafra, G.S., Duong, N.Q.K., Ozerov, A., Pérez, P.: Acoustic scene classification: an evaluation of an extremely compact feature representation. In: Conference: Detection and Classification of Acoustic Scenes and Events, vol. 35, no. 12, pp. 1340–1344. Budapest, Hungary (2016)
11. Wang, Z.H., Fang, C.: Lithology spectrum classification based on decision tree multi-classification support vector machine. J. Sun Yat-sen Univ. (Nat. Sci. Ed.) **37**(6), 93–97 + 105 (2014) (in Chinese)
12. Haritaoglu, I., Harwood, D., Davis, L.S.: Real-time surveillance of people and their activities. Planta Med. **81**(10), 847–854 (2015)
13. Girshick, R., Donahue, J., Darrell, T.: Rich feature hierarchies for accurate object detection and semantic segmentation. IEEE Trans. Comput. Vis. Pattern Recogn. 580–587 (2014)
14. Gupta, H., Gupta, D.: LPC and LPCC method of feature extraction in speech recognition system. IEEE Trans. 2016 6th International Conference (2016)
15. Abadi, M., Agarwal, A., Barham, P.: Large-scale machine learning on heterogeneous systems. Mach. Learn. (2015)

A Simplified Method of Radiator to Improve the Simulation Speed of Room Temperature Distribution

Zhenqiang Cao, Tong Niu, Haiyi Sun, and Xia Lu

Abstract Computational Fluid Dynamics (CFD) has become one of the important methods of indoor environmental analysis. Because the indoor environment of the building has the characteristics of large space and complex structure, the calculation amount of directly using CFD to simulate the indoor temperature distribution is so huge that the simulation efficiency is too low for practical application. Based on the shortcomings of the existing building environment simulation methods, Solidworks Flow Simulation and MATLAB are used to carry out research on indoor temperature distribution simulation methods. The overall treatment is equivalent to a constant heat source, which simplifies the heat dissipation model and avoids simulating complex flow patterns. Compared with traditional simulation methods, it can greatly improve simulation efficiency and model solving speed while ensuring simulation accuracy. The overall method is well supported by the simulation results.

Keywords Computational Fluid Dynamics (CFD) · Temperature distribution · Micro-segment simplified · Solidworks flow simulation

1 Introduction

Computational Fluid Dynamics (CFD) emerged in the 1960s. With the rapid development of computers after the 1990s, CFD has developed rapidly and has gradually become an important method of building environment simulation. Using a computer for CFD analysis can perform coupling simulations of multiple variables, and it can

---

T. Niu: The author contributed equally to this work.

Z. Cao · T. Niu
School of Mechanical Engineering, Shenyang Jianzhu University, Shenyang 110168, China

H. Sun (✉)
College of Science, Shenyang JianZhu University, Shenyang 110168, China
e-mail: shy_xx@163.com

X. Lu
School of Management, Shenyang Jianzhu University, Shenyang 110168, China

more clearly understand the complex process mechanism inside the system [1–5]. In the field of heat transfer simulation, CFD analysis is widely used for energy conversion of car radiators, computer radiators, fuel cells, and nanotubes [6–10]. CFD analysis is also commonly used in the field of building environment simulation. By simulating different heat source systems, the indoor temperature distribution under different conditions can be obtained [11]. In order to meet the thermal simulation of complex shapes, clustering hexahedral elements can be used to use ray intersection and ray/triangle intersection gridding methods to ensure sufficient simulation accuracy [12]. In order to improve the simulation efficiency, the mathematical model of the simulation object can be simplified. For example, indoor temperature distribution, heating inlet and outlet temperature, and shape characteristics of the radiator are analyzed after simplified modeling [13–16]. For example, when the radiator is working, the indoor air will undergo natural convection in which the heated air rises along the wall and then cools down [17]. The use of "user-defined wall function" to simplify the modeling of air convection can improve simulation efficiency. Compared with the k-ω, Shear Stress Transfer (SST), turbulence model, this model greatly reduces the number of units and improves the simulation solution speed. However, this simplified method introduces a "user-defined wall function," which makes the modeling process more complicated [18].

According to the general simulation method, both hot water and hot air must be modeled and analyzed at the same time. Because the internal structure and flow field of the heating system are complicated, the calculation amount is too large. Therefore, we simplified it by the following method.

In Sect. 2, it takes a specific room as an example. According to the thermal differential equation and various boundary conditions and the heat dissipation capacity of pipes, walls, and glass, the radiator is transformed into a constant heat source. Summing the equivalent heat dissipation power of each unit's pipeline, and its power can be regarded as the radiator power under the changed environment temperature.

When the heating power of the indoor heating is equal to the heat dissipation power, the corresponding ambient temperature is the desired equilibrium indoor temperature. In Sect. 3, it inputs the established model into Solidworks Flow Simulation to find the temperature distribution and average temperature in the room. Compared with traditional simulation methods, it can greatly improve simulation efficiency and model solving speed while ensuring simulation accuracy. The overall method is well supported by the simulation results.

2 Simplified Mathematical Model of Radiator Heat Dissipation Power

Generally, if CFD directly uses the input water temperature and flow rate as the heat source model, then it takes a long time for the simulation software to build the heat transfer model. In the transient special conduction, the solution of the temperature

Fig. 1 Schematic diagram of simplified radiator structure

field distribution over time may also get results that are inconsistent with common sense.

The calculation speed can be optimized by equating the heating pipe as a constant heat source to simplify the model. The internal structure of the coil radiator is an S-shaped coil with hot water flowing through it. The radiator can be converted into a circular heat pipe by using the principle of the same volume in Fig. 1. The volume equivalent equation is as (1).

$$\begin{cases} V = L \times D \times H \\ V = \pi \times r^2 \times l \end{cases} \tag{1}$$

where V is the radiator water capacity; L, D and H are the radiator geometry; l is the equivalent pipe length for heating; r is the inner diameter of equivalent radiator pipe.

The following will take a standard room as an example. The specific parameters are shown in Fig. 2 to verify the simplified method proposed in this paper.

When the inlet flow rate is known, the length of the radiator pipe water per second is taken as a unit length pipeline, and the heat conduction differential equation and the third type of boundary conditions are used to solve the equivalent radiator pipe heat dissipation power.

According to the continuity equation of fluid mechanics:

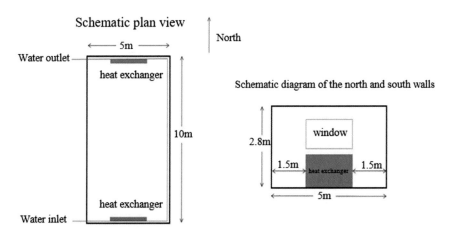

Fig. 2 Schematic diagram of the room

Fig. 3 Schematic diagram
of iron pipe heat conduction

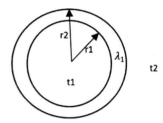

$$\begin{cases} q = Av \\ l_{\text{per}} = \frac{q}{\pi r^2 \times 3600} \end{cases} \tag{2}$$

where q is the is the liquid flow in the pipe; A is the pipe section area; v is water velocity; l_{per} is the length of a micro-segment.

The distance that the water flows per unit time is regarded as the length of the micro-segment. Each micro-segment can be regarded as a round wall radiator conduction, as shown in Fig. 3. Equation (3) is the heat conduction equation of heat transfer from the pipe to the room.

$$\begin{cases} Q_2 = 2\pi r_1 l h_w \left(t_{\text{nb}} - t_{\text{gw}}\right) \\ Q_2 = 2\pi \lambda_2 l \frac{(t_{\text{wb}} - t_{\text{nb}})}{\ln(r_1/r_2)} \\ Q_2 = 2\pi r_1 l h_q \left(t_n - t_{\text{wb}}\right) \end{cases} \tag{3}$$

where Q_2 is the micro-segment heat exchanger; r_1 is the pipe inner diameter; l is the micro-segment length; h_w is the convective heat transfer coefficient of water; t_{nb} is the pipe inner wall temperature; t_{gw} is the water temperature in pipe; λ_2 is the convection heat transfer coefficient of iron pipe; t_{wb} is the pipe outer wall temperature; r_2 is the outer diameter of pipe; h_q is the air convection heat transfer coefficient.

According to Fourier's law of heat conduction [19] and the third type of boundary conditions, the heat exchanger balance equation is established. The heat dissipation of each micro-segment can be calculated by MATLAB, and the heat of each micro-segment body can be added up. The total heat lost by the pipeline in 1 s is obtained, which is the heat generation power of the heating system in Fig. 4.

2.1 Simplified Mathematical Model of Windows and Walls Heat Dissipation Power

When analyzing the room model Fig. 5, all the energy in the room is input through the radiator, and all the energy output is completed through the exterior wall and glass. When the output energy is balanced with the input energy, the system will reach a steady state, so a heat dissipation model can be established:

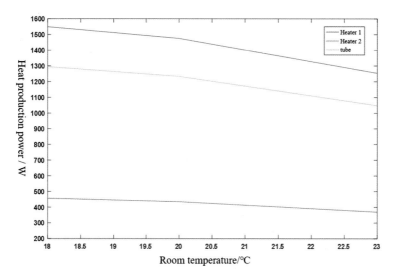

Fig. 4 Heat dissipation power diagram of each part of the pipeline

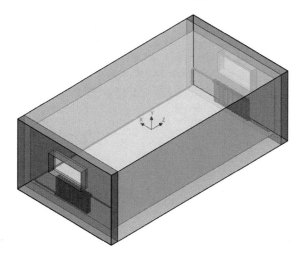

Fig. 5 3D view of the room

Fig. 6 Schematic diagram of solid heat conduction

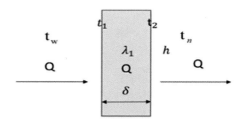

It is known that the heat in the room is dissipated through the glass Fig. 6. In order to make the room warmer, double-layer glass is usually used in the house. However, for this paper, the temperature distribution of the room is mainly considered. The coefficient is close to that of double-layer glass, which can be used in this study. The heat flux per unit time obtained by Fourier's law [19]:

$$Q_1 = \lambda_1 A \frac{t_2 - t_1}{d} \tag{4}$$

where Q_1 is the heat flux density between inner and outer glass; λ_1 is the thermal conductivity of glass; t_1 is the surface temperature of glass in contact with outdoor; t_2 is the glass and indoor contact temperature; A is the glass surface area.

$$Q_2 = hA(t_w - t_1) \tag{5}$$

where Q_2 is the heat flux density between environment and outer glass; h is the thermal convection coefficient of air; t_w is the outdoor temperature. The heat exchange between indoor air and glass can be expressed by (6):

$$Q_3 = hA(t_2 - t_n) \tag{6}$$

where Q_3 is the heat flux density between inner glass and indoor; t_2 is the indoor side glass temperature; t_n is the indoor temperature.

Since the surface temperature of the glass in contact with the outdoor and indoor sides is unknown, but the ambient temperature of the outdoor and indoor is known. So the heat dissipation power can be solved using the air convection coefficient as shown in Figs. 7 and 8.

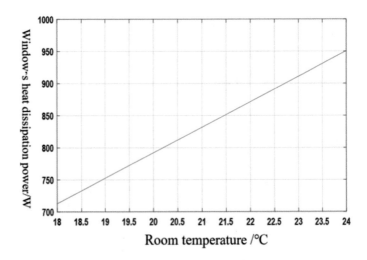

Fig. 7 Windows dissipation power

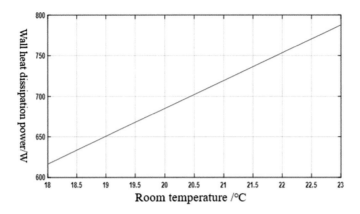

Fig. 8 Wall dissipation power

The heat dissipation power of the concrete exterior wall at a room temperature of 21 °C and an ambient temperature of 0 °C is 685.02 W.

When the dissipated power is consistent with the heat generated power, the heat will reach a dynamic balance. As shown in Fig. 9, the horizontal axis of the intersection is the power corresponding to the heating during thermal equilibrium, and the vertical axis is the average indoor temperature during thermal equilibrium.

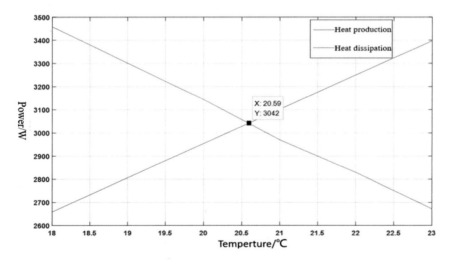

Fig. 9 System heat dissipation diagram

3 CFD Simulation

In order to improve the simulation efficiency and reduce the computer solution time, a simplified simulation method is used in this section, which is specifically as follows: treat the radiator and the internal hot water as a whole and make it equivalent to a constant heat source. When using the heat dynamic balance obtained in Fig. 9, the heating power is used as the heat generation power of the constant heat source.

Input the equivalent model into a computer, perform fluid simulation analysis, and solve the temperature distribution inside the room. At the same time, the average temperature of the room when the temperature is dynamically balanced is obtained through the simulation, and the accuracy of the heating power and the average indoor temperature corresponding to the heating power and the indoor average temperature when the heat is dynamically balanced is verified by the above method. The room temperature distribution obtained by the simulation is shown in Figs. 10, 11, and 12.

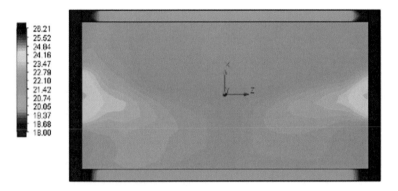

Fig. 10 Temperature distribution *XZ* plan

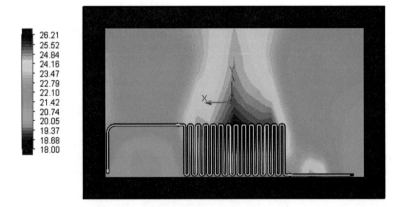

Fig. 11 Temperature distribution *XY* plan

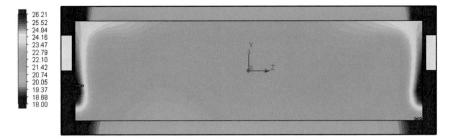

Fig. 12 Temperature distribution *YZ* plan

The average room temperature of the simulation is 20.17 °C, which is 2% error from the simplified method above, so the simplified model above is feasible. The main reason for the error is the simplified heating of complicated shapes into long straight pipes. Compared with the real heating pipes, the heat convection situation is different and the equivalent heating power is higher. Through computer simulation, the real thermal convection flow line diagram can be obtained as shown in Figs. 13 and 14.

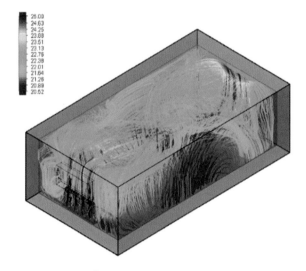

Fig. 13 Room temperature streamline

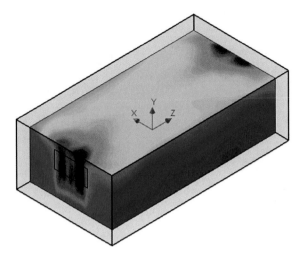

Fig. 14 Room temperature distribution

4 Conclusion

An important reference for the reasonable design of Heating, Ventilation, and Air Conditioning (HVAC) equipment is to simulate indoor plumbing equipment to optimize indoor temperature distribution and reduce energy waste. The main purpose of this paper is to propose a method to simplify the heating model of indoor plumbing equipment by avoiding the simulation of complex flow patterns, thereby increasing the solution speed and making large-scale simulation possible. Comparing the simplified simulation method with the traditional simulation effect, the same object can be simulated under the same device. The traditional method takes 1 h and 26 min to solve, while the simplified method takes 34 min. The simulation efficiency has been significantly improved. It is effective and feasible to equate the temperature and flow rate of the hot water inlet with a constant heat source. Make the indoor heat equal to the heat dissipation power at different indoor and outdoor temperatures and then adjust the inlet water temperature and flow rate, so that the indoor temperature is stable and comfortable and more convenient.

Acknowledgements This work was partially supported by Liaoning Provincial Department of Education Scientific Research Fund Project Basic Research Project (2020018), Liaoning BaiQianWan Talents Program (grant no. 2017076) and the Natural Science Foundation of Liaoning Province, China (grant no. 20170540769).

References

1. Pinghai, S., Ken, D., Thom, M.: Algae-dewatering using rotary drum vacuum filters: process modeling, simulation and techno-economics. Chem. Eng. J. **268**, 67–75 (2015)
2. Shokufe, A., Ali, G., Nima, R.: Mathematical modeling and simulation of water-alternating-gas (WAG) process by incorporating capillary pressure and hysteresis effects. Fuel **263**, 111362 (2020)
3. Baran, K., Antoni, R., Wachta H.: Thermal analysis of the factors influencing junction temperature of led panel sources. Energies **12**(20), 3941 (2019)
4. Gustavo, R.R., Garelli, L., Storti, M.: Numerical and experimental thermo-fluid dynamic analysis of a power transformer working in ONAN mode. Appl. Therm. Eng. **112**, 1271–1280 (2017)
5. Kirubagharan, R., Ramesh, C., Pragalathanl, P., Harish, N.: Geometrical analysis of automobile radiator using CFD. Materials Today, Proceedings (2020)
6. Pradeep, P.: CFD analysis of helical tube automobile radiator considering different coolants. Ind. Eng. Lett. **8**(5), 14–20 (2018)
7. Zhang, Q., Xu, L., Li, J., Ouyang, M.: Performance prediction of plate-fin radiator for low temperature preheating system of proton exchange membrane fuel cells using CFD simulation. Int. J. Hydrogen Energ. **42**(38), 24504–24516 (2017)
8. Chunhui, Z., Mesbah, U., Austin, C.: Full vehicle CFD investigations on the influence of front-end configuration on radiator performance and cooling drag. Appl. Therm. Eng. **130**, 1328–1340 (2018)
9. Luciano, G., Gustavo, R., Mario, S.: Reduced model for the thermo-fluid dynamic analysis of a power transformer radiator working in ONAF mode. Appl. Therm. Eng. **124**, 855–864 (2017)
10. Ali, K., Masoud, A.: Numerical study on thermal performance of an air-cooled heat exchanger: Effects of hybrid nanofluid, pipe arrangement and cross section. Energy Convers. Manag. **164**, 615–628 (2018)
11. Daniel, R., Mikael, R., Lars, W.: CFD modelling of radiators in buildings with user-defined wall functions. Appl. Therm. Eng. **94**, 266–273 (2016)
12. Yang, S., Pilet, T., Ordonez, J.: Volume element model for 3D dynamic building thermal modeling and simulation. Energy **148**(APR.1), 642–661 (2018)
13. Adnan, P., Sture, H.: Low-temperature ventilation pre-radiator in combination with conventional room radiators. Energ. Build. **65**, 248–259 (2013)
14. Calisir, T., Yazar, H.O., Baskaya, S.: Determination of the effects of different inlet-outlet locations and temperatures on PCCP panel radiator heat transfer and fluid flow characteristics. Int. J. Therm. Sci. **121**, 322–335 (2017)
15. Weizhen, L., Andrew, T., Alan, P.: Prediction of airflow and temperature field in a room with convective heat source. Build. Environ. **32**(6), 541–550 (1997)
16. Karthik, P., Kumaresan, V., Velraj, R.: Fanning friction (F) and colburn (J) factors of a louvered fin and flat tube compact heat exchanger. Therm. Sci. **21**(1A), 141–150 (2017)
17. El-Gendi, M.: Transient turbulent simulation of natural convection flows induced by a room radiator. Int. J. Therm. Sci. **125**, 369–380 (2018)
18. Mohammed, S., Essam, E.: CFD investigation of air flow patterns and thermal comfort in a room with diverse heating systems. Current Environ. Eng. **6**(2), 150–158 (2019)
19. Yang, S., Tao, W.: Heat Transfer, 3rd edn. Higher Education Press, China (1998)

Automation, Identification and Robotics

An Optimal Maintenance Cycle Decision of Relay Protection Device Based on Weibull Distribution Model

Qiuyu Zhuang and Meiju Liu

Abstract In view of the problem that there is no accurate optimal maintenance cycle for relay protection device, this paper is based on the Weibull distribution model. We firstly analyze the maintenance cycle data, assume and establish a distribution model, use the least square method combined with a genetic algorithm to estimate the parameters of the Weibull distribution. Then, we use MATLAB to determine the device failure rate function. Finally, we use the fuzzy decision method to get the optimal maintenance cycle for the devices. According to the optimal maintenance cycle, we can develop more appropriate state maintenance strategy and effectively support the state maintenance work.

Keywords Weibull distribution · Failure rate function · Genetic algorithm · Fuzzy decision method · Optimal maintenance cycle

1 Introduction

The state maintenance of relay protection devices has great advantages compared with traditional maintenance. First, it can avoid the breakdown caused by the failure to repair in time and reduce the workload of after-sales repair; second, it can avoid the over-repair caused by unnecessary repair and reduce the total cost of repair. The core of implementing state maintenance is to determine the optimal maintenance cycle, while the core of determining the optimal maintenance cycle is to determine the device failure rate function [1, 2]. The law of the failure rate function cannot be fully shown modeled with a basic distribution model. As the Weibull distribution is widely used in the data processing of various types of fault distribution and is mainly suitable for the modeling of the cumulative wear fault of electromechanical complex devices, we use the Weibull distribution model to get the optimal maintenance cycle in this paper [3].

Q. Zhuang (✉) · M. Liu
Faculty of Information and Control Engineering, Shenyang Jianzhu University, Shenyang, China
e-mail: 1347194056@qq.com

© The Author(s), under exclusive license to Springer Nature Singapore Pte Ltd. 2021   409
Y. Li et al. (eds.), *Advances in Simulation and Process Modelling*,
Advances in Intelligent Systems and Computing 1305,
https://doi.org/10.1007/978-981-33-4575-1_39

2 Determination of Device Failure Rate Function

For relatively complex devices, the failure rate function is an important data basis for predicting the failure rate and performing fault diagnosis [4]. In order to determine the failure rate function, we firstly analyze the maintenance cycle data to obtain the probability density function map, assuming that its failure law obeys the Weibull distribution according to the trend in the figure. Then, we establish the maintenance cycle distribution model and use the least square method combined with a genetic algorithm to estimate the parameters of Weibull distribution. Next, we use mathematical statistics to verify the hypothesis, so as to determine the cumulative distribution function, probability density function and reliability function that the maintenance cycle obeys. Finally, the failure rate function can be determined [5]. The flowchart is shown in Fig. 1.

2.1 Analysis of Maintenance Cycle Data

The maintenance cycle of the relay protection device refers to the trouble-free running time between the two adjacent failure dates. The data used in this paper was collected on 200 relay protection devices from the same manufacturer. These devices were all put into use in April 2008. All data is from the maintenance cycle collected during

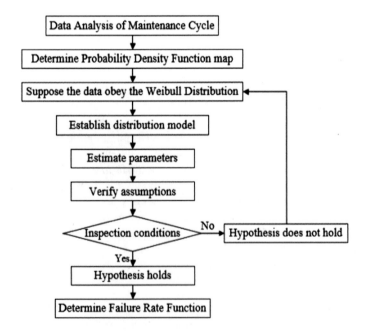

Fig. 1 Flowchart of determining device failure rate function

the 11 years from April 2008 to April 2019. The sample data was arranged and divided into 13 groups. The data is shown in Table 1. According to the frequency of each group, with the interval median as the horizontal coordinate, the frequency and cumulative frequency of each group as the vertical coordinate separately, we can obtain the cumulative distribution function map and the probability density function map. The cumulative distribution function map is shown in Fig. 2.

Table 1 Grouping data of maintenance cycle

Number	Time Interval/h	Interval median	Frequency	Cumulative frequency
1	21,900, 25,100	23,500	0.05	0.05
2	25,100, 28,300	26,700	0.06	0.11
3	28,300, 31,500	29,900	0.06	0.17
4	31,500, 34,700	33,100	0.07	0.24
5	34,700, 37,900	36,300	0.07	0.31
6	37,900, 41,100	39,500	0.07	0.38
7	41,100, 44,300	42,700	0.08	0.46
8	44,300, 47,500	45,900	0.08	0.54
9	47,500, 50,700	49,100	0.10	0.64
10	50,700, 53,900	52,300	0.09	0.73
11	53,900, 57,100	55,500	0.09	0.82
12	57,100, 60,300	58,700	0.09	0.91
13	60,300, 63,500	61,900	0.09	1.00

Fig. 2 Cumulative distribution function

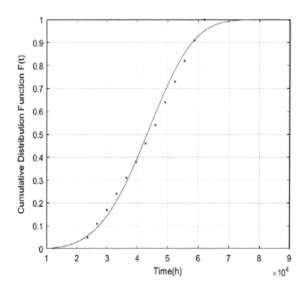

In mathematical statistics, the curve trend of the Weibull distribution is indeed shown in Fig. 2, so it can be assumed that the sample data obeys the Weibull distribution.

2.2 Weibull Parameter Estimation Based on Least Square Method

Assuming that there are several data (x_1, y_1), (x_2, y_2), ..., (x_n, y_n), then a scatter plot consisting of n points can be obtained on the plane xOy. From the figure, we can see that these points fall roughly on both sides of a straight line, so it is concluded that there is approximately a linear function between x and y. But under normal circumstances, these points are not on the same straight line. We suppose $E_i = y_i - (ax_i + b)$, it reflects the deviation between the calculated value and the actual value. Since E_i can be positive or negative, E_i^2 is used to measure the total deviation. Therefore, the method of determining the coefficients a and b which make $F(a, b)$ get the minimum is called the least square method [6, 7].

For the two-parameter Weibull distribution, the probability density function and the cumulative distribution function are as follows [8, 9].

$$f(t) = \frac{\beta}{\alpha} \left(\frac{t}{\alpha} \right)^{\beta-1} \exp\left[-\left(\frac{t}{\alpha} \right)^{\beta} \right] \tag{1}$$

$$F(t) = 1 - \exp\left[-\left(\frac{t}{\alpha} \right)^{\beta} \right] \tag{2}$$

We perform a linear transformation on Eq. (1):

$$\ln \ln \frac{1}{1 - F(t)} = -\beta \ln \alpha + \beta \ln t \tag{3}$$

Contrasted with the linear regression equation $y = ax + b$, we set up

$$y = \ln \ln \frac{1}{1 - F(t)}, x = \ln t \tag{4}$$

$$a = -\beta \ln \alpha, b = \beta \tag{5}$$

The coefficients a and b can be obtained from the simultaneous system of equations. Substituting a and b into Eq. (5), we can get the value of β and α.

The time between failures and the cumulative frequency of faults can be obtained from experimental data. x_i and y_i can be obtained according to Eq. (4), and we can finally get the value of a and b, β and α.

In the process of using MATLAB to realize the program, t_TBF[] stands for t_i, F[] stands for $F_i(t)$, x[], y[] stand for x_i, y_i. Specific steps are as follows.

First, we can get x[] according to t_TBF[]; second, we can get y[] from the Eq. (4); F[i] can be computed according to median rank, $F_i(t) \approx (i - 0.3)/(n + 0.4)$, and i stands for the sequence number of the fault data; third, we compute l_{xx} and l_{xy}; fourth, we compute $b = l_{xx}/l_{xy}$, $a = y - bx$ (x, y are the averages); $\beta = b$, $\alpha = \exp(-a/b)$.

$$l_{xx} = \sum_{i=1}^{n} (x_i - \bar{x})^2 = \sum_{i=1}^{n} x_i^2 - n\bar{x}^2 \tag{6}$$

$$l_{xy} = \sum_{i=1}^{n} (x_i - \bar{x})(y_i - \bar{y}) = \sum_{i=1}^{n} x_i y_i - n\bar{x}\bar{y} \tag{7}$$

2.3 Weibull Parameter Estimation Based on Genetic Algorithm

The basic idea of the maximum likelihood method is to select the parameter to be determined so as to maximize the probability of the sample appearing in the field of observation value and use this value as the point estimate of the unknown parameter. The basic idea of the maximum likelihood method is to select the parameter to be determined so as to maximize the probability of the sample appearing in the field of observation value, and then, we make this value as the point estimate of the unknown parameter. The likelihood function of the Weibull distribution is:

$$\ln L = \sum_{i=1}^{n} \ln[f(t_i; \beta, \alpha)] \tag{8}$$

As the process of solving the likelihood equations is relatively complicated and may result in no results due to improper initial value selection, this paper combines the idea of maximum likelihood method with genetic algorithm. Genetic algorithm is a method to search for the optimal solution by simulating natural evolution process. The main feature is to directly operate on structural objects. There is no derivation and function continuity limitation; it has better global optimization ability [10]. Therefore, the modeling process is as follows. Assume $x = [x_1, x_2]^T = [\beta, \alpha]$ as decision variable, we can get $0 < \alpha < t_n$, $0 < \beta < 5$ according to the parameters' physical meaning.

$$\min f(x) = -\sum_{i=1}^{R} \ln[f(t_i; \beta, \alpha)] \tag{9}$$

$$\text{S.T.} \begin{cases} 0 < x_1 < 5 \\ 0 < x_2 < t_n \end{cases} \tag{10}$$

In order to simplify the calculation process, we introduce the MATLAB genetic optimization toolbox method. First, we should write function M file since an M file must be written to determine the optimization objective function to use genetic algorithm in MATLAB. Second, we need to carry on the genetic algorithm optimization. In this paper, we call the genetic algorithm by commanding the function GA [11].

2.4 Hypothesis Verification of Failure Rate Distribution Model

The K-S test method is commonly used to verify whether the sample data follows the same distribution, and as the K-S test method is suitable for the verification of small assumptions and the test accuracy is high, it is used in this paper for verification.

First, we arrange the collected sample data, and then, we compute each data's hypothetical distribution function and compare it with its empirical distribution function. The maximum value of these absolute differences is regarded as the sample value D_n. Finally, we compare each sample value with the critical value, if the results satisfy Eq. (11), it proves that the assumption establishes; otherwise, it fails [12].

$$D_n = \sup_{-\infty < x < +\infty} |F_n(t) - F_0(t)| = \max\{d_i\} \leq D_{n \cdot \sigma} \tag{11}$$

$$F_n(t) = \begin{cases} 0, & t < t_1 \\ \frac{i}{n}, & t_i < t < t_{i+1} \\ 1, & t \geq t_n \end{cases} \tag{12}$$

We can get d_i according to Eqs. (11) and (12). Through the simulation experiments, the critical value $D_{n \cdot \sigma}$ is 0.3614. Each sample value is less than 0.3614, so the assumption is established and the maintenance cycle follows the Weibull distribution.

2.5 Determination of Equipment's Failure Rate Function

Based on the analysis above, the Weibull distribution obeying the scale parameter $\alpha = 47,201.4516$ and the shape parameter $\beta = 4.0683$ is obtained. The probability density function and the cumulative distribution function are shown as (13) and (14).

$$f(t) = \frac{4.0683}{47201.4516} \left(\frac{t}{47201.4516} \right)^{3.0683} \exp\left[-\left(\frac{t}{47201.4516} \right)^{4.0683} \right] \tag{13}$$

Fig. 3 Failure rate function

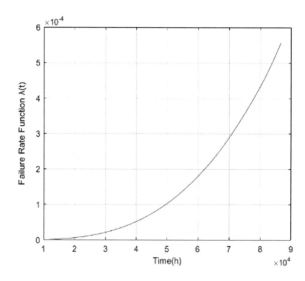

$$F(t) = 1 - \exp\left[-\left(\frac{t}{47201.4516} \right)^{4.0683} \right] \tag{14}$$

The reliability function and failure rate function can be derived from the equations above. The failure rate function map is shown in Fig. 3.

$$R(t) = \exp\left[-\left(\frac{t}{47201.4516} \right)^{4.0683} \right] \tag{15}$$

$$\lambda(t) = \frac{4.0683}{47201.4516} \left(\frac{t}{47201.4516} \right)^{3.0683} \tag{16}$$

Since the value of the shape parameter is greater than 1, it indicates that as the devices have been put into use gradually, the reliability has reduced and the probability of device failure has increased, which is consistent with the actual production.

3 Determination of the Optimal Maintenance Cycle

3.1 The Optimal Maintenance Cycle Model

When establishing the model of the optimal maintenance cycle for the relay protection device, both reliability factor and cost factor must be considered.

Reliability factor: Through analysis, we know the failure rate function is $\lambda(t) = \beta t^{\beta-1}/\alpha^{\beta}$, and we suppose $\gamma = 1/\alpha^{\beta}$ for convenience; then, we can get $\lambda(t) = \gamma t^{\beta-1}$. We integrate $\lambda(t)$ on $(0, t)$ to get the cumulative distribution function as $m(t) = \gamma t^{\beta}$. The reliability function is $R(t) = \exp(-\gamma t^{\beta})$; after the maintenance, the reliability should reach R_0 at least [13, 14]. Taking logarithms on both sides, meanwhile, we can get the minimum maintenance cycle T_1 to meet the minimum reliability requirements.

$$T_1 = \sqrt[\beta]{\frac{1}{\gamma} \ln \frac{1}{R_0}} \tag{17}$$

Cost factor: We suppose C_1 as the cost of detecting and repairing the fault during the repair process; C_2 is regarded as the cost of the fault and repair found after the devices have been put into operation $(C_2 > C_1 > 0)$. We make C_3 as the loss per unit time caused by shutdown of the devices due to maintenance $(C_3 > 0)$. T_{PS} is the average life of the device in one maintenance cycle; then, the total cost during $(0, t)$ is:

$$C(t) = C_1 \gamma t^{\beta} + C_2(\gamma T_{PS} - \gamma t^{\beta}) + C_3 t \tag{18}$$

In order to minimize the total cost, $dC(t)/dt = 0$ must be satisfied. We can get the maintenance cycle T_2 meeting the cost factor according to Eq. (18).

$$T_2 = \sqrt[\beta-1]{\frac{C_3}{(C_2 - C_1)\gamma\beta}} \tag{19}$$

3.2 Determination of Optimal Maintenance Cycle by Fuzzy Decision Method

Fuzzy decision method is a comprehensive decision-making method based on the combination of quantitative and qualitative fuzzy mathematics. Through the fuzzy mathematics method, we can analyze and evaluate the influence of fuzzy factor in a standardized way out of the overall evaluation [15].

First, we should determine the set of factors affecting the optimal maintenance cycle; second, we give the number, category of the rating expert group and the rating set of the rating expert; third, we construct the experts' scoring matrix; fourth, we construct fuzzy evaluation weight vector; finally, we can obtain the fuzzy evaluation result.

Based on the fuzzy decision method, this paper quantitatively analyzes the optimal maintenance cycle of the devices. We select 10 experts, including three personnel engaged in the relay protection device transformation, three personnel from relay protection device manufacturers and four workers with more than 10 years' working experience. Ten experts are supposed to quantitatively score the two factors, ranging from 1 to 10.

For reliability factor and cost factor, the score results are:

$$M(u_1, p_j) = \{10, 9, 10, 10, 9, 10, 10, 9, 10, 10\}, j = 1 : 10 \tag{20}$$

$$M(u_2, p_j) = \{7, 6, 4, 7, 8, 5, 6, 4, 7, 6\}, j = 1 : 10 \tag{21}$$

Based on the scoring results, the average score of the two factors is:

$$a_1 = \frac{1}{10} \sum_{j=1}^{10} M(u_1, p_j) = 9.8, a_2 = \frac{1}{10} \sum_{j=1}^{10} M(u_2, p_j) = 6 \tag{22}$$

Normalizing the average scores, we can get the influence weight of each factor:

$$w_1 = \frac{a_1}{\sum_{i=1}^{2} a_i} = 0.6203, \quad w_2 = \frac{a_2}{\sum_{i=1}^{2} a_i} = 0.3797 \tag{23}$$

The optimal maintenance cycle that meets the requirements of reliability and cost can be obtained, respectively, $T_1 = 25{,}661.4226$ h, $T_2 = 31{,}336.6751$ h. Finally, we get that the optimal maintenance cycle of the relay protection device is $T_C = 27{,}816.5818$ h.

4 Conclusion

This paper is supposed to determine the optimal maintenance cycle. After using the least square method combined with genetic algorithm to analyze the parameters, we prove that the maintenance cycle surely obeys the Weibull distribution with $\alpha = 47{,}201.4516$ and $\beta = 4.0683$. Failure rate function has been determined according to the parameters. By using the fuzzy decision method, we finally get that device optimal maintenance cycle is 3.18 years, which can be regarded as three and a half years.

Acknowledgements This research work is partially supported by Jiangsu Province Elevator Intelligent Safety Key Construction Laboratory Project, China, with Project Number: JSKLESS201707.

References

1. Gao, X.: Application technology of relay protection status maintenance. China Electric Power Press, Beijing (2008)
2. Kumm, J.J., WeBer, M.S., Schweiter, E.O.: Predicting the option routine test interval for protective relays. IEEE Trans. Power Del. **10**(2), 659–664 (1995)
3. Lai, C.D., Xie, M., Murthy, D.N.P.: A modified Weibull distribution. IEEE Trans. Reliab. **52**(1), 33–37 (2003)
4. Chen, S., Gui, W.: Statistical analysis of a lifetime distribution with a bathtub-shaped failure rate function under adaptive progressive type-II censoring. Mathematics **8**(5) (2020)
5. Zhang, Z.H.: Reliability Theory and Engineering Application. Science Press, Beijing (2012)
6. Yang, J., Lv, Z., Shi, H., Tan, S.: Performance monitoring method based on balanced partial least square and statistics pattern analysis. ISA Trans. **81**, 121–131 (2018)
7. Xu, W., Chen, W.: Feasibility study on the least square method for fitting non-gaussian noise data. Phys. A **492**, 1917–1930 (2018)
8. Wais, P.: Two and three-parameter Weibull distribution in available wind power analysis. Renew. Energ. **103**, 15–29 (2017)
9. Kim, H., Singh, C.: Power system reliability modeling with aging using thinning algorithm. IEEE Bucharest Power Tech. Conf. **1**, 1–6 (2009)
10. Dehghanian, P., Guan, Y., Kezunovic, M.: Real-time life-cycle assessment of high-voltage circuit breakers for maintenance using online condition monitoring data. IEEE Trans. Ind. Appl. **55**(2), 1135–1146 (2018)
11. Matinez-Soto, R., Rodriguez, A., Castillo, O., Aguilar, L.T.: Gain optimization for inertia wheel pendulum stabilization using particle swarm optimization and genetic algorithms. Int. J. Innovative Comput. Inf. Control. **8**(6), 4421–4430 (2012)
12. Zhang, G., Wang, X., Liang, Y.C., Liu, J.: Fast and robust spectrum sensing via Kolmogorov-Smirnov test. IEEE Trans. Commun. **58**(12), 3410–3416 (2010)
13. Savsar, M.: Realiability analysis of a flexible manufacturing cell. Reliab. Eng. Syst. Safety. **67**(2), 147–152 (2000)
14. Propst, J.E., Doan, D.R.: Improvements in modeling and evaluation of electrical power system reliability. IEEE Trans. Ind. Appl. **37**(5), 1413–1422 (2001)
15. Zamzuri, H., Zolotas, A., Goodall, R., Malzan, S.A.: Advances in tilt control design of high-speed railway vehicles: a study on fuzzy control methods. Int. J. Innovative Comput. Inf. Control. **8**(9), 6076–6080 (2012)

Hysteretic Behavior Analysis of Concrete-Filled Double-Skin Steel Tubular Column Under the Constraint of Mortise and Tenon Joint with Low-Cycle Reciprocating Load

Wei Sun, Junshan Yang, and Bing Li

Abstract The mortise and tenon (MT) joint which proposed in the early research works of authors of this article solves the difficulties in the construction of concrete-filled double-skin steel tubular (CFDST) columns effectively. The CFDST columns and steel beams constrained by MT joints can form a kind of mechanical system of the prefabricated frame, which expected to have a wide application prospect. In this paper, the finite element method was used to compare the hysteretic performance of the CFDST columns with different hollow ratio (χ) and axial compression ratio (n) under the constraints of ordinary joint and the MT joint. Through the comprehensive comparative analysis of the component stress cloud diagram, hysteresis curve and skeleton curve, it is concluded that the CFDST columns under the MT joint constraint have better energy dissipation performance and ductility than the columns under the ordinary joint constraint. The results show that the new frame system can meet the requirements of earthquake resistance.

Keywords Mortise and tenon joint · Concrete-filled double-skin steel tubular column · Hysteretic behavior analysis

1 Introduction

In recent years, there have been more and more large and complex buildings, which puts forward higher requirements on the mechanical properties of building structural components, and the combined structure formed by the combination of section steel and concrete has gradually attracted widespread attention. Because the steel–concrete composite structure can take advantage of both steel and concrete materials at the same time, especially the steel tube concrete structure, which is widely used in the field of high-rise and super high-rise buildings. Concrete-filled steel tube components can be divided into solid type and hollow sandwich type according to their structural forms. The latter is a kind of hollow structure with a smaller self-weight, while its

W. Sun (✉) · J. Yang · B. Li
School of Civil Engineering, Shenyang Jianzhu University, Shenyang, China
e-mail: lg1_315@126.com

© The Author(s), under exclusive license to Springer Nature Singapore Pte Ltd. 2021   419
Y. Li et al. (eds.), *Advances in Simulation and Process Modelling*,
Advances in Intelligent Systems and Computing 1305,
https://doi.org/10.1007/978-981-33-4575-1_40

bearing capacity is still in a high level. Therefore, it is a kind of concrete-filled steel tubular member with very reasonable mechanical properties.

Since the 1990s, the research on concrete-filled double-skin steel tubular (CFDST) columns is gradually deepened and some research results have been achieved [1–8]. However, this kind of component has not been widely used in engineering field. The reason is that it is difficult to deal with the joint for CFDST column, which brings obvious difficulties to the onsite construction. Without reasonable connection joint, CFDST columns cannot form a structural system. As a result, it has not been widely used in engineering field.

Based on the above background, literature [9] proposed a mortise and tenon (MT) joint, which effectively solves the combination problem of CFDST columns and steel beams. By the new joints, CFDST columns and steel beams can form a kind of frame structure system with prefabricated construction characteristics. At the same time, the mechanical properties of the CFDST columns constrained by the MT joints under eccentric loads are analyzed. On this basis, this paper further analyzes the low-cycle reciprocating performance of the CFDST columns constrained by the MT joints.

2   Introduction of the Mortise and Tenon Joint

The form of the MT joints is shown in Fig. 1. It consists of tenon column and pier column. They are steel pipes wrapped in concrete. Screw holes are arranged at the

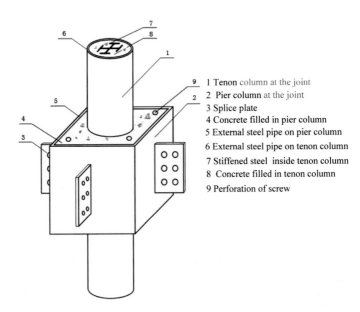

1 Tenon column at the joint
2 Pier column at the joint
3 Splice plate
4 Concrete filled in pier column
5 External steel pipe on pier column
6 External steel pipe on tenon column
7 Stiffened steel inside tenon column
8 Concrete filled in tenon column
9 Perforation of screw

Fig. 1 New node

Fig. 2 Space and
assembling form

10 Concrete filled double-skin
steel tubular column

11 End-plate

12 Screw and bolt

13 Rolled beam

14 Screw holes on the end-
plate

15 Inner cavity of inner
steel tube

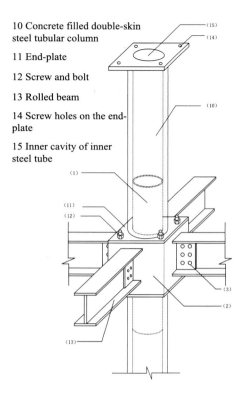

four corners of pier columns to connect the CFDST column. A splice plate is also
provided around the pier column to connect the steel structural beams. The MT joint
can connect the CFDST columns and steel beams into a kind of mechanical system
of prefabricated frame, as shown in Fig. 2. The mortise part of the MT joint inserts
into the hollow part of the CFDST column. The mortise part of the MT joint connects
with the internal cavity of the member. And then, the joint is fixed to the end plate
of the CFDST column by screws.

3   Establishment of Finite Element Model

In the process of establishing the finite element model, the axial compression ratio
n and the void ratio χ are taken as the influencing factors to research the hysteretic
behavior of CFDST columns. In this paper, the seismic performance of the members
connected by MT joint and ordinary joint under horizontal low-cycle cyclic loading is
compared and analyzed. Referring to the relevant data in reference [10], the working
conditions of the test pieces are formulated. The analysis conditions are shown in
Table 1.

Table 1 Specimen data sheet

No.	Tenon length (mm)	$D_0 \times t_0$ (mm)	$D_i \times t_i$ (mm)	χ	n
1	–	114 × 3	32 × 3	0.3	0.3
2	–	114 × 3	32 × 3	0.3	0.6
3	–	114 × 3	57 × 3	0.54	0.3
4	–	114 × 3	57 × 3	0.54	0.6
5	–	114 × 3	76 × 3	0.7	0.3
6	–	114 × 3	76 × 3	0.7	0.6
7	180	114 × 3	32 × 3	0.3	0.3
8	180	114 × 3	32 × 3	0.3	0.6
9	180	114 × 3	57 × 3	0.54	0.3
10	180	114 × 3	57 × 3	0.54	0.6
11	180	114 × 3	76 × 3	0.7	0.3
12	180	114 × 3	76 × 3	0.7	0.6

Note "D_0" is diameter of outer steel pipe. "t_0" is thickness of outer tube. "D_i" is diameter of inner steel pipe. "t_i" is thickness of inner tube. "χ" is hollow ratio. "n" is axial compression ratio. Yield strength of outer steel tube is 294.5 MPa; yield strength of inner steel tube is 374.5 MPa; compressive strength of concrete is 46.8 MPa

In the process of establishing the finite element model, the circular section CFDST column is simplified as cantilever member, and a half model is established. Its lower end is rigid restraint. The relationship curve in reference [10] is adopted for the constitutive relationship of steel tube and concrete, that is, the constitutive relation curve of steel tube is five segments, and that of concrete is compressive concrete stress–plastic strain. There are two analysis steps in the model, one is to add vertical load according to the axial compression ratio n, and the other is to add horizontal load according to Fig. 3. The shell element is used for the outer steel tube and the three-dimensional solid element is used for the concrete and the inner steel tube. The cover plates of the two joints are considered as rigid bodies. When the interaction is set, the steel pipe is the main surface and the concrete is the secondary surface. Coulomb friction is used to simulate the contact in tangent direction. The friction coefficient is 0.25, and the normal direction is set as hard contact. For the parameter setting of MT joint, the connection of cover plate and tenon and the connection of tenon and inner steel pipe also adopt this contact. The calculation model, loading mode, and boundary conditions of concrete-filled steel tube under horizontal low-cycle cyclic loading are shown in Fig. 4.

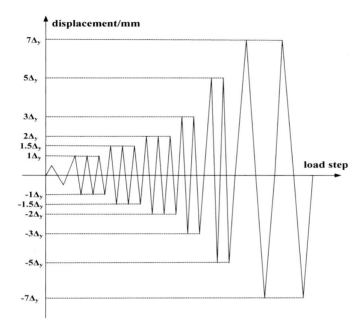

Fig. 3 Component lateral displacement loading system

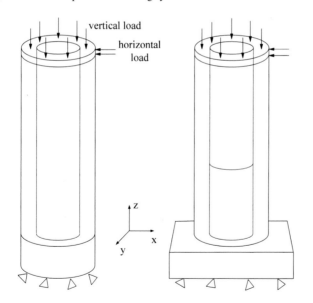

Fig. 4 Two different boundary conditions under two kind of joints

4 Analysis of Numerical Simulation Results

4.1 Comparative Analysis of Failure Characteristics

Taking the member which hollow ratio $\chi = 0.54$ and axial compression ratio $n = 0.6$ as an example, the failure characteristics of the outer steel tube, the inner steel tube, and the concrete under the two kinds of joint constraints are compared and analyzed. The stress nephogram of the outer steel tube, inner steel tube, and concrete under the two kinds of joint constraints under the horizontal low-cycle reciprocating load is shown in Fig. 5 (the left side is the ordinary joint restraint, the right side is the MT joint constraint). By comparing the stress nephogram, it can be seen that: The stress nephogram shows antisymmetric distribution, and the tensile stress at both ends of the component decreases gradually from both ends to the middle. The tensile stress value of ordinary joint restraint member is greater than that of MT joint restraint member. The tensile stress of mortise and tenon joints is uniform in the whole tenon constraint area.

With the increasing of the horizontal reciprocating load, the displacement of the column restrained by ordinary joint is also increasing, the root of the column is bulging outward, and the concrete at the position where the horizontal load is applied also appears. The tensile stress of the root concrete reaches its ultimate tensile stress value and fails. At the same time, the root of outer steel pipe exceeds the yield stress. However, due to the confinement effect of the outer steel tube, the column still maintains a certain bearing capacity. For the column constrained by MT joint, there is no drumbeat at the root, but a slight deformation at the end of the tenon. This is because the tenon at the bottom of the component has a certain restraint effect on the bottom of the component, delaying the destruction of the component root. The stress nephogram shows that the root of the column under the constraint of MT joint is uniformly stressed, and there is no stress concentration area.

Therefore, through the above analysis, it can be seen that the failure mode of components under the restraint of mortise and tenon joints is better than that of members under the restraint of ordinary joints, which can effectively improve the safety and horizontal bearing capacity of members.

4.2 Comparative Analysis of Hysteresis Curve and Skeleton Curve

The comparisons of hysteretic curves corresponding to each working condition under the two constraints are shown in Fig. 6. It can be seen from the figure that under the two kinds of joint constraints, the hysteretic curves of the columns are full, and there is no obvious pinch phenomenon. Moreover, the area of hysteretic curve of members restrained by MT joints is significantly larger than that of columns restrained by common joints, which indicates that MT joints can effectively improve the energy

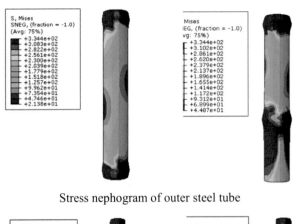

Stress nephogram of outer steel tube

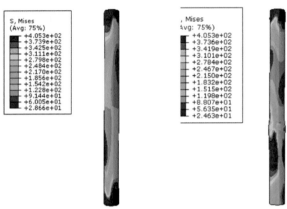

Stress nephogram of inner steel tube

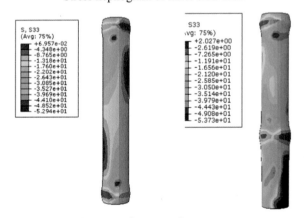

Stress nephogram of concrete

Fig. 5 Cloud comparison under two kind nodes

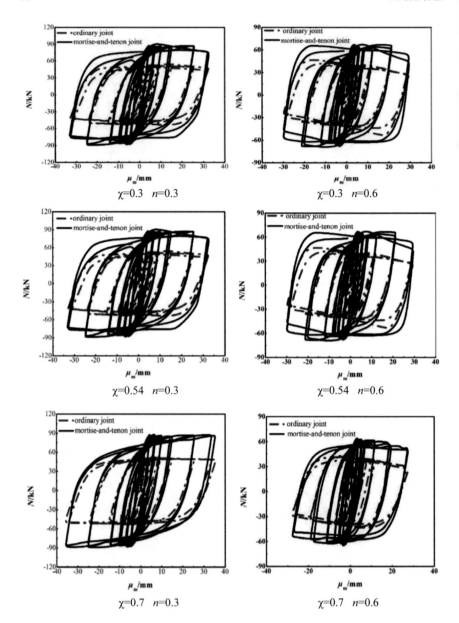

Fig. 6 Hysteresis curve of the two joints

dissipation capacity of columns. The influence of axial compression ratio n on the hysteretic behavior of the column is stronger than that of the hollow ratio χ. The larger the axial compression ratio n is, the worse the hysteretic performance of the column is under the same constraint condition and void ratio n.

The comparison of skeleton curves corresponding to each working condition under the two constraints is shown in Fig. 7. It can be seen from the figure that the bearing capacity and lateral stiffness of the members restrained by MT joints are significantly higher than those under ordinary joints. The influence of axial compression ratio n and void ratio χ on skeleton curve was compared and analyzed. It can be concluded that the bearing capacity of columns decreases with the increase of axial compression ratio n.

When $n = 0.3$, there is no obvious downward trend in the skeleton curves under the two kinds of joint constraints. In this case, the horizontal bearing capacity of the member increases with the increase of the hollow ratio χ. When $n = 0.6$, the skeleton curves of the columns under the two kinds of joint constraints show a downward trend, especially the bearing capacity of the columns under the common joint constraints decreases more obviously. Under this condition, the joint restraint

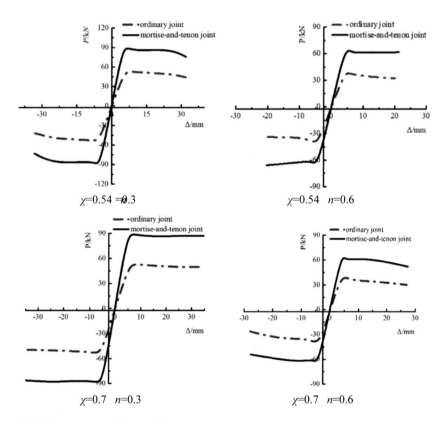

Fig. 7 Skeleton curve of the two joints

of MT joint improves the horizontal capacity more obviously, but the hollow ratio χ has little effect on it.

4.3   Seismic Coefficient Analysis

Displacement Ductility Coefficients

Ductility is an important performance indicator in the seismic resistance of engineering structures. The ductility coefficient μ is used to indicate the ductility. The displacement ductility factor is defined as:

$$\mu = \frac{\Delta u}{\Delta y} \tag{1}$$

where Δ_y—yield displacement, it is the displacement corresponding to the intersection point of the extension line of skeleton curve segment and the horizontal line passing through the peak value; Δ_u—effective ultimate displacement, it is the value of ultimate displacement is the corresponding displacement when the bearing capacity decreases to 85% of the peak bearing capacity.

The ductility coefficients of columns under the restraint of ordinary joint and MT joint are calculated and shown in Table 2. By analyzing the data in the table, the following conclusions can be drawn:

1. The ductility coefficient μ decreases with the increase of axial compression ratio n. When the axial compression ratio $n = 0.6$, the ductility coefficient μ of columns constrained by MT joint increases most obviously. The ductility coefficient μ decreases with the increase of hollow ratio χ for columns constrained by both types of joints. The main reason is that the larger hollow ratio χ is, the less the concrete part of the column section is, which lead to the bending stiffness decrease, and makes the ductility coefficient μ decrease.
2. The ductility coefficient μ of members with MT joints is larger than that with ordinary joints. Because the tenon part of the MT joint is inserted into the hollow part of the column, which changes the constraint method of the member, and the ductility of the column is improved accordingly.
3. The MT joint can significantly improve the ductility coefficient μ of columns. The curves shown in Fig. 8 shows that the larger the axial compression ratio n is, the more obvious the improvement of ductility coefficient μ by MT joint is. In this paper, when the axial compression ratio $n = 0.6$, the ductility coefficient μ of column improved by MT joint is about 24% higher than that of ordinary joint.

Stiffness

Taking the column with hollow ratio $\chi = 0.54$, as an example, the influence of joint type and axial compression ratio n on column stiffness is studied. It can be seen

Table 2 Ductility factor

Influence factors		Constraint type	Loading direction	Yield displacement (mm)	Maximum displacement (mm)	Ultimate displacement (mm)	μ
$\chi = 0.3$	$n = 0.3$	O	+	4.42	6.39	30.56	6.91
			−	−4.39	−6.47	−30.96	7.05
		MT	+	4.42	6.36	29.76	6.73
			−	−4.27	−6.54	−28.907	6.77
	$n = 0.6$	O	+	4.46	6.44	19.13	4.29
			−	−4.29	−5.44	−18.24	4.37
		MT	+	4.46	6.37	21.03	4.72
			+	−4.28	−6.25	−21.31	4.98
$\chi = 0.54$	$n = 0.3$	O	−	5.14	7.64	32.24	6.33
			+	−4.98	−7.57	−30.98	6.22
		MT	−	4.75	7.22	32.04	6.74
			+	−4.97	−6.88	−32.23	6.48
	$n = 0.6$	O	−	4.18	5.71	20.10	4.81
			+	−3.97	−5.96	−19.94	5.02
		MT	+	4.17	5.86	24.75	5.94
			+	−3.99	−5.95	−24.05	6.03
$\chi = 0.7$	$n = 0.3$	O	−	5.14	7.64	30.56	5.95
			+	−4.98	−7.57	−30.02	6.04
		MT	−	4.64	7.21	28.68	6.18
			+	−4.97	−6.88	30.96	6.21
	$n = 0.6$	O	−	3.82	5.82	19.96	5.23
			+	−3.89	−5.99	−21.56	5.54
		MT	+	3.97	5.42	25.56	6.44
			−	−3.93	−5.86	−25.96	6.61

Note: "O" is ordinary joint, " + " is forward, "−" is backward

from the curves in Fig. 9 that MT joint can significantly improve initial stiffness and residual stiffness of the columns. And the greater the axial compression ratio n is, the worse the stiffness value of the columns is.

5 Conclusions

In this paper, the numerical simulation method is used to study the mechanical properties of CFDST columns under horizontal low-cycle cyclic loading under the restraint of ordinary joint and MT joint. By comparing the stress nephogram and

Fig. 8 Ductility coefficient increase rate with the effect of χ

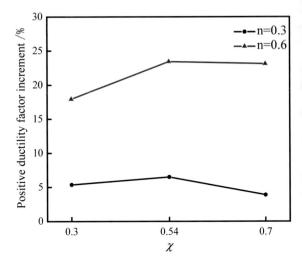

Fig. 9 Comparison of stiffness deterioration

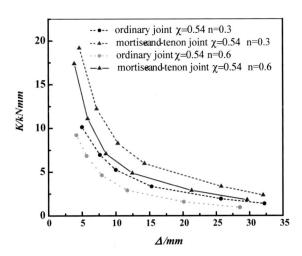

failure mode of CFDST columns under the two kinds of joints, it can be concluded that the failure position of CFDST columns under the two kinds of joints is different, MT joints can protect the root of CFDST columns from damage. Compared with the hysteretic curves and skeleton curves, the horizontal bearing capacity of CFDST columns under MT joint is significantly higher than that under ordinary joint. And with the increase of axial compression ratio n, the improvement of horizontal bearing capacity of CFDST columns under MT joint is more obvious. The CFDST columns restrained by MT joint have better energy dissipation performance and ductility than that restrained by ordinary joint.

Acknowledgements The research work in this article was carried out under the funding of the Scientific Research Project of the Education Department of Liaoning Province (Z2219013).

References

1. Zhang, Y.C., Xu, C., Lu, X.Z.: Experimental study of hysteretic behavior for concrete filled square thin-walled steel tubular columns. J. Constr. Steel Res. **63**(3), 317–325 (2007)
2. Wang, T.C., Zhang, V.: Analysis of parameter influence on seismic behaviour of concrete filled square steel tubular columns. J. Build. Struct. **34**, 339–344 (2013). (in Chinese)
3. Wang, X.N., Gao, C.C., Wang, N.N.: Research on bearing capacity of axially-loaded short columns of concrete filled double skin steel tubes. J. Hebei Univ. Eng. (Nat. Sci. Ed.) **30**(1), 20–24(2013) (in Chinese)
4. Vipulkumar, I.P., Qing, Q.L., Muhammad, N.: Numerical analysis circular concrete-filled steel tubular slender beam-columns with preload effects. Int. J. Struct. Stab. Dynam. **13**(3) (2013)
5. Kojiro, U., Hiroaki, K.: Mechanical behaviour of concrete filled double skin tubular circular deep beams. Thin-Walled Struct. **49**(2), 256–263 (2011)
6. Liu, Q., Zhang, D., Chen, B.: Pure bending of self-compacting concrete-filled double skin steel tubes in a circle. Eng. Mech. **S1**, 213–216 (2014). (in Chinese)
7. Zhang, R., Wang, C.G. Zhang, C.B.: Experimental study on seismic behaviour of recycled aggregate concrete filled square steel tube columns. J. Hefei Univ. Technol. (Nat. Sci.) **03**, 369–372 (2015) (in Chinese)
8. Huang, H., Zhu, Q., Chen, M.C.: Experimental study on concrete-filled double-skin square steel tubes under compression-bending-torsion loading conditions. China Civil Eng. J. **03**, 91–97 (2016). (in Chinese)
9. Sun, W., Zhao, Y., Yan, S.: Analysis on eccentric compression property of concrete-filled double skin steel tubular under new type mortise and tenon joint. China Concr. Cem. Prod. **226**(3), 46–49 (2014) (in Chinese).
10. Liu, W., Han, L.H.: Research on some issues of ABAQUS analysis on the behavior of axially loaded concrete-filled steel tube. J. Harb. Inst. Technol. **S1**, 157–160 (2005). (in Chinese)

Face Mask Recognition Based on MTCNN and MobileNet

Jianzhao Cao, Renning Pang, Ruwei Ma, and Yuanwei Qi

Abstract The COVID-19 can be transmitted by air droplets, aerosols, and other carriers, the spread of the virus can be effectively prevented by wearing masks in public. Therefore, it is meaningful to identify whether a mask is worn in particular places. In this paper, a method based on multi-task convolutional neural networks (MTCNN) and MobileNet algorithms is proposed to implement mask recognition on human face. Firstly, MTCNN is used to detect facial contours. Then the output image is used to train MobileNet model. By comparing the extracted facial feature data, the human with mask or not can be marked. The method has been tested in a 1.8 GHz Intel Core machine with 160×160 static images. Average accuracy rate of 94.73% and detection speed of 1.9 s are achieved.

Keywords COVID-19 · MTCNN · Face detection · Face alignment · MobileNet · Mask recognition

1 Introduction

Since December 2019, there has been an outbreak of new coronavirus pneumonia (COVID-19) in the world [1], and so far (July 17, 2020), according to the latest information from the World Health Organization, there have been 14,562,550 confirmed cases of COVID-19, including 607,781 deaths around the world.

Many researches show that COVID-19 is highly contagious, and wearing a mask can effectively reduce the probability of viral infection. Research [2] on masks also showed that masks are very helpful in preventing infection in universal community and health care settings during the COVID-19 pandemic. In this certain condition, asymptomatic infectors are difficult to detect and there is absence of vaccination, wearing masks is undoubtedly a wise choice to prevent infection [3, 4].

J. Cao · R. Pang (✉) · R. Ma · Y. Qi
Faculty of Information and Control Engineering, Shenyang Jianzhu University, Shenyang, China
e-mail: albert_pang@126.com

© The Author(s), under exclusive license to Springer Nature Singapore Pte Ltd. 2021   433
Y. Li et al. (eds.), *Advances in Simulation and Process Modelling*,
Advances in Intelligent Systems and Computing 1305,
https://doi.org/10.1007/978-981-33-4575-1_41

Nowadays, there are no specific algorithms used for mask detection. With the development of deep learning in the field of computer vision [5–7], the target detection algorithm based on neural network has been widely applied in pedestrian target detection, face detection, remote sensing image detection, medical image detection, and natural scene text detection [8–11]. In 2001, Viola and Jones proposed a face detection algorithm based on AdaBoost [12], it was the first practical boosting algorithm, and remains one of the most widely used and studied. In 2016, Chen et al. [13] proposed the RPN network and the RCNN network to address large pose variations in real-word face detection. A model called multi-task convolutional neural network (MTCNN) was designed by Zhang et al. [14]; it can exploit the inherent correlation between detection and alignment to boost up their performance, with applications in numerous fields. For better performance, convolutional neural networks are designed, from seven layers [15] to thousands of layers [16, 17]. The MobileNet [18] model is outstanding because of its lightweight and efficient performance. Through the study of relevant target detection algorithm, it is found that the deep learning model for face detection mentioned above can be applied to the mask detection. However, how to design a fast and accuracy method for mask detection is still a challenging job.

This paper aims to develop an algorithm that identifies whether pedestrians are wearing a mask, then it can be used in a particular place or public to alert people. A method is designed for mask recognition which can decrease the amount of training and optimize the loss function. MTCNN algorithm is used for face detection on account of its speed, and MobileNet structure is applied for feature extraction because of its high accuracy and strong reliability.

2   Face Detection Based on MTCNN

2.1   Algorithmic Principle

MTCNN algorithm is used in this paper, which is a rough-to-thin process, while using face detection tasks and face sub-tasks to assist face key detection. The overall framework is divided into three stages. The three stages are entered for different sizes of pictures, which are intended to detect faces of different sizes as Fig. 1.

The first stage, Proposal Network (P-Net): This network master is intended to generate candidate windows as input for stage two. On pictures of different sizes, use a fully convolutional networks (FCN) to get the border coordinates of the candidate windows, and use the border regression to correct them, and then use non-maximum value suppression to aggregate the highly overlapping candidate window.

Because P-Net uses FCN, there is no need to specify the size of the input image ($12 \times 12 \times 3$ in the input size of Fig. 1 is an example, which can be considered an anchor box). It returns two parameters, one is the coordinate of the face box, and the other is the probability of whether the face is.

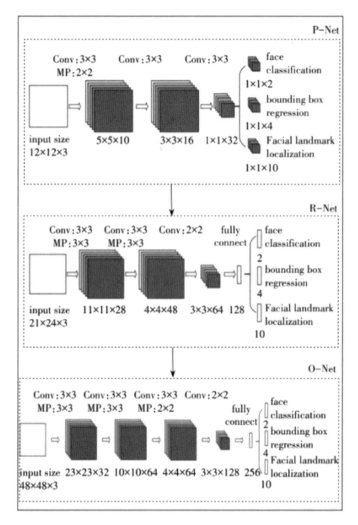

Fig. 1 MTCNN network architecture

The second stage, Refinement Network (R-Net): Use the front candidates as another input to convolutional neural networks (CNN). It further rejects the wrong candidates, corrects the border, and uses non-maximum suppression (NMS) and candidates. R-Net accepts the border of the P-Net output, the input is fixed, so the border is resized, the intermediate process is similar to R-Net, it returns two parameters: one is the coordinates of the face box, and the other is the probability of whether the face is.

The last stage, Output Network (O-Net): This network accepts the border returned by R-Net, and resizes to a fixed size, which returns three parameters: the first argument is the face box, the second argument is the probability of whether the face is face, and the third parameter is the five-key-point coordinates.

One of the most important reasons using the MTCNN algorithm for face detection is that it can independently mark the five-character features of faces, such as find the eye, nose and mouth angle that detects the face, which is essential for subsequent recognition based on sensitive features. Tag face data is used for facial feature positioning. Face feature positioning is similar to boundary regression task, as a regression problem, the European distance calculation is used to see Eq. (1).

$$\text{loss}_3 = \left\| y_t^{\text{landmarks}} - y_t^{\text{landmark}} \right\|_2^2 \tag{1}$$

where $y_i^{\text{landmarks}}$ is the network obtained face marker coordinates, y_i^{landmark} is the real face marker coordinates, a total of five groups, marked eyes, nose, mouth corner significant features.

Finally, due to the different classification regression tasks performed in different convolution network layers, different training pictures (face, non-face) are used in the learning process, in order to unify the learning goal set the task weight to balance the three sub-networks 21 as Eq. (2).

$$\min \sum_{i=1}^{N} \sum_{j \in \{1,2,3\}} \alpha_j \beta_i^j \text{loss}_j \tag{2}$$

where N represents the number of training samples, α_j represents the weight of each task, β_i^j represents the sample type indicator, and loss_j ($j = 1, 2, 3$) represents the error loss of face detection, respectively, bounding frame regression and face feature positioning.

2.2 The Process of Identification

In order to identify the contours of the face, MTCNN is used to process the images. Firstly, change the image obtained by Opencv from BGR format to RGB format for processing, and then the face image is extracted from the photo through the MTCNN face detection model. The last output image is used as the input image of the MobileNet as Fig. 2.

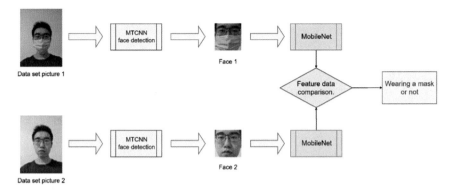

Fig. 2 Model reasoning process

3 Feature Extraction Based on MobileNet

3.1 Algorithmic Principle

MobileNet uses deep separable convolutions to build lightweight deep neural networks. It can be used in large-scale image detection and classification, and maximizes the accuracy of the model.

The depth separable convolution resolves the standard convolution into a deep convolution as shown in Fig. 3 and a 1×1 pointwise convolution as shown in Fig. 4. As shown in Fig. 5, it is supposed that the input feature map has M channels, the number of channels of the output feature map is N, and the size of the convolution kernel is $D_K \times D_K$. MobileNet convolution network reduces the network parameters of the model under the premise of ensuring network accuracy. The amount of parameters [19] of the traditional convolutional neural network convolution K, as shown in Eq. (3):

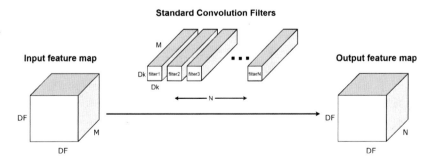

Fig. 3 Standard convolution structure

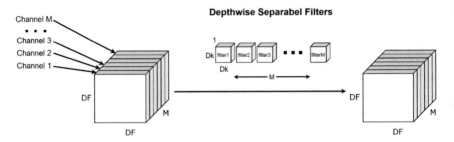

Fig. 4 Depthwise separable structure

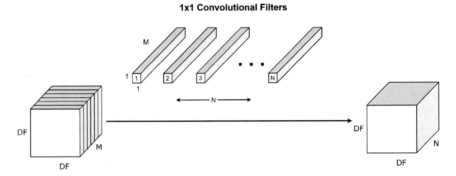

Fig. 5 Pointwise convolution structure

$$K = D_K \times D_K \times M \times N \times D_F \times D_F \qquad (3)$$

MobileNet's convolution almanac parameter [20] is shown as in Eq. (4):

$$K' = D_K \times D_K \times M \times D_F \times D_F + M \times N \times D_F \times D_F \qquad (4)$$

Compare the calculation amount of the deep convolution with the standard convolution part:

$$\frac{D_K \times D_K \times D_M \times D_M \times M + M \times N \times D_F \times D_F}{D_K \times D_K \times M \times N \times D_F \times D_F} = \frac{1}{N} + \frac{1}{D_K^2} \qquad (5)$$

The MobileNet convolution network divides traditional convolutional neural networks into depthwise convolution and pointwise convolution in the convolutional layer. Depthwise convolution is a 3×3 convolution, pointwise convolution is 1×1 convolution, through batch normalization and rectified linear unit activation functions to build the convolution layer of the network. It can be seen from the result of Eq. (5) that this operation can reduce the amount of calculation by 8 to 9 times, and improve the efficiency of calculation.

3.2   Optimization Algorithm

In this paper, the Adam [21] optimization method is used to minimize the loss function by replacing the stochastic gradient descent algorithm. The random gradient drop algorithm keeps the single learning rate (alpha) updating all the weights, and the learning rate does not change during network training. The Adam optimization algorithm designs independent adaptive learning rates by calculating the first-order moment estimation of gradient and the second-order moment estimate for different parameters. Adam algorithm has high computational efficiency and low memory requirements, and Adam algorithm gradient of the diagonal rescaling has immutability. The process of updating network parameters is represented as:

$$g_t = \frac{1}{n} \nabla\theta \sum_i L(F(Y_i, \theta), X_i) \tag{6}$$

$$m_t = u \times m_{t-1} + (1 - u) \times g_t \tag{7}$$

$$n_t = v \times n_{t-1} + (1 - v) \times g_t^2 \tag{8}$$

$$\hat{m}_t = \frac{m_t}{1 - u^t} \tag{9}$$

$$\hat{n}_t = \frac{n_t}{1 - v^t} \tag{10}$$

$$\Delta\theta_t = -\eta \times \frac{\hat{m}_t}{\sqrt{\hat{n}_t} + \varepsilon} \tag{11}$$

$$\theta_{t+1} = \theta_t + \Delta\theta_t \tag{12}$$

where g_t is the mean-square error function $L(\theta)$ to θ gradient, m_t the first-order moment $\hat{m}_t$ estimation of the gradient, n_t the second-order moment estimation of the gradient, $\hat{m}_t$ the deviation of m_t correction, $\hat{n}_t$ the deviation of n_t.

The moment estimate of the exponential decay rate u is 0.9, v is 0.99, the step η is 0.001, the numerically stable small constant ε is 10^{-8}, $\Delta\theta_t$ represents the calculated θ_t update value, θ_{t+1} the θ value of the $t + 1$ moment, and the sum of θ_t and $\Delta\theta_t$ is applied to θ_{t+1}.

Fig. 6 Example of data set pictures

4 Experiment and Analysis

4.1 The Experimental Data Set

In order to detect whether a face is wearing a mask in a complex environment, a self-built data set is used in this paper. The pictures were arbitrarily split into training and test datasets, with proportion 90% and 10%, respectively. Naturalize the size of the pictures. The data set contains two types of labels: the face wearing masks and the no mask ones shown in Fig. 6.

After getting the data set pictures, read and tag them (mask 0, no mask 1), scramble the data set, and process the labels in one-hot format before training. Due to the performance of the hardware device, each training set size is 8 pictures. Pictures in the datasets need to be tagged (mask 0, no mask 1), after which the data set will be scrambled, and the labels will be processed in one-hot format before training.

It is essential to use data enhancing, image reversing and color gamut conversing to get enhanced data, then it will be naturalized, so as to make the model more robust. In model training, pre-trained MobileNet weights are used to make training more effective. The output will be saved every three generations of training, and if the training accuracy of five successive generations does not decline, the learning rate of training will be further reduced. If the loss value does not decline after 12 generations, that means model training is completed. According the above method, the model can correctly recognize whether the face is wearing a mask in different environments.

4.2 The Experimental Results

In order to verify the effectiveness of the proposed training model for wearing masks, the experiment will compare the detection effect of YOLOv3, compare the detection performance of the two algorithms in complex cases, and draw experimental conclusions. The results of quantitative analysis of comparison between our model and YOLOv3 are presented in Table 1.

Table 1 Performance comparison of two models

Algorithm	The number of test samples	Accurate lying number	Accuracy rate (%)	Detection speed (s)
YOLOv3	707	658	93.10	2.023
MTCNN + MobileNet	707	669	94.73	1.907

The results obtained from these sequences are showed in Table 1, where the average accuracy rate of YOLOv3 is 93.10%, and the method of this paper is 94.73%. The experiments were performed with the GPU graphics card NVIDIA GeForce MX 350, and 8 GB memory on personal computer.

5 Conclusions

In this paper, a mask recognition method based on MTCNN and MobileNet is proposed, and the model shows that it has high precision and robustness in complex environments. Compared with the current popular YOLOv3, this model also demonstrates better detection efficiency. In the future, the size of the existing data set will be expanded. More in-depth research will be conducted based on hardware equipment to make contributions to social security.

Acknowledgements This research work was financially supported by the Youth Science Foundation of Liaoning Province, China.

References

1. Politis, G., Hadjileontiadis, L.: COVID-19 infection spread in Greece: ensemble forecasting models with statistically calibrated parameters and stochastic noise. medRxiv (2020)
2. MacIntyre, C.R., Chughtai, A.A.: A rapid systematic review of the efficacy of face masks and respirators against coronaviruses and other respiratory transmissible viruses for the community, healthcare workers and sick patients. Int. J. Nurs. Stud. **108**, 103629 (2020)
3. Yashavantha Rao, H.C., Jayabaskaran, C.: The emergence of a novel coronavirus (Sars-Cov-2) disease and their neuroinvasive propensity may affect in Covid-19 patients. J. Med. Virol. **92**(7), 786–790 (2020)
4. Sleator, R.D., Darby, S., Giltinan, A., Smith, N.: Covid-19: in the absence of vaccination-'mask-the-nation'. Future Med. (2020)
5. Chokkadi, S., Bhandary, A.: A study on various state of the art of the art face recognition system using deep learning techniques. arXiv:1911.08426 (2019)
6. Alawad, M., Gao, S., Qiu, J.X., Yoon, H.J., Blair Christian, J., Penberthy, L., Tourassi, G.: Automatic extraction of cancer registry reportable information from free-text pathology reports using multitask convolutional neural networks. J. Am. Med. Inform. Assoc. **27**(1), 89–98 (2020)
7. Curcic, M.: A parallel Fortran framework for neural networks and deep learning. In: ACM SIGPLAN Fortran forum, vol. 38, no. 1, pp. 4–21. ACM, New York, NY, USA (2019)

8. Ali, A.A., Abd El-Hafeez, T., Mohany, Y.K.: An accurate system for face detection and recognition. J. Adv. Math. Comput. Sci. **33**(3), 1–19 (2019)
9. Zhang, S., Zhu, X., Lei, Z., Wang, X., Shi, H., Li, S.Z.: Detecting face with densely connected face proposal network. Neurocomputing **284**, 119–127 (2018)
10. Diao, W., Sun, X., Zheng, X., Dou, F., Wang, H., Fu, K.: Efficient saliency-based object detection in remote sensing images using deep belief networks. IEEE Geosci. Remote Sens. Lett. **13**(2), 137–141 (2016)
11. Litjens, G., Kooi, T., Bejnordi, B.E., Setio, A.A.A., Ciompi, F., Ghafoorian, M., Sánchez, C.I.: A survey on deep learning in medical image analysis. Med. Image Anal. **42**, 60–88 (2017)
12. Viola, P., Jones, M.: Rapid object detection using a boosted cascade of simple features. In: Proceedings of the 2001 IEEE Computer Society Conference on Computer Vision and Pattern Recognition. CVPR 2001, vol. 1, pp. I-I. IEEE, Kauai, HI, USA (2001)
13. Chen, D., Hua, G., Wen, F., Sun, J.: Supervised transformer network for efficient face detection. In: Leibe, B., Matas, J., Sebe, N., Welling, M. (eds.) Computer Vision—ECCV 2016. Lecture Notes in Computer Science, vol. 9909. Springer, Cham (2016)
14. Zhang, K., Zhang, Z., Li, Z., Qiao, Y.: Joint face detection and alignment using multitask cascaded convolutional networks. IEEE Signal Process. Lett. **23**(10), 1499–1503 (2016)
15. Krizhevsky, A., Sutskever, I., Hinton, G.E.: ImageNet classification with deep convolutional neural networks. Commun. ACM **60**(6), 84–90 (2017)
16. He, K., Zhang, X., Ren, S., Sun, J.: Deep residual learning for image recognition. In: IEEE Conference 2016, Computer Vision and Pattern Recognition (CVPR), pp. 770–778. IEEE, Las Vegas, NV, USA (2016)
17. Huang, G., Liu, Z., Van Der Maaten, L., Weinberger, K.Q.: Densely connected convolutional networks. In: IEEE Conference 2016, Computer Vision and Pattern Recognition (CVPR), pp. 2261–2269. IEEE, Honolulu, HI, USA (2017)
18. Howard, A.G., Zhu, M., Chen, B., Kalenichenko, D., Wang, W., Weyand, T., Adam, H.: Mobilenets: efficient convolutional neural networks for mobile vision applications. arXiv preprint arXiv:1704.04861 (2017)
19. Qin, Z., Zhang, Z., Chen, X., Wang, C., Peng, Y.: Fd-Mobilenet: improved mobilenet with a fast downsampling strategy. In: 25th IEEE International Conference on Image Processing (ICIP), pp. 1363–1367. IEEE, Athens, Greece (2018)
20. Chugh, T.: An Accurate, Efficient, and Robust Fingerprint Presentation Attack Detector. Michigan State University (2020)
21. Kingma, D.P., Ba, J.: Adam: a method for stochastic optimization. arXiv preprint arXiv:1412. 6980 (2014)

Quadcopter UAV Finite Time Sliding Mode Control Based on Super-Twisting Algorithm

Jianhua Zhang, Wenbo Fei, and Yang Li

Abstract In this paper, the super-twisting sliding mode intelligent control is applied to the position and attitude of the underdrive quadcopter system. For this system, super-twisting sliding mode control has more advantages, which can solve the chattering problem of traditional sliding mode control and can converge in a finite time. Compared with other research results, this article provides a more accurate mathematical model of a quadcopter, which takes into account more influencing factors and is practical. Finally, numerical example verifies the effectiveness of the control algorithm.

Keywords Quadcopter · Finite time · Super-twisting

1 Introduction

For decades, due to its inherent characteristics, the quadcopter has been widely concerned by military, civilian, and engineering academics. The application range of small drones such as quadcopter is extremely wide, such as military reconnaissance in harsh environments, civil logistics, aerial photography, and spraying of pesticides [1–3]. In order to improve the stability and reliability of the aircraft in various scenarios, scholars have made a lot of research results in the study of intelligent control of quad-rotor drones.

The first rotorcraft came out in the twentieth century. After a miniaturized multi-rotor UAV is developed, scholars have verified the feasibility of a large number of control algorithms. At the same time, some scholars use PID combined with other algorithms to solve complex control problems [4–6]. The LQR control algorithm is also a relatively mature algorithm applied in the quadcopter control. In [7], scholars

J. Zhang (✉) · Y. Li
School of Information and Control Engineering, Qingdao University of Technology, Qingdao 266520, Shandong, China
e-mail: Jianhuazhang@aliyun.com

W. Fei
Hebei University of Science and Technology, Shijiazhuang 050018, Hebei, China

© The Author(s), under exclusive license to Springer Nature Singapore Pte Ltd. 2021   443
Y. Li et al. (eds.), *Advances in Simulation and Process Modelling*,
Advances in Intelligent Systems and Computing 1305,
https://doi.org/10.1007/978-981-33-4575-1_42

use the LQR control algorithm to study the established mathematical model, but the control performance is not ideal due to the inherent characteristics of the algorithm. However, the traditional sliding mode membrane control has a discontinuous function of high-frequency switching, which leads to chattering. Therefore, this paper proposes super-twisting sliding mode control.

The main contributions of this article are as follows:

1. Select a more accurate quadcopter mathematical model, with higher control requirements and better simulation practicality.
2. A quad-rotor aircraft based on super-twisting sliding mode control is proposed, which can converge in a finite time and reduce the chattering of traditional sliding mode.
3. The verification of the calculation example used in this paper is closer to the working logic of the actual quadcopter.

2 Mathematical Model of Quadcopter

According to the quadcopter dynamic motion law, a quadcopter dynamic model [8] is established, and the body coordinate system and the ground coordinate system are selected, as shown in Fig. 1. According to the spatial transformation of coordinate system B and coordinate system E and the Newton–Euler equation, the mathematical model of the quadcopter is calculated. The quadcopter attitude is expressed as yaw angle ψ, pitch angle θ, and roll angle ϕ.

The quadcopter position and attitude dynamics model can be obtained:

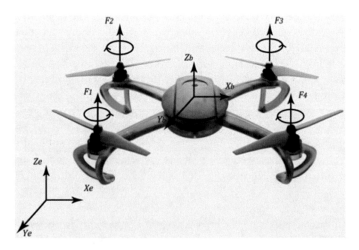

Fig. 1 Schematic of the quad-rotor system

$$
\begin{cases}
\ddot{x} = \dfrac{F_{at}}{m}(\sin\psi\sin\phi + \cos\psi\sin\theta\cos\phi) - \dfrac{\xi_x}{m}\dot{x} + d_x(\cdot) \\[3mm]
\ddot{y} = \dfrac{F_{at}}{m}(-\cos\psi\sin\phi + \sin\psi\sin\theta\cos\phi) - \dfrac{\xi_y}{m}\dot{y} + d_y(\cdot) \\[3mm]
\ddot{z} = \dfrac{F_{at}}{m}(\cos\theta\cos\phi) - g - \dfrac{\xi_z}{m}\dot{z} + d_z(\cdot) \\[3mm]
\ddot{\phi} = \dot{\theta}\dot{\psi}(I_y - I_z)/I_x + \tau_{a\phi}/I_x - \dfrac{I_r}{I_x}\dot{\theta}\varpi - \dfrac{\xi_\phi}{I_x}\dot{\phi} + d_\phi(\cdot) \\[3mm]
\ddot{\theta} = \dot{\phi}\dot{\psi}(I_z - I_x)/I_y + \tau_{a\theta}/I_y - \dfrac{I_r}{I_y}\dot{\phi}\varpi - \dfrac{\xi_\theta}{I_y}\dot{\theta} + d_\theta(\cdot)
\end{cases}
$$

$$
\begin{cases}
\ddot{\psi} = \dot{\theta}\dot{\phi}(I_x - I_y)/I_z + \tau_{a\psi}/I_z - \dfrac{\xi_\psi}{I_z}\dot{\psi} + d_\psi(\cdot)
\end{cases}
\tag{1}
$$

where lifting force F_{at}, I_r—rotor inertia, $I = \mathrm{diag}(I_x, I_y, I_z)$—three moments of inertia in the body coordinate system $\xi_k = \mathrm{diag}(\xi_\phi\ \xi_\theta\ \xi_\psi)$—air drag torque coefficient w—rotor angular velocity. $d_a(\cdot) = \mathrm{diag}[d_x(\cdot)\ d_y(\cdot)\ d_z(\cdot)]$, $d_p(\cdot) = \mathrm{diag}[d_\phi(\cdot)\ d_\theta(\cdot)\ d_\psi(\cdot)]$, $\overline{w} = w_4 + w_2 - w_3 - w_1$ is outside interference.

3 Design of Quadcopter Position and Attitude Controller

Due to the discontinuous function of the high-frequency switch under the traditional sliding mode control rate, resulting in jitter, the super-twisted sliding mode control can transfer the discrete control rate to the higher-order sliding mode surface to solve this problem [9]. And a large number of articles have proved that the algorithm can guarantee the finite time convergence [10]. The super-twist sliding mode control algorithm is composed of a continuous function of sliding mode variables and a discontinuous time difference. The algorithm is as follows:

$$
\begin{cases}
\dot{x}_1 = -k|x_1|^{\frac{1}{2}}\mathrm{sign}(x_1) + x_2 \\[2mm]
\dot{x}_2 = -k_1\mathrm{sign}(x_1) + \dot{d}_t
\end{cases}
\tag{2}
$$

where $x_i\,(i = 1, 2)$ represents the state variable, and k_1, k_2 represents the set gain, $\dot{d}_t$ is external interference. Translational Dynamics:

$$
\ddot{x} = Au_s(t) + f_1(\cdot) + d_a(x, t)
\tag{3}
$$

Expect the system position output x to be able to track x_d, so there is

$$z_1 = x - x_d; \dot{z}_1 = \dot{x} - \dot{x}_d \tag{4}$$

According to Eq. (2) super-twisting algorithm gives

$$z_2 = \dot{x} - \dot{x}_d + k|z_1|^{\frac{1}{2}} \mathrm{sgn}(z_1) \tag{5}$$

Substituting Eq. (3) into (5) gives

$$\dot{z}_2 = Au_s(t) + f_1(\cdot) + d_a(x, t) - \ddot{x}_d + \frac{k}{2} \frac{\dot{x} - \dot{x}_d}{|z_1|^{\frac{1}{2}}} \tag{6}$$

Available controller

$$u_s(t) = A^{-1} \left[-k_1 \mathrm{sgn}(z_1) + d_a(x, t) - f_1(\cdot) - d_a(x, t) + \ddot{x}_d - \frac{k}{2} \frac{\dot{x} - \dot{x}_d}{|z_1|^{\frac{1}{2}}} \right] \tag{7}$$

$$u_s(t) = A^{-1} \left[-k_1 \mathrm{sgn}(z_1) - f_1(\cdot) + \ddot{x}_d - \frac{k}{2} \frac{\dot{x} - \dot{x}_d}{|z_1|^{\frac{1}{2}}} \right] \tag{8}$$

Select the Lyapunov function

$$V(t) = \frac{1}{2} z_2^2 \tag{9}$$

$$\dot{V}(t) = z_2 \left(Au_s(t) + f_1(\cdot) + d_a(x, t) - \ddot{x}_d + \frac{k}{2} \frac{\dot{x} - \dot{x}_d}{|z_1|^{\frac{1}{2}}} \right) \tag{10}$$

$$\dot{V}(t) = \left[\dot{z}_1 + k|z_1|^{\frac{1}{2}} \mathrm{sgn}(z_1) \right] \left[-k_1 \mathrm{sgn}(z_1) \right]$$

When $z_1 > 0$, $\dot{V}(t) < 0$ is stable, and $k, k_1 > 0$, when $z_1 < 0$, $\dot{V}(t) < 0$ is stable, $k, k_1 > 0$. When is stable. Where $\mathrm{sgn}(z_1) = \begin{cases} 1 & z_1 > 0 \\ -1 & z_1 < 0 \end{cases}$.

Similarly, get the attitude controller

$$u_r(t) = B^{-1} \left[-k_1 \mathrm{sgn}(z_3) - f_2(\cdot) + \ddot{p}_d - \frac{k}{2} \frac{\dot{p} - \dot{p}_d}{|z_3|^{\frac{1}{2}}} \right]. \tag{11}$$

4 Simulation and Analysis

In this section, MATLAB simulations will be used to verify the effectiveness of the finite time control of a quadcopter based on the super-twisting algorithm.

Example 1 Consider the UAV dynamic system with the parameters as

$$m = 2, l = 0.2, g = 9.8, \xi_x = \xi_x = \xi_x = 1.2$$
$$\xi_\phi = \xi_\theta = \xi_x = 1.2, I_x = 1.25, I_y = 1.25, I_z = 2.5$$

$$k_1 = 0.1, k_2 = 0.00002, k_3 = 0.01, k_4 = 4, k_5 = 9.8, k_6 = 1.6$$

$$k_7 = 11, k_8 = 260,000, k_9 = 0.1, k_{10} = 22, k_{11} = 11, k_{12} = 0.02$$

$$\chi_d(t) = \left[5\left(1 - \cos\left(\frac{\pi}{10}t\right)\right), 5\sin\left(\frac{\pi}{10}t\right), 10\left(1 - e^{-0.3t}\right) \right]^T$$

The initial conditions of the system are selected as

$$x(0) = y(0) = z(0) = \phi(0) = \theta(0) = 0, \psi(0) = 1$$

In order to highlight the advantages of super-twisting algorithm sliding mode control compared to traditional sliding mode control, a simple comparison is made by using the traditional sliding mode control simulation of the quadcopter position.

$$\partial_d(t) = \left[10\sin\left(\frac{\pi}{10}t\right), \cos\left(\frac{\pi}{10}t\right) \right]^T$$

The initial conditions of traditional sliding mode control are selected as

$$x(0) = y(0) = z(0) = 1$$

Comparing the simulation results of position tracking control with Figs. 2 and 3 with Fig. 5, it is enough to show that compared with traditional sliding mode control, the sliding mode control based on super-twisting algorithm has more advantages. Figures 2, 3, and 4 show the verification of the reliability of the super-twisting algorithm, which can better achieve tracking control.

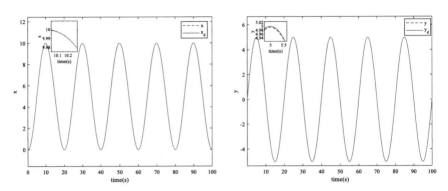

Fig. 2 Trajectories of position x, ideal position x_d, position y and ideal position y_d

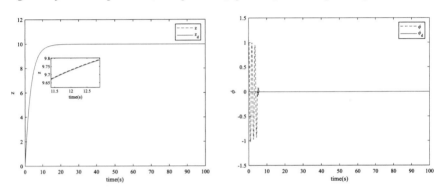

Fig. 3 Trajectories of position z, ideal position z_d, attitude ϕ and ideal position ϕ_d

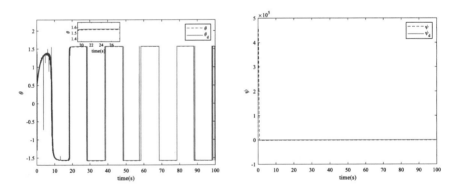

Fig. 4 Trajectories of attitude θ, ideal position θ_d, attitude ψ and ideal position ψ_d

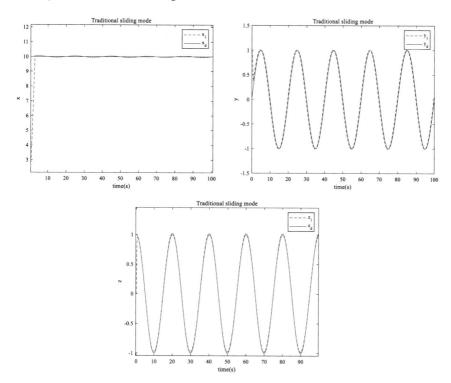

Fig. 5 Trajectories of position x_1, ideal position x_d, position y_1, ideal position y_d, position z_1, and ideal position z_d

5 Conclusions

In this paper, the super-twisting sliding mode is used to control the attitude and position of the quadcopter. Experimental simulation result shows that the controller algorithm is effective for the mathematical model of the quadcopter. Because there are uncertain nonlinear parts in the system, the next step is to verify whether the RBF neural network can be used for approximation.

References

1. Liu, J., Chen, P., Xu, X.: Estimating wheat coverage using multispectral images collected by unmanned aerial vehicles and a new sensor. In: 2018 7th International Conference on Agro-geoinformatics (Agro-geoinformatics) (2018)
2. Wang, Y., Zhang, W.: Four-rotor UAV virtual simulation system for agricultural sowing. In: 2018 IEEE 4th Information Technology and Mechatronics Engineering Conference (ITOEC) (2018)

3. Islam, S., Liu, P.X., Saddik, A.E.: Robust control of four-rotor unmanned aerial vehicle with disturbance uncertainty. IEEE Trans. Industr. Electron. **62**(3), 1563–1571 (2015)
4. Argentim, L.M., et al.: PID, LQR and LQR-PID on a quadcopter platform. In: 2013 International Conference on Informatics, Electronics and Vision (ICIEV) (2013)
5. Mie, S., Okuyama, Y., Saito, H.: Simplified quadcopter simulation model for spike-based hardware PID controller using system C-AMS. In: 2018 IEEE 12th International Symposium on Embedded Multicore/Many-core Systems-on-Chip (MCSoC) (2018)
6. Fatan, M., Sefidgari, B.L., Barenji, A.V.: An adaptive neuro PID for controlling the altitude of quadcopter robot. In: 2013 18th International Conference on Methods & Models in Automation & Robotics (MMAR) (2013)
7. Fan, B., Sun, J., Yu, Y.: A LQR controller for a quadrotor: design and experiment. In: 2016 31st Youth Academic Annual Conference of Chinese Association of Automation (YAC) (2016)
8. Song, Y., He, L., Qian, J., Fu, J.: Neuroadaptive fault-tolerant control of quadrotor UAVs: a more affordable solution. IEEE Trans. Neural Netw. Learn. Syst. **30**(7), 1975–1983 (2019)
9. Lo, J.T., Bassu, D.: Adaptive parallel identification of dynamical systems by adaptive recurrent neural networks. In: Proceedings of the International Joint Conference on Neural Networks (2003)
10. Fei, J., Feng, Z.: Fractional-order finite-time super-twisting sliding mode control of micro gyroscope based on double-loop fuzzy neural network. IEEE Trans. Syst. Man Cybern. Syst. 1–15 (2020)

Cartesian Admittance Control Based on Maxwell Model for Human Robot Interaction

Xin Wang, Jia Sun, Zhijun Gao, Languang Zhao, Jianshun Liu, and Naifeng He

Abstract Classical impedance control methods use Voigt model to describe relationship between external force and displacement. It is a kind of viscoelastic model shows elastic mechanical properties. This feature always generates a repulsive force because spring plays a major role and make user uncomfortable. By contrast, Maxwell model shows plastic mechanical properties and it shows the plastic properties. This characteristic will improve the comfortableness of human robot interaction. Therefore, we present a Maxwell model-based Cartesian admittance control that uses input of position loop to the manipulator. We compared two methods by experiment and result shows that our method can provide a good comfortableness in the case of having same effect of traditional Cartesian admittance control.

Keywords Human robot interaction · Admittance control · Industrial robot

1 Introduction

In recent years, due to the continuous improvement of robot safety [1] and the demand for various advanced tasks, the relationship between robot and human has been transformed from an isolated state to human robot interaction [2] and collaboration, which has become a hot spot in research. For the trend breakthrough the limit of structured environment, the chance of interaction between human and robot will increase rapidly. Therefore, robots can cope with uncertainty and complexity, especially to prevent damage to the robot itself and the environment and have compliance becomes one of several prerequisites for robots.

In general, there are two kinds of ways make a robot have compliance. One is called passive compliance, which use mechanisms such as remote center compliance

X. Wang · J. Sun · Z. Gao (✉) · L. Zhao · J. Liu
Faculty of Information and Control Engineering, Shenyang Jianzhu University, Shenyang, China
e-mail: gzj@sjzu.edu.cn

N. He
XCMG Construction Machinery, Xuzhou, China

device, which is also called RCC device [3]. Other mechanisms can also achieve same effect such as series elastic actuator [4]. The other way is servo control or active compliance control. Typical compliance control methods are hybrid position/force control [5] and impedance control [6]. When robots interact with environment, the touch force and displacement are coupled. By adjusting a mass, spring constant and viscosity coefficient of impedance control, impedance control can achieve a dynamic relationship between displacement and force close the real situation and admittance control is a mirror of impedance control. In recent years, several different methods of impedance have been proposed from multiple perspectives. Ott et al. proposed unified impedance and admittance control [7] can switch between impedance control and admittance control to combine the capacity of two control laws. Scherzinger et al. proposed forward dynamic compliance control [8] which has excellent performance and stability in singularities. Lee et al. proposed task space control [9] improved e accuracy, robustness, and stability of forward dynamics control and solve the problem when manipulator move through singularities, velocity of manipulator change fast. Maciej et al. [10] proposed impedance control based on MPC which behaves exactly as impedance control and can add limits of position, velocity, energy, and so on.

All of impedance control above use Voigt model that can be described as a connect which a spring and damper connected parallel. It is a kind of viscoelastic model [11] shows elastic mechanical properties. When apply a force on it, it always generates a repulsive force because spring plays a major role. This may influence the human comfortableness. By contrast, there is another model called Maxwell model shows plastic mechanical properties. It often is used as fluid description [12]. Compare with Voigt model, it is described as a connect that spring and damper is in series, it shows the plastic properties. When apply a force on it, it will have a plastic deformation not return elastic force. This characteristic will improve the comfortableness of human robot interaction. Fu and Zhao proposed an impedance control framework [13], and a plastic deformation behavior is produced under unexcepted contact force. Then, they proposed a general Maxwell model-based Cartesian admittance control [14] that uses input of velocity to the manipulator, and it will decrease the accuracy of Cartesian trajectory.

Therefore, we present a Maxwell model-based Cartesian admittance control that uses input of position loop to the manipulator.

This paper is organized as follows. An introduction and comparison of two models are present in Sect. 2. A Maxwell model-based Cartesian admittance control is proposed in Sect. 3. Simulation study is given in Sects. 4 and 5 concludes this paper.

2 Voigt Model and Maxwell Model

$M, B, K,$ and F_{ext} denote the mass of lump object, dump coefficient, spring constant, and external force apply, respectively, in every case of two models. x_d represents the desired position and x represents the actual position. $x_e = x - x_d$ is the displacement of end-effector of manipulator. Two models are shown in Fig. 1.

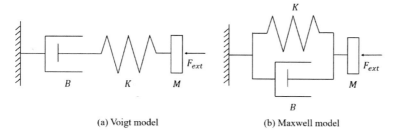

(a) Voigt model (b) Maxwell model

Fig. 1 Two viscoelastic models: **a** Voigt model **b** Maxwell model

2.1 Voigt Model

The relationship between external force F_{ext} and displacement x_e can be illustrated as Eq. (1), and it is a second order differential equation. This equation is used widely in most impedance control laws. It is common to know that when external force vanished, the displacement of end-effector will come to zero eventually. This deformation is called elastic deformation.

$$M\ddot{x}_e + B\dot{x}_e + Kx_e = F_{ext} \tag{1}$$

It is easy to know overdamped condition of Voigt model is inequality (2).

$$B^2 > 4MK \tag{2}$$

2.2 Maxwell Model

Due to the different arrange of spring and damper, Maxwell model has a more different deformed feature and dynamics. Dynamic equation of Maxwell model can be described as inequality (2). When external force vanished, the displacement of end-effector will be a constant. This feature is called plastic deformation.

$$M\ddot{x}_e + KB^{-1}M\dot{x}_e + Kx_e = F_{ext} + KB^{-1}\int F_{ext} \tag{3}$$

Its overdamped condition is shown as inequality (4).

$$4B^2 < MK \tag{4}$$

Maxwell model also can be transformed to a Voigt model, as shown in Fig. 2. Due to the integral term of external force, it has a plastic deformation.

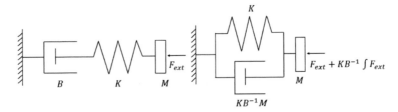

Fig. 2 Equivalent transformation of the Maxwell model to Voigt model

3 Cartesian Maxwell-Based Admittance Control

The basic idea of admittance control is that the control system adopts the inner loop based on position control and the outer loop based on force control.

This control law enables the system to show the impedance characteristics of Voigt model, which does not require the dynamic model of manipulator. This method is especially suitable for servo control system with good position control effect. The result of admittance control is the desired position of the joint, which is easily realized by traditional industrial robots. The application of human robot interaction is based on admittance control.

In conventional situation, relationship between external force and displacement is described as follows:

$$M_d \ddot{x}_m + B_d \dot{x}_e + K_d x_e = F_{\text{ext}} \tag{5}$$

We can get the modified acceleration of manipulator by the following Eq. (6)

$$\ddot{x}_m = M_d^{-1}(F_{\text{ext}} - B_d \dot{x}_e - K_d x_e) \tag{6}$$

External force is described on base frame of manipulator.

For calculating the modified acceleration, the current position and velocity of end-effector need to be used. The position can be solved using the forward kinematics of the robot, while the velocity can be solved using the following Eq. (7)

$$\dot{x} = J\dot{q} \tag{7}$$

The control law makes the control system equivalent to a Voigt model and makes it equivalent to a Maxwell model by replace Eqs. (5) and (6) with (8) and (9).

$$M_d \ddot{x}_m + K_d B_d^{-1} M_d \dot{x}_e + K_d x_e = F_{\text{ext}} + K_d B_d^{-1} \int F_{\text{ext}} \tag{8}$$

$$\ddot{x}_m = M_d^{-1}\left(F_{\text{ext}} + K_d B_d^{-1} \int F_{\text{ext}} - K_d B_d^{-1} M_d \dot{x}_e - K_d x_e \right) \tag{9}$$

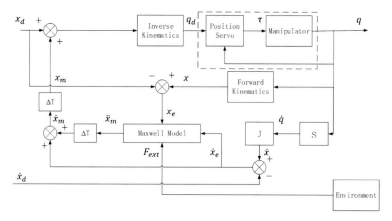

Fig. 3 The complete control structure of Cartesian admittance control based on Maxwell model

The complete control structure of Cartesian admittance control is shown in Fig. 3.

4 Simulation Study

The manipulator used in our experiments is based on DOREP [15]. We use MDH method to calculate kinematics and kinematic parameters are available in the manuals.

We presented an experiment of Maxwell model-based Cartesian admittance control and Voigt model-based Cartesian admittance control. The experiment is to verify the effectiveness of method proposed in Sect. 3 and demonstrate the differences between two methods based on the Maxwell model and Voigt model. During the following experiments, the desired viscoelastic coefficients of the Maxwell model are listed in Table 1. For comparing the result, we translate the viscoelastic coefficients into equivalent coefficients of Voigt model and ignore integral part of external force.

Table 1 Desired viscoelastic coefficients in experiment

Viscoelastic coefficients	Desired values
M_{pd}	diag [20, 20, 20] N s^2/m
B_{pd}	diag [30, 30, 30] N s/m
K_{pd}	diag [300, 300, 300] N/m
M_{od}	diag [20, 20, 20] N m s^2/rad
B_{od}	diag [30, 30, 30] N m s/rad
K_{od}	diag [300, 300, 300] N m/rad

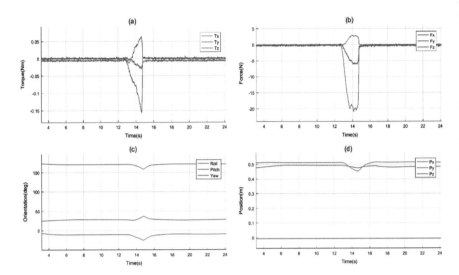

Fig. 4 Robot reaction of typical Maxwell model-based Cartesian compliance control when five successive external forces applied on the end effector. **a** External torques during human interaction. **b** External forces during human interaction. **c** Orientation of end effector during human interaction. **d** Position of end effector during human interaction

In the test, we keep the manipulator desired position and orientation static. Then, we perform a force on the force sensor and record the position and orientation data of flange and external force and torque applied on the sensor. We use two control methods perform the same experiment, and the figure of data is shown in Figs. 4 and 5.

We can draw the following conclusions from Figs. 4 to 5. When force sensor detected external force and torque, two methods can both have a position and orientation displacement and after human interaction disappeared, two methods can both make manipulator move back to the initial position and orientation. Two methods can both avoid the influence of signal noise of sensor. When we translate the viscoelastic coefficients into equivalent coefficients of Voigt model and ignore integral part of external force, two methods have same effect and Maxwell model-based Cartesian admittance control have an ignorable plastic deformation which shows the Maxwell model-based Cartesian admittance control can replace the traditional Cartesian admittance control and have a good comfortableness.

5 Conclusion

This paper presents a novel Maxwell model-based Cartesian admittance control which use position loop control the manipulator. We compare two methods and results shows effectiveness of our control method. Maxwell model-based Cartesian

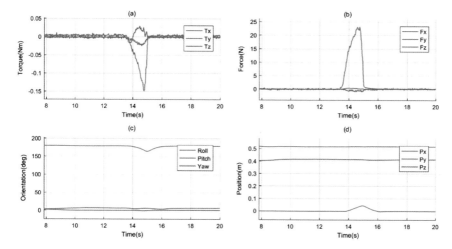

Fig. 5 Robot reaction of typical Maxwell model-based Cartesian compliance control when five successive external forces applied on the end effector. **a** External torques during human interaction. **b** External forces. **c** Orientation of end effector. **d** Position of end effector

admittance control uses position loop control the manipulator can be used on most kind of manipulators and provide a good comfortableness in the case of having same effect of traditional Cartesian admittance control.

Acknowledgements This work was supported by the National key R&D plan tasks 2018YFF0300304-04, National Natural Science Foundation of China under contact 61903357, Scientific research project of Liaoning Provincial Department of Education LNJC201912, Liaoning Provincial Natural Science Foundation of China 2020-MS-032 and China Postdoctoral Science Foundation 2020M672600.

References

1. Dhillon, B.S.: Robot Reliability and Safety. Springer Science & Business Media, Berlin (2012)
2. Goodrich, M.A., Schultz, A.C.: Human-Robot Interaction: A Survey. Now Publishers Inc. (2008)
3. Qiao, H., Dalay, B.S., Parkin, R.M.: Precise robotic chamferless peg-hole insertion operation without force feedback and remote centre compliance (RCC). J. Mech. Eng. Sci. **208**(2), 89–104 (1994)
4. Robinson, D.W., Pratt, J.E., Paluska, D.J., et al.: Series elastic actuator development for a biomimetic walking robot. In: 1999 IEEE/ASME International Conference on Advanced Intelligent Mechatronics, pp. 561–568. IEEE, Atlanta (1999)
5. Raibert, M.H., Craig, J.J.: Hybrid position/force control of manipulators. J. Dyn. Syst. Meas. Control **102**, 126–133 (1981)
6. Hogan, N.: Impedance control: an approach to manipulation. I: Theory. J. Dyn. Syst. Meas. Control **107**(1), 1–7 (1985)

7. Ott, C., Mukherjee, R., Nakamura, Y.: Unified impedance and admittance control. In: 2010 IEEE International Conference on Robotics and Automation, pp. 554–561. IEEE, Anchorage (2010)
8. Scherzinger, S., Roennau, A., Dillmann, R.: Forward dynamics compliance control (FDCC): a new approach to Cartesian compliance for robotic manipulators. In: 2017 IEEE/RSJ International Conference on Intelligent Robots and Systems (IROS), pp. 4568–4575. IEEE, Vancouver (2017)
9. Lee, D., Lee, W., Park, J., et al.: Task space control of articulated robot near kinematic singularity: forward dynamics approach. IEEE Robot. Autom. Lett. 5(2), 752–759 (2020)
10. Maciej, B., Hassan, O., Bernard, B.: Model predictive impedance control. In: 2020 IEEE International Conference on Robotics and Automation (ICRA), pp. 4702–4708. IEEE, Paris (2020)
11. Spriggs, T.W., Huppler, J.D., Bird, R.B.: An experimental appraisal of viscoelastic models. Trans. Soc. Rheol. 10(1), 191–213 (1966)
12. Ehlers, W., Markert, B.: On the viscoelastic behaviour of fluid saturated porous materials. Granular Matter 2(3), 153–161 (2000)
13. Fu, L., Zhao, J.: Design and implementation of plastic deformation behavior by Cartesian impedance control based on Maxwell model. Complexity 2018(1), 1–9 (2018)
14. Fu, L., Zhao, J.: Maxwell model-based compliance control for human robot friendly interaction. IEEE Trans. Cogn. Dev. Syst. (2020)
15. Liu, G., Han, B., Li, Q., et al.: DOREP: an educational experiment platform for robot control based on MATLAB and the real-time controller. In: International Conference on Intelligent Robotics and Applications, pp. 555–565. Springer, Shenyang (2019)

Bearing Fault Detection Method Based on Improved Convolution Network

Pengyu Cheng, Binbin Li, and Bin Jiao

Abstract Bearing is the key component of rotating machinery. It is very important for the normal operation of the equipment to carry out accurate fault detection and analysis of the bearing. Aiming at this problem, this paper proposes a fault diagnosis model based on residual network (RESNET). By transforming one-dimensional vibration data into two-dimensional data, after successive operations such as convolution, pooling, activation, and the residual network module is used to deepen the network while effectively correcting the parameters behind, and finally realizes the improvement of the fault recognition ability. The experimental results prove that this method is compared with other methods the accuracy rate is greatly improved.

Keywords Deep learning · Rolling bearing · Fault diagnosis · Convolution neural network · Residual network

1 Introduction

Rolling bearing is a key component in rotating machinery equipment. Because of its poor ability to withstand impact, 30% of mechanical failures are caused by rolling bearing failures [1]. Therefore, accurate fault detection and analysis of bearing is of great significance for the normal operation of the equipment. 80% of the rolling bearing fault diagnosis methods at home and abroad are based on the vibration signals of rolling bearings [2], but the vibration signals collected by them have non-stationary and non-linear characteristics [3]. In addition, it is difficult to fully characterize the fault characteristics with the features collected from the time domain or frequency domain alone. Therefore, some scholars try to combine the two methods to analyze the fault, such as wavelet decomposition [4], empirical mode decomposition (EMD) [5] and resonance sparse decomposition [6]. With the development of artificial intelligence technology, scholars gradually use back-propagation neural network (BPNN)

P. Cheng (✉) · B. Li · B. Jiao
School of Electrical Engineering, Shanghai Dianji University, Shanghai, China
e-mail: libb@sdju.edu.cn

© The Author(s), under exclusive license to Springer Nature Singapore Pte Ltd. 2021
Y. Li et al. (eds.), *Advances in Simulation and Process Modelling*,
Advances in Intelligent Systems and Computing 1305,
https://doi.org/10.1007/978-981-33-4575-1_44

[7], support vector machine (SVM) [8], and hidden Markov [9] to diagnose bearing fault.

The concept of deep learning is a qualitative leap in the field of feature extraction and pattern recognition, which can obtain deep-level feature representation and avoid the complexity of manual feature selection and the dimension disaster of high-dimensional data [10]. Some scholars try to use the deep neural network to study the fault diagnosis of rolling bearing [11], but because the potential of deep learning has not been fully exploited, to a certain extent, it is still unable to get rid of the interference of complex signal processing technology [12], and it is still in the initial stage [13]. In bearing fault diagnosis, there are deep belief network (DBN) [14], deep auto encoder (DAE) [15], convolutional neural network (CNN) [16, 17], and recurrent neural network (RNN) [18]. In this context, a bearing fault detection method based on residual network (RESNET) is proposed. By inputting the original data into the network, the fault features are extracted and then the fault diagnosis is carried out. The results show that this method will greatly improve the accuracy of identification.

2 Theoretical Basis

2.1 Convolution Neural Network

Convolutional neural networks have three advantages, they are local connection, weight sharing, and pooling. First of all, the hidden layer nodes of the general deep neural network are fully connected to each pixel of an image, while the convolutional neural network connects each hidden node only to a certain local area of the image to reduce the number of training parameters. Secondly, in the convolutional layer of the convolutional neural network, the weights corresponding to the neurons are the same. Because the weights are the same, the amount of training parameters can be reduced. Finally, because the images to be processed are often relatively large, and in the actual process, there is no need to analyze the original image. The most important thing is to effectively obtain the characteristics of the image. Therefore, the idea similar to image compression can be used to after convolution, a down-sampling process is used to adjust the image size.

1. Convolution layer: In the convolution neural network, each convolution layer is composed of several convolution units, and the parameters of each convolution unit are optimized by the back-propagation algorithm. The purpose of convolution operation is to extract different features of input. The first convolution layer may only extract some low-level features. With the deepening of network, more complex features can be extracted from low-level features.

$$S(i, j) = (I * K)(i, j) = \sum \sum I(i, j) K(m, n) + b \qquad (1)$$

where m and n is the size of the convolution kernel and b is the offset value.

$$a = f(s(i, j)) \tag{2}$$

In order to reduce the gradient loss, ReLU function is chosen as the activation function.

2. Pooling layer: In convolution neural network, after feature extraction of input image in convolution layer, output image will be sent to pooling layer for feature selection and filtering some unimportant information. The pooling layer contains pooling function, whose function is to select the important features in the feature map. Generally, we have maximum pooling and average pooling. The calculation formula is as follows:

$$p = \max(a) = \max(f(s(i, j))) \tag{3}$$

$$P = \frac{1}{n} \sum f(s(i, j)) \tag{4}$$

where n is the size of the convolution.

3. Full connection layer: In convolution neural network, after convolution operation of convolution layer and pooling operation of pooling layer, relevant important features have been selected. The role of full connection layer is to assemble the extracted local features into a complete graph through weight matrix, which is convenient for classification research.

2.2 Residual Network

Deep learning emphasizes a deep word, but it is found that with the increasing number of neural network layers, the accuracy of the model is constantly improved. When the network deepens to a certain extent, the training accuracy and test accuracy decrease rapidly, which shows that when the neural network deepens, it becomes more difficult to train. With the deepening of the network, the gradient problem becomes more obvious, failure to effectively modify the parameters has greatly reduced network performance and has not reached the expected level.

The core idea of the residual network is to use identity mapping. Imagine a fully trained neural network, add several identity maps to the output, which deepens the depth of the network without increasing errors. The idea of using the identity mapping to directly transmit the output of the previous layer to the rear mentioned here promotes the development of deep residual networks.

The residual network block is shown in Fig. 1. The original input image is x, and the residual network has two paths. The first is the main path of the neural network, that is, the output $F(x)$ obtained from the traditional forward propagation path. There is also a shortcut that $H(x) = F(x) + x$ is obtained by superposition of the original

Fig. 1 Basic structure of
residual network

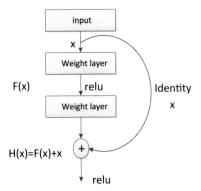

input image data and the output $F(x)$. Finally, $H(x)$ is imported into the activation function to obtain the final output. By using the identity mapping in residual network, the normal propagation of gradient can be ensured while deepening the network, so that the parameters of the back network can be effectively modified, and finally, the efficiency of the network can be improved.

3 Experimental Process

3.1 Experimental Steps

1. The vibration signals under different loads are collected by bearing data of Western Reserve University.
2. The one-dimensional vibration signal is randomly inserted into a two-dimensional characteristic atlas.
3. Set the small batch size, and divide the 2D feature atlas into small batch sample groups.
4. Set the initial parameters of the network according to the size of the input image, and train the model based on RESNET.
5. Use Gaussian initialization to initialize the weights in the fault diagnosis model.
6. Select a small sample of faulty bearing data.
7. The forward propagation of convolution neural network is carried out, and the parameters are continuously optimized by using the random gradient descent method.
8. The experimental results were analyzed.
9. Compared with other models.
10. Draw a conclusion.

3.2 Experimental Process

This experiment uses bearing vibration data provided by Western Reserve University as the experimental data. The experimental device consists of a 1.5 kW electric motor, torque sensor/decoder, power tester, and electronic controller. Use electric spark discharge to process single-point damage on the outer ring, inner ring and rolling elements of the bearing, and place an acceleration sensor above the bearing seat of the motor drive end to collect the vibration acceleration signal of the faulty bearing, and finally pass the 16-channel data recorder collect vibration signals.

The experiment is to diagnose the normal bearing, outer ring failure, inner ring failure, and rolling element failure under different loads. Record the normal state as N and the inner ring fault, outer ring fault, and rolling element fault with a fault diameter of 0.1778 mm as IR007, OR007 and B007, respectively. The load of the motor is recorded as load 0, load 1, load 2, and load 3.

The experimental data is shown in Table 1. Firstly, import the data of Western Reserve University, and intercept 480,000 data under normal state and fault state under different loads. Then, 10 * 10 two-dimensional input images are formed by randomly inserting samples. Each load constitutes 4800 sample sets. After scrambling them, 2880 samples are selected as training samples and 1920 as test samples.

Input the selected two-dimensional data into the fault diagnosis model shown in Fig. 2. The input image is first filled with zeros to expand the dimension of the input data. This method is filled with three layers, and the output goes through the convolution operation of the convolution layer. The normalization operation of the batch norm layer, the activation operation of the activation layer using the ReLU function as the activation function, and the pooling operation of the maximum pooling

Table 1 Experimental data

Sample state	Training sample set				Test sample set			
	Load 0	Load 1	Load 2	Load 3	Load 0	Load 1	Load 2	Load 3
B007	2880	2880	2880	2880	1920	1920	1920	1920
IR007	2880	2880	2880	2880	1920	1920	1920	1920
OR007	2880	2880	2880	2880	1920	1920	1920	1920
N	2880	2880	2880	2880	1920	1920	1920	1920

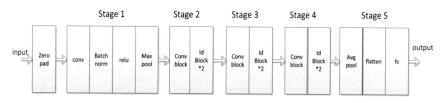

Fig. 2 Fault diagnosis model

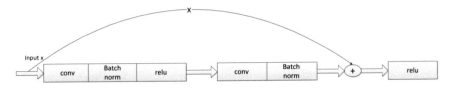

Fig. 3 Standard identity block

layer to extract some features, and then input the output into the second to fourth parts of model, where id block is the standard identity block as shown in Fig. 3. The shortcut route of the standard identity block in this article is two layers, id block*2 means two standard identity blocks, the obtained output is input to the average pooling layer and expanded into one-dimensional data. The extracted local features are assembled through the weight matrix through the fully connected layer, and finally, the softmax function is used for efficient classification.

3.3 Analysis and Comparison of Experimental Results

The experimental results are shown in Table 2 and Fig. 4. This chart shows the impact of the size of different small batch sample sets on the accuracy of the model under different loads. It can be seen from the figure that load 1 has the best adaptability to the small batch sample set. In the case of different small batch sample sets, there is a higher accuracy rate (Table 2 and Fig. 4).

From a horizontal perspective, the experimental model has the highest adaptability to the small batch sample set of 32. The model has a higher accuracy rate under normal and fault data of different loads. At the same time, when the small batch sample set is 8, the experimental results are not too ideal, the accuracy rate for load 3 is only 68.9%.

It can be seen from Table 3 that the accuracy of this method is greatly improved compared with other methods. The accuracy of the BPNN + EMD method is about 30–40%; the accuracy of the SVN + EMD method is about 20%. The accuracy of the Lenet-5 method has just reached 90%, and the accuracy of this method has reached about 99%. Because the parameters can be effectively corrected with the help of the

Table 2 Experimental results

	8	16	32	64	128
Load 0	77.9	99	98.8	93.9	94.1
Load 1	98.7	97.8	98.8	99.6	99.7
Load 2	89.9	89.3	99.3	98.6	98.6
Load 3	68.9	92.3	99.3	84.7	89.1

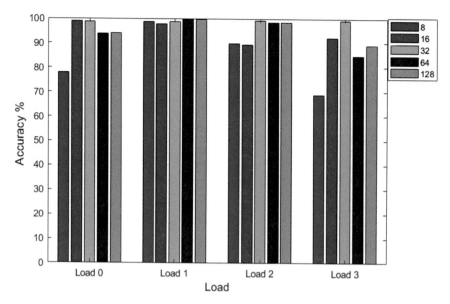

Fig. 4 Experimental results

Table 3 Fault diagnosis accuracy of different methods %

Method	Load 0	Load 1	Load 2	Load 3
BPNN + EMD	34.73	37.53	37.81	39.20
SVM + EMD	19.07	23.76	22.54	19.60
Lenet-5	90.33	92.40	92.55	88.64
This research method	99	99.7	99.3	99.3

residual network, which improves the overall performance of the deep convolutional network.

4   Conclusions

The innovation of this method lies in the combination of convolution and residual blocks, and the establishment of fault diagnosis models based on RESNET to qualitatively classify bearing faults. By adding a standard identity module behind the convolutional neural network, the parameters of the back network can be effectively corrected while the network is deepened, so as to improve the ability of fault identification. Experimental results show that this research method has higher accuracy in diagnosing bearing faults than previous methods.

References

1. Gan, M., Wang, C., Zhu, C.A.: Construction of hierarchical diagnosis network based on deep learning and its application in the fault pattern recognition of rolling element bearings. Mech. Syst. Signal Process. **72**(73), 92–104 (2016)
2. Cerrada, M., Sanchez, R., Li, C., et al.: A review on data-driven fault severity assessment in rolling bearings. Mech. Syst. Signal Process. **99**, 169–196 (2018)
3. Zheng, J.D., Dai, J.X., Zhu, X.L., Pan, H.Y., Pan, Z.W.: A rolling bearing fault diagnosis approach based on improved multiscale fuzzy entropy. J. Vibr. Meas. Diagn. **38**(05), 929–934 (2018) (in Chinese)
4. Zhu, X.Y., Wang, Y.J., Zhang, Y.Q., Yuan, J.Y.: Method of incipient fault diagnosis of bearing based on adaptive optimal Morlet wavelet. J. Vibr. Meas. Diagn. **38**(05), 1021–1029 (2018) (in Chinese)
5. Li, Y., Xu, M., Huang, W., et al.: An improved EMD method for fault diagnosis of rolling bearing. In: Prognostics and System Health Management Conference, Chengdu, China, pp 1–5. IEEE (2017)
6. Sun, Z.L.: Research on Fault Diagnosis Method of Rolling Bearing Based on Resonance Sparse Decomposition. Beijing Jiaotong University, Beijing (2017)
7. Zhang, L.P., Liu, H.M., Lu, C.: Fault diagnosis technology of rolling bearing based on LMD and BP neural network. In: Intelligent Control and Automation, pp. 1327–1331. Guilin, China. IEEE (2016)
8. Li, Y.B., Xu, M.Q., Zhao, H.Y., Huang, W.H.: A study on rolling bearing fault diagnosis method based on hierarchical fuzzy entropy and ISVM-BT. J. Vibr. Eng. **29**(01), 184–192 (2016) (in Chinese)
9. Georgoulas, G., Mustafa, M.O., Tsoumas, I.P., et al.: Principal component analysis of the start-up transient and hidden Markov modeling for broken rotor bar fault diagnosis in asynchronous machines. Expert Syst. Appl. Int. J. **40**(17), 7024–7033 (2013)
10. Duan, Y.J., Lv, Y.S., Zhang, J., Zhao, X.L., Wang, F.Y.: Deep learning for control the state of the art and prospects. Acta Autom. Sinica **42**(05), 643–654 (2016) (in Chinese)
11. Deng, S., Cheng, Z., Li, C., et al.: Rolling bearing fault diagnosis based on deep Boltzmann machines. In: Prognostics & System Health Management Conference, pp. 1–6. Chengdu, China. IEEE (2017)
12. Guo, X.J., Chen, L., Shen, C.Q.: Hierarchical adaptive deep convolution neural network and its application to bearing fault diagnosis. Measurement **93**, 490–502 (2016)
13. Wang, Y.J., Na, X.D., Kang, S.Q., et al.: State recognition method of a rolling bearing based on EEMD-Hilbert envelope spectrum and DBN under variable load. Proc. CSEE **37**(23), 6943–6950 (2017) (in Chinese)
14. Shao, H.D., Jiang, H.K., Wang, F., Wang, Y.N.: Rolling bearing fault diagnosis using adaptive deep belief network with dual-tree complex wavelet packet. ISA Trans. **69**, 187–201 (2017)
15. Shao, H.D., Jiang, H.K., Li, X.Q., Wu, S.P.: Intelligent fault diagnosis of rolling bearing using deep wavelet auto-encoder with extreme learning machine. Knowl. Based Syst. **140**, 1–14 (2018)
16. Zhang, W., Li, C.H., Peng, G.L., Chen, Y.H., Zhang, Z.J.: A deep convolutional neural network with new training methods for bearing fault diagnosis under noisy environment and different working load. Mech. Syst. Sig. Process. **100**, 439–453 (2018)
17. Lu, C., Wang, Z.Y., Zhou, B.: Intelligent fault diagnosis of rolling bearing using hierarchical convolutional network based health state classification. Adv. Eng. Inform. **32**, 139–151 (2017)
18. Cui, Q., Li, Z., Yang, J., et al.: Rolling bearing fault prognosis using recurrent neural network. In: Control & Decision Conference, pp. 1196–1201. Chongqing, China, IEEE (2017)

Path Planning of AFM-Based Manipulation Using Virtual Nano-hand

Shuai Yuan, Tianshu Chu, and Jing Hou

Abstract During performing AFM-based nano-manipulations, traditional method is unstable, and the manipulation efficiency is low due to the uncertainty of the AFM tip position. As for these problems, this paper refers to the macro-robot caging strategy, proposes to plan the tip maneuvering trajectory using the "Z-shape" path to form a virtual nano-hand. Then, the model parameters are discussed and optimized through simulation. Meanwhile, the Monte Carlo method is used to illustrate the effectiveness of the optimization. The simulation result indicates that the optimized parameters can make the manipulation more stable and efficient. Finally, the AFM experiment with optimized parameters is carried out to verify the effectiveness and stability of the virtual nano-hand method.

Keywords Virtual nano-hand · "Z-shape" path planning · Caging strategy · AFM

1 Introduction

Atomic force microscopy (AFM) has been widely used in electronic devices [1], conductive materials [2], atomic imaging [3], etc. However, the tip position has uncertainty during nano-manipulation due to the nonlinearity of the PZT ($PbZrTiO_3$) driver and the system thermal drift. In addition, the single tip of AFM can only exert a point force on the maneuvered object, which has an adverse effect on nano-manipulation. The tip uncertainty and unstable manipulation limit the further application of AFM.

In order to improve the stability and efficiency, some researchers use kinematic model to predict the position of the maneuvered object [4–6]. References [7, 8] constructed a dual-tip parallel manipulation AFM system. Reference [9] uses depth from defocus for achieving automatic approximation of the AFM tip to the sample. Some researchers have improved the traditional AFM manipulation methods, such as sequential parallel pushing [10], continuous directional pushing [11], parallel pushing strategy [12] and so on. As for the system uncertainty, Ref. [13] proposed

S. Yuan (✉) · T. Chu · J. Hou
Faculty of Information and Control Engineering, Shenyang Jianzhu University, Shenyang, China
e-mail: reidyuan@163.com

© The Author(s), under exclusive license to Springer Nature Singapore Pte Ltd. 2021
Y. Li et al. (eds.), *Advances in Simulation and Process Modelling*,
Advances in Intelligent Systems and Computing 1305,
https://doi.org/10.1007/978-981-33-4575-1_45

a random method to estimate the tip position, which improves the tip localization accuracy. These studies have solved the existing problems of AFM manipulation to a certain extent.

As for unstable problems in AFM nano-manipulation, virtual nano-hand method is proposed to plan the tip to move according to the "Z-shape" path, which can simulate the parallel manipulation of the dual tip to realize fixed-posture nano-manipulation. Then, the driving parameters are optimized by simulations, and an AFM experiment is carried out to construct two nanoparticle structures, which can further illustrate the effectiveness of the virtual nano-hand method and increase the stability of nano-manipulation.

2 Virtual Nano-hand Method

As shown in Fig. 1a, in traditional nano-manipulation, the tip position in task space has uncertainty, and only point force can be applied between the tip and the nano-object. The nanoparticle position will distribute within a certain region after manipulation. Therefore, it is difficult to achieve stable nano-manipulation with fixed posture, and the maneuvered result is lack of stability and low efficiency.

In order to solve the problems in traditional nano-manipulation method, this paper refers to the caging strategy of macroscopic multi-robot collaboration as shown in Fig. 1b [14]. This strategy uses multiple small robots to move the maneuvered object in collaboration, which is similar to use multiple fingers to clamp an object and move it from one position to another. The tip adopts the manipulation mode of multiple rapid movements and multiple contact points to form a virtual nano-hand (the blue area in Fig. 1c) for pushing the nanoparticle horizontally to the left side. Compared with the traditional manipulation method, the position error of the nanoparticle becomes smaller, and the manipulation result is more stable.

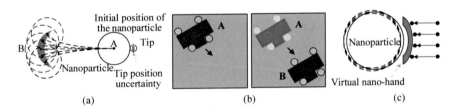

Fig. 1 Diagram of traditional manipulation method and virtual nano-hand method **a** traditional manipulation method **b** caging strategy **c** virtual nano-hand method

3 Path Planning of the Tip Manipulation Based on Virtual Nano-hand Method

3.1 Particle Motion Model Using Least Action Principle

Based on the Ref. [6], the particle kinematics model based on the least action principle is furtherly discussed. It is represented that the nanoparticle driven by AFM tip makes a uniform circular motion, and its rotation center locates on the center line that passes through the particle center perpendicular to the pushing direction, the contact plane between the nanoparticle and the substrate is integrated as infinitesimal element: an equivalent ring.

By analyzing the force and torque of the contact plane of the substrate and the pushing direction, a kinematic model is established. The detail formula derivation process can refer to Ref. [6].

$$F = \frac{\int_0^{2\pi} \int_0^R (f_c + cL\omega)\sqrt{(s + r\sin\theta)^2 + r^2\cos^2\theta}\,dr\,d\theta}{D + s} \tag{1}$$

where s is the instantaneous center position of the particle, and ω is the rotation angle velocity, which is an unknown quantity, here the relationship between s and ω is discussed as following.

When pushing the nanoparticle, the nanoparticle is mainly subjected to the propulsive force of the tip and the friction of the substrate. In addition, the friction between the tip and the particle will change the direction of the propulsive force. As shown in Fig. 2, the tip contacts the particle at point C and pushes the particle with velocity V, and the driving force is F_c, the friction between the tip and the particle is f_t, and the resultant force is F. The particle rotates with an angular velocity of ω centered on point O. V_c is the velocity of the contact point C, and the component of V and V_c in the direction of F_c are equal, so the following formulas can be obtained:

$$V_c \cos(\theta_1 + \alpha) = V \cos\theta_2 \tag{2}$$

Fig. 2 Force analysis of the tip pushing nanoparticles

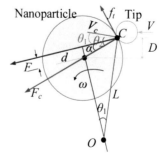

$$V_c = L\omega \tag{3}$$

$$\omega = \frac{V\cos\theta_2}{L\cos(\theta_1 + \alpha)} \tag{4}$$

$$\cos\theta_2 = \frac{\sqrt{R^2 - D^2}}{R} \tag{5}$$

$$\tan\alpha = f_t/F_c = \mu \tag{6}$$

$$\cos\theta_1 = \frac{d + s}{L} \tag{7}$$

Through using the above-mentioned formulas, the relationship between s and ω can be obtained as

$$\omega = \frac{V\sqrt{R^2 - D^2}}{(d + s)R\cos\alpha - R^2\cos\alpha\sin\alpha} \tag{8}$$

The method of calibrating the coefficient μ is the same as Ref. [6], then as long as s is calculated, the angular velocity ω can be obtained, and the position of the nanoparticle can be calculated.

3.2 Analysis of the Tip Movement Process

When the tip moves horizontally, due to the tip position uncertainty, it is difficult to push the nanoparticle with pushing direction via the nanoparticle center. The nanoparticle will rotate by the action of the tip, and the vertical distance between the tip center and the particle center is defined as d_i. The distance is defined as d_p that the tip moves horizontally to push the particles from point A to point B. Then, the particle will have offsets d_x and d_y in the horizontal and vertical directions, as shown in Fig. 3.

Fig. 3 Diagram of the tip movement process

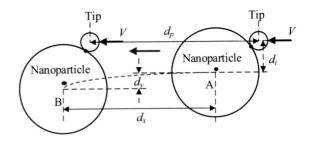

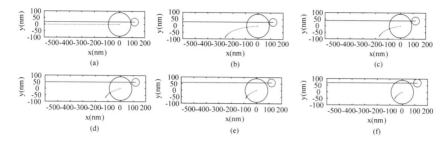

Fig. 4 Maneuvering trajectory of the particle and the tip using traditional manipulation method

Table 1 Final position coordinate of the particle using traditional manipulation method

Number	d_i (nm)	x-axis (nm)	y-axis (nm)
(a)	20	−498.0859	−1.3476
(b)	30	−261.9300	−90
(c)	40	−169.9914	−80
(d)	50	−119.5348	−70
(e)	60	−85.2799	−60
(f)	70	−60.0883	−50

It can be seen from Fig. 3 that d_i and d_p determine the maneuvering trajectory of the particle. In order to illustrate the influence of d_i on the manipulation result, a simulation is carried out with different d_i. Figure 4a–f represents that d_i increases from 20 to 70 nm with d_p value 500 nm, and the coordinate statistics are shown in Table 1.

It can be seen from Fig. 7 that the nanoparticle and the tip will detach from each other with the increase of d_i, which greatly reduces the stability and accuracy. In the same way, the larger the pushing distance d_p is, the more serious the vertical deflection of the particles will be, which makes the stability of the manipulation become worse.

3.3 The Tip "Z-Shape" Path Planning

In order to simplify the modeling complexity and further abstract the manipulation process, a "Z-shape" maneuvering trajectory is proposed. As shown in Fig. 5, the tip moves one step forward on the top, then gets back and moves to the bottom for next manipulation, the "Z-shape" path planning procedure is shown in Fig. 5. The path for the first and second push of the tip is $P_1 \rightarrow P_2 \rightarrow P_3 \rightarrow P_4$. The small circle represents the tip, and the points in the circle represent the tip position uncertainty.

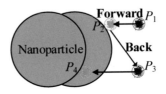

Fig. 5 "Z-shape" path planning procedure

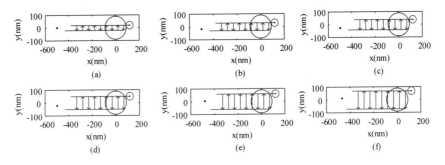

Fig. 6 Final position of the particle using "Z-shape" path

Table 2 Final position using "Z-shape" path

Number	d_i (nm)	x-axis (nm)	y-axis (nm)
(a)	20	−498.3	0.2812
(b)	30	−499.3	−17.28
(c)	40	−499.3	−26.64
(d)	50	−496.7	−21.93
(e)	60	−481.1	4.682
(f)	70	−468.2	11.37

Figure 6 shows the simulation result using "Z-shape" maneuvering trajectory with different d_i. Each set of arrows represents the completion of a "Z-shape" manipulation. In the simulation, d_p is 50 nm with continuously pushing five times. Figure 6a–f represents that d_i increases from 20 to 70 nm. The statistical results of the particle center are shown in Table 2. Comparing the results of Tables 1 and 2, it is obvious that after using the "Z-shape" maneuvering trajectory, the result can be more stable.

4 Model Parameters Optimization

Due to the tip position uncertainty, the nanoparticle has multiple possible positions after being pushed. In order to limit the nanoparticle between the top and bottom of

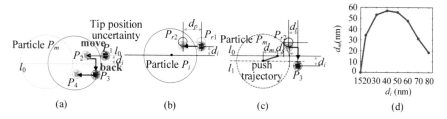

Fig. 7 Optimization for maximum moving distance d_m

pushing direction (this constraint is used in the following step 2), the offset distance d_i, pushing distance d_p needs to be optimized. The estimation of optimal d_i and d_p is as follows:

1. As shown in Fig. 7a, it is necessary to ensure the nanoparticle moves between P_1P_2 and P_3P_4 during the manipulation. As for the first manipulation $P_1 \rightarrow P_2$, it is assumed that the tip is at the upper boundary P_{r1} of the possible area P_1, the tip will reach P_{r2} after moving a distance d_p, as shown in Fig. 7b. While the nanoparticle is moving forward, it rotates down at the same time.
2. The tip moves back to area P_3, in order to ensure the constraint, it is supposed that l_1 passes through the upper boundary of P_3. As shown in Fig. 7c, the nanoparticle should be on the upper side of l_1 after manipulation. Therefore, the maximum advance distance of the nanoparticle can be defined as d_m.
3. 500 sample points are randomly selected by using Monte Carlo method within the distribution region to represent the possible tip positions, and it is assumed that these 500 random points satisfy the Gaussian distribution with $\sigma = 5$ nm. As for different d_i, d_m is calculated separately.

The optimized curve of relation between d_i and d_m is shown in Fig. 7d, and the maximum value of d_m is about 57 nm. At this time, the corresponding offset distance d_i and pushing distance d_p are 40 nm and 102 nm, respectively. Then, these two values are used as the pushing parameters for simulation. Figure 8a–f represents that d_p increases from 70 to 150 nm. The statistical results are shown in Table 3. It can be seen that as for a certain d_i with the same of the pushing time, the manipulation result obtained by using optimal parameters can keep the particle within its constraint, and simultaneously, the nanoparticle moves the farthest distance. When d_p is greater than the optimized value, the nanoparticle will exceed the constraint area, causing the divergent result.

5 AFM Experiment

In order to verify the virtual nano-hand method, an experiment is carried out using AFM. The experimental sample is the polystyrene nanosphere with a radius of about

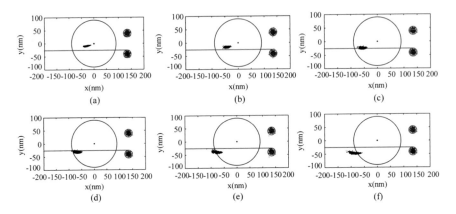

Fig. 8 Simulation using optimal parameters

Table 3 Final position using optimal parameters

Number	d_p (nm)	x-axis (nm)	y-axis (nm)
(a)	60	−11.35	−4.74
(b)	80	−25.84	−12.01
(c)	102	−56.73	−23.86
(d)	120	−67.81	−30.98
(e)	140	−77.99	−39.29
(f)	160	−86.52	−7.02

100 nm. In the experiment, the parameters are set as follows: The pushing speed is 1 μm/s, the offset distance d_i and the pushing distance d_p are set according to the optimized value, which are ±40 nm and 102 nm with continuously pushing five times. Figure 9 shows the experimental results, the vertical displacements of the nanoparticles are 902 nm and 916 nm, and the horizontal displacements are 12 nm and 19 nm, respectively. Finally, two experiments of constructing nanoparticle structures are

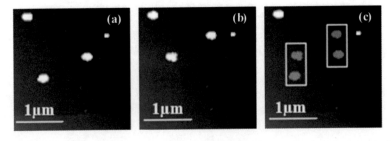

Fig. 9 Experimental results of virtual nano-hand manipulation method **a** before manipulation **b** after manipulation **c** superimposed image

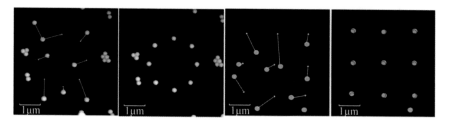

Fig. 10 Nanoparticle structure constructed using virtual nano-hand manipulation method

carried out. The experimental results are shown in Fig. 10. The experimental results furtherly illustrate that the proposed method can effectively reduce the impact of tip uncertainty and promote the efficiency, reliability and stability of the manipulation.

6 Conclusion

As for the problems of low controllability and uncertainty in AFM nano-manipulation, a virtual nano-hand method referring to the caging strategy is proposed to achieve stable manipulation. On the foundation of the precise tip localization, a nano-manipulation model using the least action principle is established, and the "Z-shape" maneuvering trajectory is designed to perform nano-manipulation. Then, the driven parameters are optimized, and a more stable result is obtained through simulation. Finally, the AFM experiment is carried out to construct two nanoparticle structures, and the experiment results furtherly illustrate that the virtual nano-hand method can not only effectively reduce the influence of the tip uncertainty, but also improve the stability, reliability and efficiency, which is of great significance to AFM-based nano-manipulation.

References

1. Buckwell, M., Zarudnyi, K., Montesi, L., et al.: Conductive AFM topography of intrinsic conductivity variations in silica based dielectrics for memory applications. ECS Trans. **75**(5), 3–9 (2016)
2. Zhang, D., Wang, X., Song, W., et al.: Analysis of crystallization property of LDPE/Fe$_3$O$_4$ nano-dielectrics based on AFM measurements. J. Mater. Sci. Mater. Electron. **28**(4), 3495–3499 (2017)
3. Boneschanscher, M., Van Der Lit, J., Sun, Z.X., et al.: Quantitative atomic resolution force imaging on epitaxial graphene with reactive and nonreactive AFM probes. ACS Nano **6**(11), 10216–10221 (2012)
4. Hou, J., Wu, C.D., Liu, L.Q., et al.: Research on the kinematics model of nanoparticles based on AFM. Chin. J. Sci. Instrum. **32**(8), 1851–1857 (2011) (in Chinese)

5. Korayem, M.H., Homayooni, A., Hefzabad, R.N.: Non-classic multiscale modeling of manipulation based on AFM, in aqueous and humid ambient. Surf. Sci. **671**(671), 27–35 (2018)
6. Liu, T.B., Liu, Y., Yuan, S., et al.: Research on nanoparticle operation modeling based on the principle of least action. High Technol. Lett. **25**(7), 725–733 (2015) (in Chinese)
7. Xie, H., Régnier, S.: High-efficiency automated nanomanipulation with parallel imaging manipulation force microscopy. IEEE Trans. Nanotechnol. **11**(1), 21–33 (2012)
8. Loganathan, M., Al-Ogaidi, A., Bristow, D.A.: Design and control of a dual-probe atomic force microscope. IEEE/ASME Trans. Mechatron. **23**(1), 424–433 (2018)
9. Liu, J., Ma, J.C., Yu, P., et al.: Research on AFM automatic approximation method based on image focus positioning. Chin. J. Sci. Instrum. **39**(1), 58–67 (2018) (in Chinese)
10. Xu, K.M., Kalantari, A., Qian, X.P.: Efficient AFM based nanoparticle manipulation via sequential parallel pushing. IEEE Trans. Nanotechnol. **11**(4), 666–675 (2012)
11. Zhao, W., Xu, K.M., Qian, X.P., et al.: Tip based nanomanipulation through successive directional push. J. Manuf. Sci. Eng. **132**(3), 311–322 (2010)
12. Liu, H.Z., Wu, S., Zhang, J.M., et al.: Strategies for the AFM-based manipulation of silver nanowires on a flat surface. Nanotechnology **28**(36), 5301–5312 (2017)
13. Yuan, S., Yao, X., Luan, F.J., et al.: Research on optimal estimation of AFM probe position based on stochastic method. Chin. J. Sci. Instrum. **38**(9), 2120–2129 (2017) (in Chinese)
14. Wan, W., Rui, F.: Efficient planar caging test using space mapping. IEEE Trans. Autom. Sci. Eng. **15**(1), 278–289 (2018)

Automatic Reading Algorithm of Pointer Water Meter Based on Deep Learning and Double Centroid Method

Hongqing Li, Juan Wang, Bing Bai, and Chuang Lu

Abstract Aiming at the problem of the reading of the eight-pointer water meter, it is proposed, in this paper, an automatic recognize reading algorithm based on deep learning and double centroid method. The YOLOv3 neural network is used to detect the object of the small dial. Based on the centers of small dial, the gesture of large dial is corrected by perspective transformation. Then, the whole region of pointers is segmented by U-net neural network. According to the segmented region, a new double centroid method of getting the angle of the pointer region is proposed. The angle of the pointer region can accurately be obtained by this method. Finally, in terms of people's meter-reading method, a new reading rule is established. A large amount of experimental results show the proposed algorithm can guarantee the satisfying effect even when the dial is unclear or the shooting angle is sloped.

Keywords Pointer water meter · Automatic reading · Deep learning · Double centroid method

1 Introduction

In daily life, various meters have become an indispensable part. Among them, pointer meters are widely used in human production and life. But problems have appeared in the use of pointer meters. For example, the water affairs company needs staff to read meters and monitor meters manually, which consumes manpower and brings trouble to users. Some engineers have proposed to use image processing technology to recognize the reading of the pointer meter. At present, LRCD [1] and Hough transform method [2] are used to recognize the reading. These are all based on

H. Li · J. Wang (✉)
Faculty of Information and Control Engineering,
Shenyang Jianzhu University, Shenyang 110168, China
e-mail: wangjuanneu@163.com

B. Bai · C. Lu
Liaoning Academy of Agricultural Sciences,Shenyang 110161, China

© The Author(s), under exclusive license to Springer Nature Singapore Pte Ltd. 2021   477
Y. Li et al. (eds.), *Advances in Simulation and Process Modelling*,
Advances in Intelligent Systems and Computing 1305,
https://doi.org/10.1007/978-981-33-4575-1_46

traditional machine vision methods [3, 4]. The traditional machine vision algorithms are fast and flexible, but they have some disadvantages in recognition accuracy and poor robustness. In recent years, deep learning has shown many advantages and has been fully developed in the fields of object detection and semantic segmentation. In fact, the method of deep learning has more accurate, robust and less flexible feature.

Taking the eight-pointer water meter as an example, the proposed algorithm is fast, accurate and robust. This algorithm combining the advantages of both deep learning and traditional machine vision algorithms achieves satisfying result.

2 Preprocessing Based on YOLOv3

The YOLOv3 is a type of ResNet. It is a object detection neural network model. It is the third generation of the YOLO series. In this part, the YOLOv3 is used to find the region of the small dial and get the center of the small dial. According to the accurate and positive region of the small dial, the large dial is corrected by perspective transformation.

2.1 Small Dial Detection Based on YOLOv3

The YOLOv3 model is a neural network model with the advantage of rapidity and accuracy. The backbone network has been sufficiently improved from the previous generation of darknet-19 to the current darknet-53. After YOLOv3 has undergone multiple residual network convolutions, it adopts three-layer output and upsamplings the output of previous two layers. Then, the output of layer is concated with the upsamplings of previous layers. Three anchors of different sizes are used in the output of each layer, and finally, non-Maximum suppression is used to control multiple detections of objects of the same type.

In fact, due to the shooting angle and lighting problems, traditional threshold segmentation or other traditional methods are used, which cause excessive noise and unstable threshold [1]. So, the YOLOv3 model is used to find the minimum bounding rectangle of the small dial. As shown in Fig. 1, even though the image of the dial is sloped, the positions of the small dial are still accurately marked.

2.2 Perspective Transformation of the Large Dial

When the image is captured, the problem of shooting angle usually appears. Therefore, the angle of the dial cannot directly be used in the reading of the small dial. In addition to the positive shooting angle, the shooting angle should be as vertical as possible to the dial. If it is not vertical, there will be a visual error. So, dial must be corrected before recognizing [5]. The angle of the pointer can be used for reading

Fig. 1 Small dial after YOLOv3

after the dial has been corrected. Then, the center of four small dials corresponding to two large dials are chosen to calculating the transformation matrix for perspective transformation. Finally, perspective transformation is performed in the large dial.

The general equation for perspective transformation [6] is as follows:

$$\begin{bmatrix} x \\ y \\ z \end{bmatrix} = \begin{bmatrix} m_{11} & m_{12} & m_{13} \\ m_{21} & m_{22} & m_{23} \\ m_{31} & m_{32} & m_{33} \end{bmatrix} \begin{bmatrix} u \\ v \\ 1 \end{bmatrix} \tag{1}$$

$$x' = \frac{x}{z} = \frac{m_{11}u + a_{12}v + m_{13}}{a_{31}u + m_{32}v + m_{33}} \tag{2}$$

$$y' = \frac{y}{z} = \frac{m_{21}u + m_{22}v + m_{23}}{m_{31}u + m_{32}v + m_{33}} \tag{3}$$

where x, y and z are image coordinates in three-dimensional space, matrix M is transformation matrix, u and v are original image coordinates, x' and y' are new image coordinates.

Finally, as shown in Fig. 2, we can see the corrected dial.

Fig. 2 Large dial after
perspective transformation

3 Automatic Reading Algorithm Based on U-Net and Double Centroid Method

After preprocessing, we get a corrected dial. In order to obtain the region of the pointer, U-net [7] is used to segment the small dial. In the case of a single background, the U-net network has better feature reduction. We can use this advantage to extract small pointer and use the double centroid method to obtain the angle of the pointer.

3.1 Semantic Segmentation of Small Pointer Based on U-Net

The U-net network is relatively simple. The first half is used for feature extraction, while the second half is used for feature upsampling. The U-net adopts a completely different feature fusion method: concat and uses the feature concat together in the channel dimension to form thicker features, which make the texture of the image easier to be segmented. After the dial is rotated, the small dial is resized to the same specification. The trained U-net model can be used to segment the shape of the dial pointer. Moreover, since the dial at this time is positive and the angle of view is

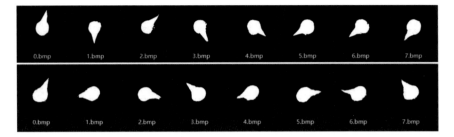

Fig. 3 Pointer after semantic segmentation

vertical, the angle can be used as reading. By network of semantic segmentation, the parameters to be adjusted are more less traditional algorithms [8], and the robustness is enhanced.

The result of semantic segmentation is shown in Fig. 3.

3.2 Measuring Angle of the Pointer Based on Double Centroid Method

In practical applications, we tried a series of methods such as convex hull, template matching [9] and Hough circle [10]. The effect of these methods is still not satisfying. Even though the pointer has been segmented by U-net, the small pointer is still incomplete due to stains, rust and reflections. Therefore, an new method that uses double centroids to measure the angle is proposed. In the previous step, all the regions of small dial are resized to the same size. The purpose is to obtain the angle of the small dial using the double centroid method.

The centroid of the small pointer can be obtained by Eq. (4). Then, a solid circle centered on this centroid is drawn to cover the region of pointer. The circle should be slightly larger than the base circle of pointer. At the same time, in order to prevent noise interference, the largest region would be choose in the remaining regions. In this way, the small pointer head can be retained. Next the centroid of the small pointer head can be obtained by Eq. (4). Finally, the angle of the pointer can be obtained by these two centroids. As shown in Fig. 4, this method is far more accurate than existing methods. At the same time, it minimizes the influence of rust or light. Even though the image such as the 582.jpg in Fig. 4 has rust or light, the effect is still satisfying.

$$x_c = \frac{\sum_{i=1}^{n} p_i x_i}{\sum_{i=1}^{n} p_i} \quad y_c = \frac{\sum_{i=1}^{n} p_i y_i}{\sum_{i=1}^{n} p_i} \tag{4}$$

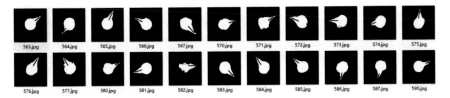

Fig. 4 Double centroid method

Table 1 Relationship between angle and reading

Angle	Reading	Angle	Reading
[0, 36)	0	[180, 216)	5
[36, 72)	1	[216, 252)	6
[72, 108)	2	[252, 288)	7
[108, 144)	3	[252, 288)	8
[144, 180)	4	[324, 360)	9

Table 2 Carry rule table

Current pointer angle difference	Last pointer angle	
	Angle < 0	Angle > 0
Angle difference <5	−1	0
Angle difference >31	0	+1

3.3 Pointer Reading Recognition

Firstly, it is roughly divided into ten classes according to the angle. A relationship table is established as shown in Table 1.

Now, the problem is that if there is an angle in the middle of the two classes, it is very difficult to estimate the reading. At this time, its previous dial should be observed. If its previous reading is judged to be more than 0, the reading can be carried or keep the current state. If the previous digit is judged to be less than 0, then the reading would keep the current state or abdicated. A rule table is established as shown in Table 2.

4 Datasets and Experimental Configurations

The image datasets of the pointer water meter used in this paper come from Shenyang Water Affairs Group Limited, with a total of 2000 images.

4.1 Dataset Description and Model Parameters

The main problem of the YOLOv3 model is to detect small dials in any direction. In order to enhance the generalization performance of the model and maintain the balance of samples in all directions, rotation data augmentation is used. After three rotation data augmentations, we still have 8000 samples. We take 7200 of them as the training set, the remaining 400 images as the validation set and 400 images as the test set. During the training of the YOLOv3 model, the initial learning rate is 0.001, the momentum parameter is 0.9, the Gaussian distribution is used to initialize the convolution kernel, and the weight decay is 0.0005. Three different gradient descent methods are used to train the network to make the loss function converge to a minimum value.

The datasets of the U-net model are marked with the small dial area processed by YOLOv3. A total of 1228 small dials are marked. The 1104 images are used as the training set, 62 images are used as the validation set, and 62 images as the test set. The initial learning rate is 0.0001, the weight decay is 0.0001, and the momentum parameter is 0.5. The Gaussian distribution is used to initialize the convolution kernel, and the gradient descent method is used to train the model to make the loss function converge to a minimum.

4.2 Hardware and Software Configuration

The experimental hardware configuration includes AMD R9 3900X cpu, NVIDIA GeForce GTX2080s gpu, and the software system includes windows10 system, cuda10.0, cudnn7.4.1.5, tensorflow-gpu 1.13.2 and keras2.1.5.

5 Results

The results show that the algorithm is better than the other algorithms. Even though the performance of the algorithm is affected by some factor, but a satisfying effect can still be achieved.

5.1 Training Results of Neural Network Model

The YOLOv3 model based on the kares framework and the U-net model based on the tensorflow framework are used to recognize the readings of the water meter.

For YOLO, we evaluate the performance of the model by using mean average precision (mAP).

Table 3 Deep learning net model

network model	mean IOU	mAP
YOLOv3		0.995
U-net	0.987	

For U-net, we evaluate the performance of the model by using mean intersection over union (IOU).

The results of the test set are shown in Table 3.

In the additional test of YOLOv3 model, three images are not recognized. The phenomenon attracts our attention. The accuracy of the whole algorithm is based on the YOLOv3 recognition. In the later study, we found that mistakes are caused by overexposure.

5.2 Results of the Algorithm

Fisrt of all, the end-to-end neural network model [11] was tried. It is obvious that algorithm proposed in this paper is more flexible in reading stage. So, the model is divided two parts that include object detection and semantic segmentation.

After that, based on references [3, 4], a lot of traditional machine vision algorithms are studied such as template matching method to correct the dials and obtain reading. Compared with the traditional algorithm, the algorithm proposed in this paper is more robust, and a few parameters need to be adjusted.

But in the YOLOv3 model, overexposure is not considered. This leads to error in the algorithm, if YOLOv3 cannot find eight dials. The YOLOv3 model still needs to improve the recognition performance with supplementing overexposed datasets or other preprocessing methods. In the U-net model, too much rust and light can also lead to faulty recognization.

Since this algorithm is not an end-to-end deep learning algorithm, we carry out the final reading test and compare it with other algorithms at the end. In the test, 100 images are taken for testing. As shown in Table 4, among other algorithms, although a little faster, the accuracy is much lower than the algorithm in this paper. In the algorithm of this paper, only six readings of the small dials are wrong. The accuracy is 99.3%. It only takes 25 seconds on a computer with a gpu of NVIDIA GTX2060.

Table 4 Algorithm performance evaluation and comparison

Algorithm	Time (s)	Accuracy (%)
LRCD [1]	0.16	65.4
SVM [9]	0.20	82.4
This paper	0.25	99.3

6 Conclusion

This paper proposes the algorithm based on deep learning and double centroid method to achieve the intelligent recognize reading. As the results show, this algorithm achieves better results than other algorithm mentioned in the references and give much more consistent predictions in reading tasks in our dataset. At the same time, it has the advantages of fastness and robustness. Therefore, this paper has great significance for instrument recognition.

Acknowledgements This work was supported by the National Natural Science Foundation of China under Grant 61703288 and Natural Science Foundation of Liaonning Province of China under Grant 2019-BS-195. The study has been conducted with ethics approval obtained from Shenyang Water Affairs Group Limited.

References

1. Zhang, Z.J., Yuan, L.I., Zhou, C.B., Yuan, W.Q.: Image recognition based automatic reading method for multi-pointer meter. J. Shenyang Univ. Technol. **33**(5), 550–555 (2011)
2. Li, Z.W.: Study on a new recognition method of pointer meters. Microcomput. Inf. **23**(11), 113–114 (2007)
3. Kaehler, A., Bradski, G.: Learning OpenCV: Computer Vision in C++ with the OpenCV Library. O'Reilly Media Inc, Sebastopol, CA (2013)
4. Gonzalez, R.C., Woods, R.E.: Digital Image Processing. 3rd edn. Prentice-Hall, Inc. Division of Simon and Schuster One Lake Street Upper Saddle River, NJ, United States (2008)
5. Xing, H.Q., Du, Z.Q., Su, B.: Detection and recognition method for pointer-type meter in transformer substation. Chine. J. Sci. Instrum. **38**(11), 2813–2821 (2017)
6. Zhu, Y.W.: Perspective transformation and perspective projection. J. Eng. Graph. **1**(9), 1–11 (1988)
7. Ronneberger, O., Fischer, P., Brox, T.: U-net: Convolutional networks for biomedical image segmentation. In: Navab, N,. Hornegger, J., Wells, W.M., Frangi, A.F. (eds.) Medical Image Computing and Computer-Assisted Intervention 2015, LNCS, vol. 9351, pp. 234–241. Springer (2015)
8. Song, W., Zhang, W.J., Zhang, J.Q., Wang, Y.P., Zhou, Q., Shi, W.R.: Meter reading recognition method via the pointer region feature. Chine. J. Sci. Instrum. **35**(12), 20–28 (2014)
9. Zhang, Y.Q., Ding, M.L., Fu, W.Y.F., Li, Y.Q.: Reading recognition of pointer meter based on pattern recognition and dynamic three-points on a line. In: Verikas, A., Radeva, P., Nikolaev, D.P., Zhang, W., Zhou, J. (eds.) Ninth International Conference on Machine Vision (ICMV 2016), vol. 10341, pp. 101–106. SPIE (2017)
10. Gao, J.L., Guo, L., Lv, Y.Y., Wu, Q.W., Mu, D.Q.: Pointer meter reading method based on improved orb and hough algorithm. Comput. Eng. Appl. **54**(23), 252–258 (2018)
11. Liu, J.L., Wu, H.Y., Chen, Z.H.: Automatic identification method of pointer meter under complex environment. In: International Conference on Machine Learning and Computing, vol. 12, pp. 276–282. ACM(2020)

Author Index

B
Bai, Bing, 477
Bai, Shan, 3
Bai, Zhu, 193
Bai, Shan, 3
Bai, Zhu, 193
Bao, Longsheng, 243

C
Cao, Jianzhao, 29, 433
Cao, Yi, 233
Cao, Zhenqiang, 395
Chen, Shizhong, 103
Chen, Yifan, 263, 277
Cheng, Lei, 341
Cheng, Pengyu, 459
Chu, Tianshu, 467

D
Dai, Guanghui, 253
Deng, Yuanyuan, 21

F
Fan, Rang-Lin, 119
Fan, Shanshan, 205
Fei, Wenbo, 443
Feng, Jin, 333
Fu, Baochuan, 315

G
Gan, Yulin, 325

Gao, Shushida, 223
Gao, Zhijun, 21, 39, 451
Ge, Qi, 139
Gu, Fan, 71, 351, 365
Guan, Feng, 223, 233
Guang, Yongxing, 83
Guo, Chao, 303
Guo, Lingzhong, 263, 277
Guo, Mingze, 159

H
Han, Guojing, 13
He, Naifeng, 451
He, Yang, 325
Hou, Jing, 467
Huang, Na, 325
Huang, Zhe, 159

J
Jiang, Baoping, 315
Jiang, Yuhan, 71
Jiao, Bin, 459

L
Li, Binbin, 459
Li, Bing, 419
Li, Donghua, 375
Li, Guang, 341
Li, Guochang, 303
Li, Huanlin, 341
Li, Manwen, 59
Li, Ning, 159

Li, Peng, 83
Li, Qinghe, 351, 365
Li, Songhua, 103, 181
Li, Tao, 341
Li, Xiaohu, 159
Li, Xuefeng, 131
Li, Yahui, 93
Li, Yang, 443
Li, Yu Peng, 287
Li, Ziyang, 315
Li, Zongze, 205
Li,Hongqing, 477
Liang, Jiahua, 59
Liu, Fei, 287
Liu, Jian, 385
Liu, Jianshun, 451
Liu, Junjie, 243
Liu, Maohua, 59
Liu, Meiju, 409
Liu, Qiang, 325
Liu, Tao, 263, 277
Liu, Tianqi, 47
Liu, Weidong, 205
Liu, Zijin, 103
Lu, Chuang, 477
Lu, Shengliang, 13
Lu, Tianyi, 149
Lu, Zhengran, 303
Lu, Xia, 395
Lv, Lianjie, 375

M
Ma, Chi, 13
Ma, Ruwei, 29, 433
Ma, Yue, 333

N
Ning, Shuyan, 385
Niu, Tong, 395

P
Pang, Renning, 29, 433
Peng, Xuezhao, 223, 233

Q
Qi, Song-Qiang, 119
Qi, Yuanwei, 29, 433
Qiao, Feng, 263, 277
Qiao, Tianling, 83
Qu, Qiuxia, 149

R
Ren, Shuwen, 103

S
Sha, Shuya, 149
Shao, Qianqian, 139, 169, 213, 253
Song, Lijuan, 131
Sun, Haiyi, 159, 395
Sun, Jia, 451
Sun, Jian, 181
Sun, Jun, 375
Sun, Liangliang, 149
Sun, Wei, 419

T
Tan, Youchen, 325
Tan, Youchen, 325

W
Wang, Bing, 287
Wang, Chi, 47
Wang, Juan, 477
Wang, Kechong, 181
Wang, Sanmu, 385
Wang, Shaofan, 13
Wang, Xin, 39, 451
Wang, Yang, 243
Wang, Yongbao, 233
Wang, Yonghua, 103
Wang, Zhanfei, 333
Wang, Zhongpu, 223
Wang, Zhuhan, 21
Wu, Nana, 325
Wu, Yanjie, 341
Wu, Zhengtian, 315

X
Xia, Zhongxian, 103
Xie, Chenlei, 47
Xing, Yan, 205
Xing, Yingwen, 205
Xu, Qicheng, 243

Y
Yang, Jingxuan, 243
Yang, Junshan, 419
Yang, Lijian, 315
Yang, Xudong, 71, 351, 365
Yao, Fang-Hua, 119

Yi, Jiarui, 385
Yuan, Baolong, 149
Yuan, Shuai, 467

Z
Zhang, Jianhua, 443
Zhang, Lan, 83
Zhang, Lei, 193
Zhang, Licheng, 325
Zhang, Qiang, 333

Zhang, Siqi, 139, 169, 213, 253
Zhang, Xiaojing, 169
Zhang, Yi, 47
Zhang, Yunfeng, 139, 169, 213, 253
Zhang, Yuyu, 71
Zhao, Languang, 451
Zhao, Mingrui, 385
Zhao, Shengkai, 159
Zheng, Yingcheng, 213
Zhuang, Qiuyu, 409
Zhu, Wanying, 149

Printed in the United States
by Baker & Taylor Publisher Services